DICTIONARY OF ANALYSIS, CALCULUS, AND DIFFERENTIAL EQUATIONS

COMPREHENSIVE DICTIONARY OF MATHEMATICS

Stan Gibilisco
Editorial Advisor

FORTHCOMING AND PUBLISHED VOLUMES

Algebra, Arithmetic and Trigonometry
Steven Krantz

Classical & Theoretical Mathematics
Catherine Cavagnaro and Will Haight

Applied Mathematics for Engineers and Scientists
Emma Previato

Probability & Statistics
To be determined

The Comprehensive Dictionary of Mathematics
Stan Gibilisco

A VOLUME IN THE
COMPREHENSIVE DICTIONARY
OF MATHEMATICS

DICTIONARY OF ANALYSIS, CALCULUS, AND DIFFERENTIAL EQUATIONS

Douglas N. Clark

University of Georgia
Athens, Georgia

CRC Press is an imprint of the
Taylor & Francis Group, an **informa** business

Library of Congress Cataloging-in-Publication Data

Dictionary of analysis, calculus, and differential equations / [edited by] Douglas N. Clark.
p. cm. — (Comprehensive dictionary of mathematics)
ISBN 0-8493-0320-6 (alk. paper)
1. Mathematical analysis—Dictionaries. 2. Calculus—Dictionaries. 3. Differential equations—Dictionaries. I. Clark, Douglas N. (Douglas Napier), 1944– II. Series.

QA5 .D53 1999
515'.03—dc21 99-087759

Direct all inquiries to CRC Press LLC, 2000 N.W. Corporate Blvd., Boca Raton, Florida 33431.

International Standard Book Number 0-8493-0320-6
Library of Congress Card Number 99-087759

Preface

Book 1 of the **CRC Press Comprehensive Dictionary of Mathematics** covers analysis, calculus, and differential equations broadly, with overlap into differential geometry, algebraic geometry, topology, and other related fields. The authorship is by 15 mathematicians, active in teaching and research, including the editor.

Because it is a dictionary and not an encyclopedia, definitions are only occasionally accompanied by a discussion or example. Because it is a dictionary of mathematics, the primary goal has been to define each term rigorously. The derivation of a term is almost never attempted.

The dictionary is written to be a useful reference for a readership which includes students, scientists, and engineers with a wide range of backgrounds, as well as specialists in areas of analysis and differential equations and mathematicians in related fields. Therefore, the definitions are intended to be accessible, as well as rigorous. To be sure, the degree of accessibility may depend upon the individual term, in a dictionary with terms ranging from Albanese variety to z intercept.

Occasionally a term must be omitted because it is archaic. Care was taken when such circumstances arose because an archaic term may not be obsolete. An example of an archaic term deemed to be obsolete, and hence not included, is right line. This term was used throughout a turn-of-the-century analytic geometry textbook we needed to consult, but it was not defined there. Finally, reference to a contemporary English language dictionary yielded straight line as a synonym for right line.

The authors are grateful to the series editor, Stanley Gibilisco, for dealing with our seemingly endless procedural questions and to Nora Konopka, for always acting efficiently and cheerfully with CRC Press liaison matters.

Douglas N. Clark
Editor-in-Chief

Contributors

Gholamreza Akbari Estahbanati
University of Minnesota, Morris

Ioannis K. Argyros
Cameron University

Douglas N. Clark
University of Georgia

John M. Davis
Auburn University

Lifeng Ding
Georgia State University

Johnny L. Henderson
Auburn University

Amy Hoffman
University of Georgia

Alan Hopenwasser
University of Alabama

Arthur R. Lubin
Illinois Institute of Technology

Brian W. McEnnis
Ohio State University, Marion

Judith H. Morrel
Butler University

Giampiero Pecelli
University of Massachusetts, Lowell

David S. Protas
California State University, Northridge

David A. Stegenga
University of Hawaii

Derming Wang
California State University, Long Beach

a.e. *See* almost everywhere.

Abel summability A series $\sum_{j=0}^{\infty} a_j$ is *Abel summable* to A if the power series

$$f(z) = \sum_{j=0}^{\infty} a_j z^j$$

converges for $|z| < 1$ and

$$\lim_{x \to 1-0} f(x) = A.$$

Abel's Continuity Theorem *See* Abel's Theorem.

Abel's integral equation The equation

$$\int_a^x \frac{u(t)}{(x-t)^\alpha} dt = f(x),$$

where $0 < \alpha < 1, a \le x \le b$ and the given function $f(x)$ is $\mathbf{C}^1$ with $f(a) = 0$. A continuous solution $u(x)$ is sought.

Abel's problem A wire is bent into a planar curve and a bead of mass m slides down the wire from initial point (x, y). Let $T(y)$ denote the time of descent, as a function of the initial height y. *Abel's mechanical problem* is to determine the shape of the wire, given $T(y)$. The problem leads to *Abel's integral equation:*

$$\frac{1}{\sqrt{2g}} \int_0^y \frac{f(v)}{\sqrt{y-v}} dv = T(y).$$

The special case where $T(y)$ is constant leads to the *tautochrone.*

Abel's Theorem Suppose the power series $\sum_{j=0}^{\infty} a_j x^j$ has radius of convergence R and that $\sum_{j=0}^{\infty} a_j R^j < \infty$, then the original series converges uniformly on $[0, R]$.

A consequence is that convergence of the series $\sum a_j$ to the limit L implies *Abel summability* of the series to L.

Abelian differential An assignment of a meromorphic function f to each local coordinate z on a Riemann surface, such that $f(z)dz$ is invariantly defined. Also *meromorphic differential.*

Sometimes, analytic differentials are called *Abelian differentials of the first kind,* meromorphic differentials with only singularities of order ≥ 2 are called *Abelian differentials of the second kind,* and the term *Abelian differential of the third kind* is used for all other Abelian differentials.

Abelian function An inverse function of an Abelian integral. *Abelian functions* have two variables and four periods. They are a generalization of elliptic functions, and are also called *hyperelliptic functions. See also* Abelian integral, elliptic function.

Abelian integral (**1.**) An integral of the form

$$\int_0^x \frac{dt}{\sqrt{P(t)}},$$

where $P(t)$ is a polynomial of degree > 4. They are also called hyperelliptic integrals.

See also Abelian function, elliptic integral of the first kind.

(**2.**) An integral of the form $\int R(x, y)dx$, where $R(x, y)$ is a rational function and where y is one of the roots of the equation $F(x, y) = 0$, of an algebraic curve.

Abelian theorems Any theorems stating that convergence of a series or integral implies summability, with respect to some summability method. *See* Abel's Theorem, for example.

abscissa The first or x-coordinate, when a point in the plane is written in rectangular coordinates. The second or y-coordinate is called the *ordinate.* Thus, for the point (x, y), x is the abscissa and y is the ordinate. The *abscissa* is the horizontal distance of a

0-8493-0320-6/00/$0.00+$.50

point from the y-axis and the ordinate is the vertical distance from the x-axis.

abscissa of absolute convergence The unique real number σ_a such that the Dirichlet series

$$\sum_{j=1}^{\infty} a_j e^{-\lambda_j s}$$

(where $0 < \lambda_1 < \lambda_2 \cdots \to \infty$) converges absolutely for $\Re s > \sigma_a$, and fails to converge absolutely for $\Re s < \sigma_a$. If the Dirichlet series converges for all s, then the abscissa of absolute convergence $\sigma_a = -\infty$ and if the Dirichlet series never converges absolutely, $\sigma_a = \infty$. The vertical line $\Re s = \sigma_a$ is called the *axis of absolute convergence.*

abscissa of boundedness The unique real number σ_b such that the sum $f(s)$ of the Dirichlet series

$$f(s) = \sum_{j=1}^{\infty} a_j e^{-\lambda_j s}$$

(where $0 < \lambda_1 < \lambda_2 \cdots \to \infty$) is bounded for $\Re s \geq \sigma_b + \delta$ but not for $\Re s \geq \sigma_b - \delta$, for every $\delta > 0$.

abscissa of convergence (1.) The unique real number σ_c such that the Dirichlet series

$$\sum_{j=1}^{\infty} a_j e^{-\lambda_j s}$$

(where $0 < \lambda_1 < \lambda_2 \cdots \to \infty$) converges for $\Re s > \sigma_c$ and diverges for $\Re s < \sigma_c$. If the Dirichlet series converges for all s, then the abscissa of convergence $\sigma_c = -\infty$, and if the Dirichlet series never converges, $\sigma_c = \infty$. The vertical line $\Re s = \sigma_c$ is called the *axis of convergence.*

(2.) A number σ such that the Laplace transform of a measure converges for $\Re z > \sigma$ and does not converge in $\Re z > \sigma - \epsilon$, for any $\epsilon > 0$. The line $\Re z = \sigma$ is called the *axis of convergence.*

abscissa of regularity The greatest lower bound σ_r of the real numbers σ' such that the function $f(s)$ represented by the Dirichlet series

$$f(s) = \sum_{j=1}^{\infty} a_j e^{-\lambda_j s}$$

(where $0 < \lambda_1 < \lambda_2 \cdots \to \infty$) is regular in the half plane $\Re s > \sigma'$. Also called *abscissa of holomorphy.* The vertical line $\Re s = \sigma_r$ is called the *axis of regularity.* It is possible that the abscissa of regularity is actually less than the abscissa of convergence. This is true, for example, for the Dirichlet series $\sum (-1)^j j^{-s}$, which converges only for $\Re s > 0$; but the corresponding function $f(s)$ is entire.

abscissa of uniform convergence The unique real number σ_u such that the Dirichlet series

$$\sum_{j=1}^{\infty} a_j e^{-\lambda_j s}$$

(where $0 < \lambda_1 < \lambda_2 \cdots \to \infty$) converges uniformly for $\Re s \geq \sigma_u + \delta$ but not for $\Re s \geq \sigma_u - \delta$, for every $\delta > 0$.

absolute continuity (1.) For a real valued function $f(x)$ on an interval $[a, b]$, the property that, for every $\epsilon > 0$, there is a $\delta > 0$ such that, if $\{(a_j, b_j)\}$ are intervals contained in $[a, b]$, with $\sum (b_j - a_j) < \delta$ then $\sum |f(b_j) - f(a_j)| < \epsilon$.

(2.) For two measures μ and ν, absolute continuity of μ with respect to ν (written $\mu << \nu$) means that whenever E is a ν-measurable set with $\nu(E) = 0$, E is μ-measurable and $\mu(E) = 0$.

absolute continuity in the restricted sense Let $E \subset \mathbf{R}$, let $F(x)$ be a real-valued function whose domain contains E. We say that F is *absolutely continuous in the restricted sense* on E if, for every $\epsilon > 0$ there is a $\delta > 0$ such that for every sequence $\{[a_n, b_n]\}$ of non-overlapping intervals whose endpoints belong to E, $\sum_n (b_n - a_n) < \delta$ implies that $\sum_n O\{F; [a_n, b_n]\} < \epsilon$. Here, $O\{F; [a_n, b_n]\}$ denotes the oscillation of the function F in $[a_n, b_n]$, i.e., the difference between the least upper bound and the greatest lower bound of the values assumed by $F(x)$ on $[a_n, b_n]$.

absolute convergence (**1.**) For an infinite series $\sum_{n=1}^{\infty} a_j$, the finiteness of $\sum_{j=1}^{\infty} |a_j|$. (**2.**) For an integral

$$\int_S f(x)dx,$$

the finiteness of

$$\int_S |f(x)|dx.$$

absolute curvature The absolute value

$$|k| = \left|\frac{d^2r}{ds^2}\right| = +\sqrt{\left|g_{ik}\frac{D}{ds}\left(\frac{dx^i}{ds}\right)\frac{D}{ds}\left(\frac{dx^k}{ds}\right)\right|}$$

of the first curvature vector $\frac{d^2r}{ds^2}$ is the *absolute curvature* (*first*, or *absolute geodesic curvature*) of the regular arc C described by n parametric equations

$$x^i = x^i(t) \quad (t_1 \leq t \leq t_2)$$

at the point $(x^1, x^2, \ldots, x^n)$.

absolute maximum A number M, in the image of a function $f(x)$ on a set S, such that $f(x) \leq M$, for all $x \in S$.

absolute minimum A number m, in the image of a function $f(x)$ on a set S, such that $f(x) \geq m$ for all $x \in S$.

absolute value For a real number a, the absolute value is $|a| = a$, if $a \geq 0$ and $|a| = -a$ if $a < 0$. For a complex number $\zeta = a + bi$, $|\zeta| = \sqrt{a^2 + b^2}$. Geometrically, it represents the distance from $0 \in \mathbf{C}$. Also called *amplitude, modulus*.

absolutely continuous spectrum *See* spectral theorem.

absolutely convex set A subset of a vector space over $\mathbf{R}$ or $\mathbf{C}$ that is both convex and balanced. *See* convex set, balanced set.

absolutely integrable function *See* absolute convergence (for integrals).

absorb For two subsets A, B of a topological vector space X, A is said to *absorb* B if, for some nonzero scalar α,

$$B \subset \alpha A = \{\alpha x : x \in A\}.$$

absorbing A subset M of a topological vector space X over $\mathbf{R}$ or $\mathbf{C}$, such that, for any $x \in X$, $\alpha x \in M$, for some $\alpha > 0$.

abstract Cauchy problem Given a closed unbounded operator T and a vector v in the domain of T, the *abstract Cauchy problem* is to find a function f mapping $[0, \infty)$ into the domain of T such that $f'(t) = Tf$ and $f(0) = v$.

abstract space A formal system defined in terms of geometric axioms. Objects in the space, such as lines and points, are left undefined. Examples include abstract vector spaces, Euclidean and non-Euclidean spaces, and topological spaces.

acceleration Let $p(t)$ denote the position of a particle in space, as a function of time. Let

$$s(t) = \int_0^t \left(\left(\frac{dp}{dt}, \frac{dp}{dt}\right)\right)^{\frac{1}{2}} dt$$

be the length of path from time $t = 0$ to t. The *speed* of the particle is

$$\frac{ds}{dt} = \left(\left(\frac{dp}{dt}, \frac{dp}{dt}\right)\right) = \|\frac{dp}{dt}\|,$$

the *velocity* $\mathbf{v}(t)$ is

$$\mathbf{v}(t) = \frac{dp}{dt} = \frac{dp}{ds}\frac{ds}{dt}$$

and the *acceleration* $\mathbf{a}(t)$ is

$$\mathbf{a}(t) = \frac{d^2p}{dt^2} = \frac{dT}{ds}\left(\frac{ds}{dt}\right)^2 + T\frac{d^2s}{dt^2},$$

where T is the unit tangent vector.

accretive operator A linear operator T on a domain D in a Hilbert space H such that $\Re(Tx, x) \geq 0$, for $x \in D$. By definition, T is accretive if and only if $-T$ is *dissipative.*

accumulation point Let S be a subset of a topological space X. A point $x \in X$ is an *accumulation point* of S if every neighborhood of x contains infinitely many points of $E\backslash\{x\}$.

Sometimes the definition is modified, by replacing "infinitely many points" by "a point."

addition formula A functional equation involving the sum of functions or variables. For example, the property of the exponential function:

$$e^a \cdot e^b = e^{a+b}.$$

additivity for contours If an arc γ is subdivided into finitely many subarcs, $\gamma = \gamma_1 + \ldots + \gamma_n$, then the contour integral of a function $f(z)$ over γ satisfies

$$\int_\gamma f(z)dz = \int_{\gamma_1} f(z)dz + \ldots + \int_{\gamma_n} f(z)dz.$$

adjoint differential equation Let

$$L = a_0 \frac{d^n}{dt^n} + a_1 \frac{d^{n-1}}{dt^{n-1}} + \ldots + a_n$$

be a differential operator, where $\{a_j\}$ are continuous functions. The adjoint differential operator is

$$L^+ = (-1)^n \left(\frac{d^n}{dt^n}\right) M_{\bar{a}_0} + (-1)^{n-1} \left(\frac{d^{n-1}}{dt^{n-1}}\right) M_{\bar{a}_1} + \ldots + M_{\bar{a}_n}$$

where M_g is the operator of multiplication by g. The *adjoint differential equation* of $Lf = 0$ is, therefore, $L^+ f = 0$.

For a system of differential equations, the functions $\{a_j\}$ are replaced by matrices of functions and each $\bar{a}_j$ above is replaced by the conjugate-transpose matrix.

adjoint operator For a linear operator T on a domain D in a Hilbert space H, the adjoint domain is the set $D^* \subset H$ of all $y \in H$ such that there exists $z \in H$ satisfying

$$(Tx, y) = (x, z),$$

for all $x \in D$. The *adjoint operator* T^* of T is the linear operator, with domain D^*, defined by $T^* y = z$, for $y \in D^*$, as above.

adjoint system *See* adjoint differential equation.

admissible Baire function A function belonging to the class on which a functional is to be minimized (in the calculus of variations).

AF algebra A **C*** algebra $\mathcal{A}$ which has an increasing sequence $\{A_n\}$ of finite-dimensional **C*** subalgebras, such that the union $\cup_n A_n$ is dense in $\mathcal{A}$.

affine arc length (**1.**) For a plane curve $\mathbf{x} = \mathbf{x}(t)$, with

$$\left(\frac{d\mathbf{x}}{dt}, \frac{d^2\mathbf{x}}{dt^2}\right) \neq 0,$$

the quantity

$$s = \int \left(\frac{d\mathbf{x}}{dt}, \frac{d^2\mathbf{x}}{dt^2}\right).$$

(**2.**) For a curve $\mathbf{x}(p) = \{x_1(p), x_2(p), x_3(p)\}$ in 3-dimensional affine space, the quantity

$$s = \int \det \begin{pmatrix} x_1 & x_2 & x_3 \\ x_1' & x_2' & x_3' \\ x_1'' & x_2'' & x_3'' \end{pmatrix}^{\frac{1}{6}} dt.$$

affine connection Let B be the bundle of frames on a differentiable manifold M of dimension n. An *affine connection* is a connection on B, that is, a choice $\{H_b\}_{b \in B}$, of

subspaces $H_b \subset B_b$, for every $b \in B$, such that
(i.) $B_b = H_b + V_b$ (direct sum) where V_b is the tangent space at b to the fiber through b;
(ii.) $H_{bg} = g_*(H_b)$, for $g \in GL(n, \mathbf{R})$; and
(iii.) H_b depends differentiably on b.

affine coordinates Projective space P^n is the set of lines in $\mathbf{C}^{n+1}$ passing through the origin. *Affine coordinates* in P^n can be chosen in each patch $U_j = \{[(x_0, x_1, \ldots, x_n)] : x_j \neq 0\}$ (where $[(x_1, x_2, \ldots, x_n)]$ denotes the line through 0, containing the point $(x_0, x_1, \ldots, x_n)$). If $z = [(z_0, \ldots, z_n)]$, with $z_j \neq 0$, the affine coordinates of z are $(z_0/z_j, \ldots, z_{j-1}/z_j, z_{j+1}/z_j, \ldots, z_n/z_j)$. Also called *nonhomogeneous coordinates.*

affine curvature (1.) For a plane curve $\mathbf{x} = \mathbf{x}(t)$, the quantity

$$\kappa = (\mathbf{x}'', \mathbf{x}''')$$

where $' = \frac{d}{ds}$, (arc length derivative).
(2.) For a space curve $\mathbf{x}(p) = \{x_1(p), x_2(p), x_3(p)\}$, the quantity

$$\kappa = \det \begin{Bmatrix} x_1^{(4)} & x_2^{(4)} & x_3^{(4)} \\ x_1' & x_2' & x_3' \\ x_1''' & x_2''' & x_3''' \end{Bmatrix},$$

where derivatives are with respect to affine arc length.

One also has the first and second affine curvatures, given by

$$\kappa_1 = -\frac{\kappa}{4}, \kappa_2 = \frac{\kappa'}{4} - \tau,$$

where τ is the affine torsion. *See* affine torsion.

affine diffeomorphism A diffeomorphism q of n-dimensional manifolds induces maps of their tangent spaces and, thereby, a $GL(n, \mathbf{R})$-equivariant diffeomorphism of their frame bundles. If each frame bundle carries a connection and the induced map of frame bundles carries one connection to the other, then q is called an *affine diffeomorphism,* relative to the given connections.

affine differential geometry The study of properties invariant under the group of affine transformations. (The general linear group.)

affine length Let X be an affine space, V a singular metric vector space and k a field of characteristic different from 2. Then (X, V, k) is a metric affine space with metric defined as follows. If x and y are points in X, the unique vector A of V such that $Ax = y$ is denoted by $\overrightarrow{x, y}$. The square affine length (distance) between points x and y of X is the scalar $\overrightarrow{x, y}^{\,2}$.

If (X, V, R) is Euclidean space, $\overrightarrow{x, y}^{\,2} \geq 0$ and the Euclidean distance between the points x and y is the nonnegative square root $\sqrt{\overrightarrow{x, y}^{\,2}}$. In this case, the square distance is the square of the Euclidean distance. One always prefers to work with the distance itself rather than the square distance, but this is rarely possible. For instance, in the Lorentz plane $\overrightarrow{x, y}^{\,2}$ may be negative and, therefore, there is no real number whose square is $\overrightarrow{x, y}^{\,2}$.

affine minimal surface The extremal surface of the variational problem $\delta\Omega = 0$, where Ω is affine surface area. It is characterized by the condition that its affine mean curvature should be identically 0.

affine normal (1.) For a plane curve $\mathbf{x} = \mathbf{x}(t)$, the vector $\mathbf{x}' = \frac{d\mathbf{x}}{ds}$, where s is affine arc length.
(2.) For a surface $(\mathbf{x})$, the vector $\mathbf{y} = \frac{1}{2}\Delta\mathbf{x}$, where Δ is the second Beltrami operator.

affine principal normal vector For a plane curve $\mathbf{x} = \mathbf{x}(t)$, the vector $\mathbf{x}'' = \frac{d^2\mathbf{x}}{ds^2}$, where s is affine arc length.

affine surface area Let (x_1, x_2, x_3) denote the points on a surface and set

$$L = \frac{\partial^2 x_3}{\partial x_1^2}, M = \frac{\partial^2 x_3}{\partial x_1 \partial x_2}, N = \frac{\partial^2 x_3}{\partial x_2^2}.$$

The *affine surface area* is

$$\Omega = \int\int |LN - M^2|^{\frac{1}{4}} du dv.$$

affine symmetric space A complete, connected, simply connected, n-dimensional manifold M having a connection on the frame bundle such that, for every $x \in M$, the geodesic symmetry $exp_x(Z) \to exp_x(-Z)$ is the restriction to $exp_x(M_x)$ of an affine diffeomorphism of M. *See* affine diffeomorphism.

affine torsion For a space curve $\mathbf{x}(p) = \{x_1(p), x_2(p), x_3(p)\}$, the quantity

$$\tau = -\det \begin{Bmatrix} x_1^{(4)} & x_2^{(4)} & x_3^{(4)} \\ x_1'' & x_2'' & x_3'' \\ x_1''' & x_2''' & x_3''' \end{Bmatrix},$$

where derivatives are with respect to affine arc length.

affine transformation (1.) A function of the form $f(x) = ax + b$, where a and b are constants and x is a real or complex variable. (2.) Members of the general linear group (invertible transformations of the form $(az + b)/(cz + d)$).

Ahlfors function *See* analytic capacity.

Ahlfors' Five Disk Theorem Let $f(z)$ be a transcendental meromorphic function, and let $A_1, A_2, \ldots, A_5$ be five simply connected domains in $\mathbf{C}$ with disjoint closures. There exists $j \in \{1, 2, \ldots, 5\}$ and, for any $R > 0$, a simply connected domain $D \subset \{z \in \mathbf{C} : |z| > R\}$ such that $f(z)$ is a conformal map of D onto A_j. If $f(z)$ has a finite number of poles, then 5 may be replaced by 3.

See also meromorphic function, transcendental function.

Albanese variety Let R be a Riemann surface, $H^{1,0}$ the holomorphic 1, 0 forms on R, $H^{0,1*}$ its complex dual, and let a curve γ in R act on $H^{0,1}$ by integration:

$$w \to I(\gamma) = \int_\gamma w.$$

The *Albanese variety*, $Alb(R)$ of R is

$$Alb(R) = H^{0,1*}/I(H_1(Z)).$$

See also Picard variety.

Alexandrov compactification For a topological space X, the set $\hat{X} = X \cup \{x\}$, for some point $x \notin X$, topologized so that the closed sets in $\hat{X}$ are (i.) the compact sets in X, and (ii.) all sets of the form $E \cup \{x\}$ where E is closed in X.

$\hat{X}$ is also called the *one point compactification* of X.

algebra of differential forms Let M be a differentiable manifold of class $\mathbf{C}^r$ ($r \geq 1$), $T_p(M)$ its tangent space, $T_p^*(M) = T_p(M)^*$ the dual vector space (the linear mappings from $T_p(M)$ into $\mathbf{R}$) and $T^*(M) = \cup_{p \in M} T_p^*(M)$. The *bundle of i-forms* is

$$\wedge^i(T^*(M)) = \cup_{p \in M} \wedge^i (T_p^*(M)),$$

where, for any linear map $f : V \to W$, between two vector spaces, the linear map

$$\wedge^i f : \wedge^i V \to \wedge^i W$$

is defined by $(\wedge^i f)(v_1 \wedge \cdots \wedge v_k) = f(v_1) \wedge \cdots \wedge f(v_k)$. The bundle projection is defined by $\pi(z) = p$, for $z \in \wedge^i(T_p^*(M))$.

A *differential i-form* or *differential form of degree i* is a section of the bundle of i-forms; that is, a continuous map

$$s : M \to \wedge^i(T^*(M))$$

with $\pi(s(p)) = p$. If $D^i(M)$ denotes the vector space of differential forms of degree i, the *algebra of differential forms* on M is

$$D^*(M) = \sum_{i \geq 0} \oplus D^i(M).$$

It is a graded, anticommutative algebra over $\mathbf{R}$.

algebra of sets A collection $\mathcal{F}$ of subsets of a set S such that if $E, F \in \mathcal{F}$, then (i.) $E \cup F \in \mathcal{F}$, (ii.) $E \backslash F \in \mathcal{F}$, and (iii.) $S \backslash F \in \mathcal{F}$. If $\mathcal{F}$ is also closed under the taking of countable unions, then $\mathcal{F}$ is called a σ-algebra. Algebras and σ-algebras of sets are sometimes called *fields* and *σ-fields* of sets.

algebraic analysis The study of mathematical objects which, while of an analytic nature, involve manipulations and characterizations which are algebraic, as opposed to inequalities and estimates. An example is the study of algebras of operators on a Hilbert space.

algebraic function A function $y = f(z)$ of a complex (or real) variable, which satisfies a polynomial equation

$$a_n(z)y^n + a_{n-1}(z)y^{n-1} + \ldots + a_0(z) = 0,$$

where $a_0(z), \ldots, a_n(z)$ are polynomials.

algebraic singularity *See* branch.

algebroidal function An analytic function $f(z)$ satisfying the irreducible algebraic equation

$$A_0(z)f^k + A_1(z)f^{k-1} + \cdots + A_k(z) = 0$$

with single-valued meromorphic functions $A_j(z)$ in a complex domain G is called *k-algebroidal* in G.

almost complex manifold A smooth manifold M with a field of endomorphisms J on $T(M)$ such that $J^2 = J \circ J = -I$, where I is the identity endomorphism. The field of endomorphisms is called an *almost complex structure* on M.

almost complex structure *See* almost complex manifold.

almost contact manifold An odd dimensional differentiable manifold M which admits a tensor field ϕ of type $(1, 1)$, a vector field ζ and a 1-form ω such that

$$\phi^2 X = -X + \omega(X)\zeta, \quad \omega(\zeta) = 1,$$

for X an arbitrary vector field on M. The triple (ϕ, ζ, ω) is called an *almost contact structure* on M.

almost contact structure *See* almost contact manifold.

almost everywhere Except on a set of measure 0 (applying to the truth of a proposition about points in a measure space). For example, a sequence of functions $\{f_n(x)\}$ converges *almost everywhere* to $f(x)$, provided that $f_n(x) \to f(x)$ for $x \in E$, where the complement of E has measure 0. Abbreviations are *a.e.* and *p.p.* (from the French *presque partout*).

almost periodic function in the sense of Bohr A continuous function $f(x)$ on $(-\infty, \infty)$ such that, for every $\epsilon > 0$, there is a $p = p(\epsilon) > 0$ such that, in every interval of the form $(t, t + p)$, there is at least one number τ such that $|f(x + \tau) - f(x)| \leq \epsilon$, for $-\infty < x < \infty$.

almost periodic function on a group For a complex-valued function $f(g)$ on a group G, let $f_s : G \times G \to \mathbf{C}$ be defined by $f_s(g, h) = f(gsh)$. Then f is said to be *almost periodic* if the family of functions $\{f_s(g, h) : s \in G\}$ is totally bounded with respect to the uniform norm on the complex-valued functions on $G \times G$.

almost periodic function on a topological group On a (locally compact, Abelian) group G, the uniform limit of trigonometric polynomials on G. A trigonometric polynomial is a finite linear combination of characters (i.e., homomorphisms into the multiplicative group of complex numbers of modulus 1) on G.

alpha capacity A financial measure giving the difference between a fund's actual return and its expected level of performance, given its level of risk (as measured by the beta capacity). A positive alpha capacity indicates that the fund has performed better than expected based on its beta capacity whereas a negative alpha indicates poorer performance.

alternating mapping The mapping $\mathcal{A}$, generally acting on the space of covariant tensors on a vector space, and satisfying

$$\mathcal{A}\Phi(\mathbf{v}_1, \ldots, \mathbf{v}_r) = \frac{1}{r!}\sum_{\sigma} \operatorname{sgn}\sigma\,\Phi(\mathbf{v}_{\sigma(1)}, \ldots, \mathbf{v}_{\sigma(r)}),$$

where the sum is over all permutations σ of $\{1, \ldots, r\}$.

alternating multilinear mapping A mapping $\Phi : V \times \cdots \times V \to W$, where V and W are vector spaces, such that $\Phi(v_1, \ldots, v_n)$ is linear in each variable and satisfies

$$\Phi(v_1, \ldots, v_i, \ldots, v_j, \ldots, v_n) = -\Phi(v_1, \ldots, v_j, \ldots, v_i, \ldots, v_n).$$

alternating series A formal sum $\sum a_j$ of real numbers, where $(-1)^j a_j \geq 0$ or $(-1)^j a_{j+1} \geq 0$; i.e., the terms alternate in sign.

alternating tensor *See* antisymmetric tensor.

alternizer *See* alternating mapping.

amenable group A locally compact group G for which there is a left invariant mean on $L^\infty(G)$.

Ampere's transformation A transformation of the surface $z = f(x, y)$, defined by coordinates X, Y, Z, given by

$$X = \frac{\partial f}{\partial x}, Y = \frac{\partial f}{\partial y}, Z = \frac{\partial f}{\partial x}x + \frac{\partial f}{\partial y}y - z.$$

amplitude function For a normal lattice, let e_1, e_2, e_3 denote the stationary values of the Weierstrass $\wp$-function and, for $i = 1, 2, 3$, let $\mathbf{f}_i(u)$ be the square root of $\wp - e_i$, whose leading term at the origin is u^{-1}. Two of the Jacobi-Glaisher functions are

$$\mathrm{cs}u = \mathbf{f}_1, \mathrm{sn}u = 1/\mathbf{f}_2,$$

which are labeled in analogy with the trigonometric functions, on account of the relation $\mathrm{sn}^2u+\mathrm{cs}^2u = 1$. As a further part of the analogy, the *amplitude*, am u, of u, is defined to be the angle whose sine and cosine are sn$u$ and cs$u$.

amplitude in polar coordinates In polar coordinates, a point in the plane $\mathbf{R}^2$ is written (r, θ), where r is the distance from the origin and $\theta \in [0, 2\pi)$ is the angle the line segment (from the origin to the point) makes with the positive real axis. The angle θ is called the *amplitude.*

amplitude of complex number *See* argument of complex number.

amplitude of periodic function The absolute maximum of the function. For example, for the function $f(x) = A\sin(\omega x - \phi)$, the number A is the amplitude.

analysis A branch of mathematics that can be considered the foundation of calculus, arising out of the work of mathematicians such as Cauchy and Riemann to formalize the differential and integral calculus of Newton and Leibniz. *Analysis* encompasses such topics as limits, continuity, differentiation, integration, measure theory, and approximation by sequences and series, in the context of metric or more general topological spaces. Branches of analysis include real analysis, complex analysis, and functional analysis.

analysis on locally compact Abelian groups The study of the properties (inversion, etc.) of the Fourier transform, defined by

$$\hat{f}(\gamma) = \int_G f(x)(-x, \gamma)dx,$$

with respect to Haar measure on a locally compact, Abelian group G. Here $f \in L^1(G)$

and γ is a homomorphism from G to the multiplicative group of complex numbers of modulus 1. The classical theory of the Fourier transform extends with elegance to this setting.

analytic *See* analytic function.

analytic automorphism A mapping from a field with absolute value to itself, that preserves the absolute value.

See also analytic isomorphism.

analytic capacity For a compact planar set K, let $\Omega(K) = K_1 \cup \{\infty\}$, where K_1 is the unbounded component of the complement of K. Let $\mathcal{A}(K)$ denote the set of functions f, analytic on $\Omega(K)$, such that $f(\infty) = 0$ and $\|f\|_{\Omega(K)} \leq 1$. If K is not compact, $\mathcal{A}(K)$ is the union of $\mathcal{A}(E)$ for E compact and $E \subset K$. The *analytic capacity* of a planar set E is

$$\gamma(E) = \sup_{f \in \mathcal{A}(E)} |f'(\infty)|.$$

If K is compact, there is a unique function $f \in \mathcal{A}(K)$ such that $f'(\infty) = \gamma(K)$. This function f is called the *Ahlfors function* of K.

analytic continuation A function $f(z)$, analytic on an open disk $A \subset \mathbf{C}$, is a *direct analytic continuation* of a function $g(z)$, analytic on an open disk B, provided the disks A and B have nonempty intersection and $f(z) = g(z)$ in $A \cap B$.

We say $f(z)$ is an *analytic continuation* of $g(z)$ if there is a finite sequence of functions $f_1, f_2, \ldots, f_n$, analytic in disks $A_1, A_2, \ldots, A_n$, respectively, such that $f_1(z) = f(z)$ in $A \cap A_1$, $f_n(z) = g(z)$ in $A_n \cap B$ and, for $j = 1, \ldots, n-1$, $f_{j+1}(z)$ is a direct analytic continuation of $f_j(z)$.

analytic continuation along a curve Suppose $f(z)$ is a function, analytic in a disk D, centered at z_0, $g(z)$ is analytic in a disk E, centered at z_1, and C is a curve with endpoints z_0 and z_1. We say that g is an analytic continuation of f along C, provided there is a sequence of disks $D_1, \ldots, D_n$, with centers on C and an analytic function $f_j(z)$ analytic in D_j, $j = 1, \ldots, n$, such that $f_1(z) = f(z)$ in $D = D_1$, $f_n(z) = g(z)$ in $D_n = E$ and, for $j = 1, \ldots, n-1$, $f_{j+1}(z)$ is a direct analytic continuation of $f_j(z)$. *See* analytic continuation.

analytic curve A curve $\alpha : I \to M$ from a real interval I into an analytic manifold M such that, for any point $p_0 = \alpha(t_0)$, the chart (U_{p_0}, ϕ_{p_0}) has the property that $\phi_{p_0}(\alpha(t))$ is an analytic function of t, in the sense that $\phi_{p_0}(\alpha(t)) = \sum_{j=0}^{\infty} a_j(t-t_0)^j$ has a nonzero radius of convergence, and $a_1 \neq 0$.

analytic disk A nonconstant, holomorphic mapping $\phi : D \to \mathbf{C}^n$, were D is the unit disk in $\mathbf{C}^1$, or the image of such a map.

analytic function (1.) A real-valued function $f(x)$ of a real variable, is *(real) analytic* at a point $x = a$ provided $f(x)$ has an expansion in power series

$$f(x) = \sum_{j=0}^{\infty} c_j(x-a)^j,$$

convergent in some neighborhood $(a-h, a+h)$ of $x = a$.
(2.) A complex valued function $f(z)$ of a complex variable is *analytic* at $z = z_0$ provided

$$f'(w) = \lim_{z \to w} \frac{f(w) - f(z)}{w - z}$$

exists in a neighborhood of z_0. *Analytic in a domain* $D \subseteq \mathbf{C}$ means analytic at each point of D. Also *holomorphic, regular, regular-analytic.*
(3.) For a complex-valued function $f(z_1, \ldots, z_n)$ of n complex variables, analytic in each variable separately.

analytic functional A bounded linear functional on $\mathcal{O}(U)$, the Fréchet space of analytic functions on an open set $U \subset \mathbf{C}^n$, with the topology of uniform convergence on compact subsets of U.

analytic geometry The study of shapes and figures, in 2 or more dimensions, with the aid of a coordinate system.

Analytic Implicit Function Theorem Suppose $F(x, y)$ is a function with a convergent power series expansion

$$F(x, y) = \sum_{j,k=0}^{\infty} a_{jk}(x - x_0)^j (y - y_0)^k,$$

where $a_{00} = 0$ and $a_{01} \neq 0$. Then there is a unique function $y = f(x)$ such that
(i.) $F(x, f(x)) = 0$ in a neighborhood of $x = x_0$;
(ii.) $f(x_0) = y_0$; and
(iii.) $f(x)$ can be expanded in a power series

$$f(x) = \sum_{j=0}^{\infty} b_j (x - x_0)^j,$$

convergent in a neighborhood of $x = x_0$.

analytic isomorphism A mapping between fields with absolute values that preserves the absolute value.

See also analytic automorphism.

analytic manifold A topological manifold with an atlas, where compatibility of two charts (U_p, ϕ_p), (U_q, ϕ_q) means that the composition $\phi_p \circ \phi_q^{-1}$ is analytic, whenever $U_p \cap U_q \neq \emptyset$. *See* atlas.

analytic neighborhood Let P be a polyhedron in the PL (piecewise linear) n-manifold M. Then an *analytic neighborhood* of P in M is a polyhedron N such that (1) N is a closed neighborhood of P in M, (2) N is a PL n-manifold, and (3) $N \downarrow P$.

analytic polyhedron Let W be an open set in $\mathbf{C}^n$ that is homeomorphic to a ball and let $f_1, \ldots, f_k$ be holomorphic on W. If the set

$$\Omega = \{z \in W : |f_j(z)| < 1, j = 1, \ldots, k\}$$

has its closure contained in W, then Ω is called an *analytic polyhedron.*

analytic set A subset A of a Polish space X such that $A = f(Z)$, for some Polish space Z and some continuous function $f : Z \to X$.

Complements of analytic sets are called *co-analytic sets.*

analytic space A topological space X (the underlying space) together with a sheaf $\mathcal{S}$, where X is locally the zero set Z of a finite set of analytic functions on an open set $D \subset \mathbf{C}^n$ and where the sections of $\mathcal{S}$ are the analytic functions on Z. Here *analytic functions on* Z (if, for example, D is a polydisk) means functions that extend to be analytic on D.

The term *complex space* is used by some authors as a synonym for *analytic space.* But sometimes, it allows a bigger class of functions as the sections of $\mathcal{S}$. Thus, while the sections of $\mathcal{S}$ are $\mathcal{H}(Z) = \mathcal{H}(D)/\mathcal{I}(Z)$ (the holomorphic functions on D modulo the ideal of functions vanishing on Z) for an analytic space, $\mathcal{H}(Z)$ may be replaced by $\hat{\mathcal{H}}(Z) = \mathcal{H}(D)/\hat{\mathcal{I}}$, for a complex space, where $\hat{\mathcal{I}}$ is some other ideal of $\mathcal{H}(D)$ with zero set Z.

angle between curves The angle between the tangents of two curves. *See* tangent line.

angular derivative Let $f(z)$ be analytic in the unit disk $D = \{z : |z| < 1\}$. Then f has an *angular derivative* $f'(\zeta)$ at $\zeta \in \partial D$ provided

$$f'(\zeta) = \lim_{r \to 1-} f'(r\zeta).$$

antiderivative A function $F(x)$ is an *antiderivative* of $f(x)$ on a set $S \subset \mathbf{R}$, provided F is differentiable and $F'(x) = f(x)$, on S. Any two antiderivatives of $f(x)$ must differ by a constant (if S is connected) and so, if $F(x)$ is one antiderivative of f, then any antiderivative has the form $F(x) + C$, for some real constant C. The usual notation for the most general antiderivative of f is

$$\int f(x)dx = F(x) + C.$$

antiholomorphic mapping A mapping whose complex conjugate, or adjoint, is analytic.

antisymmetric tensor A covariant tensor Φ of order r is *antisymmetric* if, for each $i, j, 1 \leq i, j \leq r$, we have

$$\Phi(\mathbf{v}_1, \ldots, \mathbf{v}_i, \ldots, \mathbf{v}_j, \ldots, \mathbf{v}_r) \\ = -\Phi(\mathbf{v}_1, \ldots, \mathbf{v}_j, \ldots, \mathbf{v}_i, \ldots, \mathbf{v}_r).$$

Also called an *alternating*, or *skew* tensor, or an *exterior form*.

Appell hypergeometric function An extension of the hypergeometric function to two variables, resulting in four kinds of functions (Appell 1925):

$$G_1(a; b, c; d; x, y) \\ = \sum_{m=0}^{\infty}\sum_{n=0}^{\infty} \frac{(a)_{m+n}(b)_m(c)_n}{m!n!(d)_{m+n}} x^m y^n$$

$$G_2(a; b, c; d, d'; x, y) \\ = \sum_{m=0}^{\infty}\sum_{n=0}^{\infty} \frac{(a)_{m+n}(b)_m(c)_n}{m!n!(d)_m(d')_n} x^m y^n$$

$$G_3(a, a'; b, c'; d; x, y) \\ = \sum_{m=0}^{\infty}\sum_{n=0}^{\infty} \frac{(a)_m(a')_n(b)_m(c)_n}{m!n!(d)_{m+n}} x^m y^n$$

$$G_4(a; b; d, d'; x, y) \\ = \sum_{m=0}^{\infty}\sum_{n=0}^{\infty} \frac{(a)_{m+n}(b)_{m+n}}{m!n!(d)_m(d')_n} x^m y^n.$$

Appell defined these functions in 1880, and Picard showed in 1881 that they can be expressed by integrals of the form

$$\int_0^1 u^a(1-u)^b(1-xu)^d(1-yu)^q\, du.$$

approximate derivative *See* approximately differentiable function.

approximate identity On $[-\pi, \pi]$, a sequence of functions $\{e_j\}$ such that
(i.) $e_j \geq 0, j = 1, 2, \ldots$;
(ii.) $1/2\pi \int_{-\pi}^{\pi} e_j(t)dt = 1$;
(iii.) for every ϵ with $\pi > \epsilon > 0$,

$$\lim_{j\to\infty} \int_{-\epsilon}^{\epsilon} e_j(t)dt = 0.$$

approximately differentiable function A function $F : [a, b] \to \mathbf{R}$ (at a point $c \in [a, b]$) such that there exists a measurable set $E \subseteq [a, b]$ such that $c \in E$ and is a density point of E and $F|_E$ is differentiable at c. The *approximate derivative* of F at c is the derivative of $F|_E$ at c.

approximation (1.) An approximation to a number x is a number that is close to x. More precisely, given an $\epsilon > 0$, an approximation to x is a number y such that $|x - y| < \epsilon$. We usually seek an approximation to x from a specific class of numbers. For example, we may seek an approximation of a real number from the class of rational numbers.
(2.) An approximation to a function f is a function that is close to f in some appropriate measure. More precisely, given an $\epsilon > 0$, an approximation to f is a function g such that $\|f - g\| < \epsilon$ for some norm $\|\cdot\|$. We usually seek an approximation to f from a specific class of functions. For example, for a continuous function f defined on a closed interval I we may seek a polynomial g such that $\sup_{x\in I} |f(x) - g(x)| < \epsilon$.

arc length (1.) For the graph of a differentiable function $y = f(x)$, from $x = a$ to $x = b$, in the plane, the integral

$$\int_a^b \sqrt{1 + (\frac{dy}{dy})^2} dx.$$

(2.) For a curve $t \to p(t), a \leq t \leq b$, of class $\mathbf{C}^1$, on a Riemannian manifold with inner product $\Phi(X_p, Y_p)$ on its tangent space at p, the integral

$$\int_a^b \left(\Phi(\frac{dp}{dt}, \frac{dp}{dt})\right)^{\frac{1}{2}} dt.$$

Argand diagram The representation $z = re^{i\theta}$ of a complex number z.

argument function The function $\arg(z) = \theta$, where z is a complex number with the representation $z = re^{i\theta}$, with r real and non-negative. The choice of θ is, of course, not unique and so $\arg(z)$ is not a function without further restrictions such as $-\pi < \arg(z) \le \pi$ (principal argument) or the requirement that it be continuous, together with a specification of the value at some point.

argument of complex number The angle θ in the representation $z = re^{i\theta}$ of a complex number z. Also *amplitude.*

argument of function The domain variable; so that if $y = f(x)$ is the function assigning the value y to a given x, then x is the *argument* of the function f. Also *independent variable.*

argument principle Let $f(z)$ be analytic on and inside a simple closed curve $C \subset \mathbf{C}$, except for a finite number of poles inside C, and suppose $f(z) \neq 0$ on C. Then $\Delta \arg f$, the net change in the argument of f, as z traverses C, satisfies $\Delta \arg f = N - P$, the number of zeros minus the number of poles of f inside C.

arithmetic mean For n real numbers, $a_1, a_2, \dots, a_n$, the number $\frac{a_1+a_2+\dots+a_n}{n}$. For a real number r, the *arithmetic mean of order* r is

$$\frac{\sum_{j=1}^{n}(r+1)\cdots(r+n-j)a_j/(n-j)!}{\sum_{j=1}^{n}(r+1)\cdots(r+n-j)/(n-j)!}.$$

arithmetic progression A sequence $\{a_j\}$ where a_j is a linear function of $j : a_j = cj + r$, with c and r independent of j.

arithmetic-geometric mean The *arithmetic-geometric mean* (AGM) $M(a, b)$ of two numbers a and b is defined by starting with $a_0 \equiv a$ and $b_0 \equiv b$, then iterating

$$a_{n+1} = \tfrac{1}{2}(a_n + b_n) \quad b_{n+1} = \sqrt{a_n b_n}$$

until $a_n = b_n$. The sequences a_n and b_n converge toward each other, since

$$\begin{aligned} a_{n+1} - b_{n+1} &= \tfrac{1}{2}(a_n + b_n) - \sqrt{a_n b_n} \\ &= \frac{a_n - 2\sqrt{a_n b_n} + b_n}{2}. \end{aligned}$$

But $\sqrt{b_n} < \sqrt{a_n}$, so

$$2b_n < 2\sqrt{a_n b_n}.$$

Now, add $a_n - b_n - 2\sqrt{a_n b_n}$ so each side

$$a_n + b_n - 2\sqrt{a_n b_n} < a_n - b_n,$$

so

$$a_{n+1} - b_{n+1} < \tfrac{1}{2}(a_n - b_n).$$

The AGM is useful in computing the values of complete elliptic integrals and can also be used for finding the inverse tangent. The special value $1/M(1, \sqrt{2})$ is called Gauss's constant.

The AGM has the properties

$$\begin{aligned} \lambda M(a, b) &= M(\lambda a, \lambda b) \\ M(a, b) &= M\left(\tfrac{1}{2}(a+b), \sqrt{ab}\right) \\ M(1, \sqrt{1-x^2}) &= M(1+x, 1-x) \\ M(1, b) &= \frac{1+b}{2} M\left(1, \frac{2\sqrt{b}}{1+b}\right). \end{aligned}$$

The Legendre form is given by

$$M(1, x) = \prod_{n=0}^{\infty} \tfrac{1}{2}(1 + k_n),$$

where $k_0 \equiv x$ and

$$k_{n+1} \equiv \frac{2\sqrt{k_n}}{1 + k_n}.$$

Solutions to the differential equation

$$(x^3 - x)\frac{d^2y}{dx^2} + (3x^2 - 1)\frac{dy}{dx} + xy = 0$$

are given by $[M(1 + x, 1 - x)]^{-1}$ and $[M(1, x)]^{-1}$.

A generalization of the arithmetic-geometric mean is

$$I_p(a,b) = \int_0^\infty \frac{x^{p-2}dx}{(x^p+a^p)^{1/p}(x^p+b^p)^{(p-1)/p}},$$

which is related to solutions of the differential equation

$$x(1-x^p)Y'' + [1-(p+1)x^p]Y' - (p-1)x^{p-1}Y = 0.$$

When $p = 2$ or $p = 3$, there is a modular transformation for the solutions of the above equation that are bounded as $x \to 0$. Letting $J_p(x)$ be one of these solutions, the transformation takes the form

$$J_p(\lambda) = \mu J_p(x),$$

where

$$\lambda = \frac{1-u}{1+(p-1)u} \quad \mu = \frac{1+(p-1)u}{p}$$

and

$$x^p + u^p = 1.$$

The case $p = 2$ gives the arithmetic-geometric mean, and $p = 3$ gives a cubic relative discussed by Borwein and Borwein (1990, 1991) and Borwein (1996) in which, for $a, b > 0$ and $I(a,b)$ defined by

$$I(a,b) = \int_0^\infty \frac{t\,dt}{[(a^3+t^3)(b^3+t^3)^2]^{1/3}},$$

$$I(a,b) = I\left(\frac{a+2b}{3}, \left[\frac{b}{3}(a^2+ab+b^2)\right]\right).$$

For iteration with $a_0 = a$ and $b_0 = b$ and

$$a_{n+1} = \frac{a_n+2b_n}{3}$$

$$b_{n+1} = \frac{b_n}{3}(a_n^2 + a_nb_n + b_n^2),$$

$$\lim_{n\to\infty} a_n = \lim_{n\to\infty} b_n = \frac{I(1,1)}{I(a,b)}.$$

Modular transformations are known when $p = 4$ and $p = 6$, but they do not give identities for $p = 6$ (Borwein 1996).

See also arithmetic-harmonic mean.

arithmetic-harmonic mean For two given numbers a, b, the number $A(a,b)$, obtained by setting $a_0 = a, b_0 = b$, and, for $n \geq 0$, $a_{n+1} = \frac{1}{2}(a_n + b_n)$, $b_{n+1} = 2a_nb_n/(a_n + b_n)$ and $A(a,b) = \lim_{n\to\infty} a_n$. The sequences a_n and b_n converge to a common value, since $a_n - b_n \leq \frac{1}{2}(a_{n-1} - b_{n-1})$, if a, b are nonnegative, and we have $A(a_0, b_0) = \lim_{n\to\infty} a_n = \lim b_n = \sqrt{ab}$, which is just the geometric mean.

Arzela-Ascoli Theorem The theorem consists of two theorems:

Propagation Theorem. If $\{f_n(x)\}$ is an equicontinuous sequence of functions on $[a,b]$ such that $\lim_{n\to\infty} f_n(x)$ exists on a dense subset of $[a,b]$, then $\{f_n\}$ is uniformly convergent on $[a,b]$.

Selection Theorem. If $\{f_n(x)\}$ is a uniformly bounded, equicontinuous sequence on $[a,b]$, then there is a subsequence which is uniformly convergent on $[a,b]$.

associated radii of convergence Consider a power series in n complex variables: $\sum a_{i_1i_2...i_n} z_1^{i_1} z_2^{i_2} \ldots z_n^{i_n}$. Suppose $r_1, r_2, \ldots, r_n$ are such that the series converges for $|z_1| < r_1, |z_2| < r_2, \ldots, |z_n| < r_n$ and diverges for $|z_1| > r_1, |z_2| > r_2, \ldots, |z_n| > r_n$. Then $r_1, r_2, \ldots, r_n$ are called *associated radii of convergence.*

astroid A hypocycloid of four cusps, having the parametric equations

$$x = 4a\cos^3 t, y = 4a\sin^3 t.$$

$(-\pi \leq t \leq \pi)$. The Cartesian equation is

$$x^{\frac{2}{3}} + y^{\frac{2}{3}} = a^{\frac{2}{3}}.$$

asymptote For the graph of a function $y = f(x)$, either (i.) a *vertical asymptote:* a vertical line $x = a$, where $\lim_{x\to a} f(x) = \infty$; (ii.) a *horizontal asymptote:* a horizontal line $y = a$ such that $\lim_{x\to\infty} f(x) = a$; or (iii.)

an *oblique asymptote:* a line $y = mx + b$ such that $\lim_{x\to\infty}[f(x) - mx - b] = 0$.

asymptotic curve Given a regular surface M, an *asymptotic curve* is formally defined as a curve $\mathbf{x}(t)$ on M such that the normal curvature is 0 in the direction $\mathbf{x}'(t)$ for all t in the domain of $\mathbf{x}$. The differential equation for the parametric representation of an asymptotic curve is

$$eu'^2 + 2fu'v' + gv'^2 = 0,$$

where e, f, and g are second fundamental forms. The differential equation for asymptotic curves on a Monge patch $(u, v, h(u, v))$ is

$$h_{uu}u'^2 + 2h_{uu}u'v' + h_{vv}v'^2 = 0,$$

and on a polar patch $(r\cos\theta, 4\sin\theta, h(r))$ is

$$h''(r)r'^2 + h'(r)r\theta'^2 = 0.$$

asymptotic direction A unit vector X_p in the tangent space at a point p of a Riemannian manifold M such that $(S(X_p), X_p) = 0$, where S is the shape operator on $T_p(M)$: $S(X_p) = -(d\mathbf{N}/dt)_{t=0}$.

asymptotic expansion A divergent series, typically one of the form

$$\sum_{j=0}^{\infty} \frac{A_j}{z^j},$$

is an *asymptotic expansion* of a function $f(z)$ for a certain range of z, provided the remainder $R_n(z) = z^n[f(z) - s_n(z)]$, where $s_n(z)$ is the sum of the first $n + 1$ terms of the above divergent series, satisfies

$$\lim_{|z|\to\infty} R_n(z) = 0$$

(n fixed) although

$$\lim_{n\to\infty} |R_n(z)| = \infty$$

(z fixed).

asymptotic path A path is a continuous curve. *See also* asymptotic curve.

asymptotic power series *See* asymptotic series.

asymptotic rays Let M be a complete, open Riemannian manifold of dimension ≥ 2. A geodesic $\gamma : [0, \infty) \to M$, emanating from p and parameterized by arc length, is called a *ray emanating from p* if $d(\gamma(t), \gamma(s)) = |t - s|$, for $t, s \in [0, \infty)$. Two rays, γ, γ' are *asymptotic* if $d(\gamma(t), \gamma'(t)) \leq |t - s|$ for all $t \geq 0$.

asymptotic sequence Let R be a subset of $\mathbf{R}$ or $\mathbf{C}$ and c a limit point of R. A sequence of functions $\{f_j(z)\}$, defined on R, is called an *asymptotic sequence* or *scale* provided

$$f_{j+1}(z) = o(f_j(z))$$

as $z \to c$ in R, in which case we write the asymptotic series

$$f(z) \sim \sum_{j=0}^{\infty} a_j f_j(z) \qquad (z \to c, \text{ in } R)$$

for a function $f(z)$, whenever, for each n,

$$f(z) = \sum_{j=0}^{n-1} a_j f_j(z) + O(f_n(z)),$$

as $z \to c$ in R.

asymptotic series *See* asymptotic sequence.

asymptotic stability Given an autonomous differential system $y' = f(y)$, where $f(y)$ is defined on a set containing $y = 0$ and satisfies $f(0) = 0$, we say the solution $y \equiv 0$ is *asymptotically stable, in the sense of Lyapunov*, if

(i.) for every $\epsilon > 0$, there is a $\delta_\epsilon > 0$ such that, if $|y_0| < \delta_\epsilon$, then there is a solution $y(t)$ satisfying $y(0) = y_0$ and $|y(t)| < \epsilon$, for $t \geq 0$; and

(ii.) $y(t) \to 0$, as $t \to \infty$.

Whenever (i.) is satisfied, the solution $y \equiv 0$ is said to be *stable, in the sense of Lyapunov.*

asymptotic tangent line A direction of the tangent space $T_p(S)$ (where S is a regular surface and $p \in S$) for which the normal curvature is zero.

See also asymptotic curve, asymptotic path.

Atiyah-Singer Index Theorem A theorem which states that the analytic and topological indices are equal for any elliptic differential operator on an n-dimensional compact differentiable $\mathbf{C}^\infty$ boundaryless manifold.

atlas By definition, a topological space M is a differentiable [resp., $\mathbf{C}^\infty$, analytic] manifold if, for every point $p \in M$, there is a neighborhood U_p and a homeomorphism ϕ_p from U_p into $\mathbf{R}^n$. The neighborhood U_p or, sometimes, the pair (U_p, ϕ_p), is called a *chart.* Two charts U_p, U_q are required to be *compatible*; i.e., if $U_p \cap U_q \neq \emptyset$ then the functions $\phi_p \circ \phi_q^{-1}$ and $\phi_q \circ \phi_p^{-1}$ are differentiable [resp, $\mathbf{C}^\infty$, analytic]. The set of all charts is called an *atlas*. An atlas $\mathcal{A}$ is *complete* if it is maximal in the sense that if a pair U, ϕ is compatible with one of the U_p, ϕ_p in $\mathcal{A}$, then U belongs to $\mathcal{A}$.

In the case of a differentiable [resp., $\mathbf{C}^\infty$, analytic] manifold *with boundary,* the maps ϕ_p may map from U_p to either $\mathbf{R}^n$ or $\mathbf{R}^n_+ = \{(x_1, \dots, x_n) : x_j \geq 0, \text{ for } j = 1, \dots, n\}$.

atom For a measure μ on a set X, a point $x \in X$ such that $\mu(x) > 0$.

automorphic form Let G be a Kleinian group acting on a domain $D \subset \mathbf{C}$ and q a positive integer. A measurable function $\sigma : D \to \mathbf{C}$ is a *measurable automorphic form of weight* $-2q$ *for* G if

$$(\sigma \circ g)(g')^q = \sigma$$

almost everywhere on D, for all $g \in G$.

automorphic function A meromorphic function $f(z)$ satisfying $f(Tz) = f(z)$ for T belonging to some group of linear fractional transformations (that is, transformations of the form $Tz = (az+b)/(cz+d)$). When the linear fractional transformations come from a subgroup of the modular group, f is called a *modular* function.

autonomous linear system *See* autonomous system.

autonomous system A system of differential equations $\frac{d\mathbf{y}}{dt} = \mathbf{f}(\mathbf{y})$, where $\mathbf{y}$ and $\mathbf{f}$ are column vectors, and $\mathbf{f}$ is independent of t.

auxiliary circle Suppose a central conic has center of symmetry P and foci F and F', each at distance a from P. The circle of radius a, centered at P, is called the *auxiliary circle.*

axiom of continuity One of several axioms defining the real number system uniquely: Let $\{x_j\}$ be a sequence of real numbers such that $x_1 \leq x_2 \leq \dots$ and $x_j \leq M$ for some M and all j. Then there is a number $L \leq M$ such that $x_j \to L, j \to \infty$ and $x_j \leq L, j = 1, 2, \dots$.

This axiom, together with axioms determining addition, multiplication, and ordering serves to define the real numbers uniquely.

axis (1.) The Cartesian coordinates of a point in a plane are the directed distances of the point from a pair of intersecting lines, each of which is referred to as an axis. In three-dimensional space, the coordinates are the directed distances from coordinate planes; an *axis* is the intersection of a pair of coordinate planes.

(2.) If a curve is symmetric about a line, then that line is known as an *axis* of the curve. For example, an ellipse has two axes: the major axis, on which the foci lie, and a minor axis, perpendicular to the major axis through the center of the ellipse.

(3.) The *axis* of a surface is a line of sym-

metry for that surface. For example, the axis of a right circular conical surface is the line through the vertex and the center of the base. The axis of a circular cylinder is the line through the centers of the two bases.
(4.) In polar coordinates (r, θ), the polar axis is the ray that is the initial side of the angle θ.

axis of absolute convergence *See* abscissa of absolute convergence.

axis of convergence *See* abscissa of convergence.

axis of regularity *See* abscissa of regularity.

axis of rotation A surface of revolution is obtained by rotating a curve in the plane about a line in the plane that has the curve on one side of it. This line is referred to as the *axis of rotation* of the surface.

B

Baire σ-algebra The smallest σ-algebra on a compact Hausdorff space X making all the functions in $\mathbf{C}(X)$ measurable. The sets belonging to the Baire σ-algebra are called the *Baire subsets* of X.

Baire Category Theorem A nonempty, complete metric space is of the second category. That is, it cannot be written as the countable union of nowhere dense subsets.

Baire function A function that is measurable with respect to the ring of Baire sets. Also *Baire measurable function*.

Baire measurable function *See* Baire function.

Baire measure A measure on a Hausdorff space X, for which all the Baire subsets of X are measurable and which is finite on the compact G_δ sets.

Baire property A subset A of a topological space has the *Baire property* if there is a set B of the first category such that $(A\backslash B) \cup (B\backslash A)$ is open.

Baire set *See* Baire σ-algebra.

balanced set A subset M of a vector space V over $\mathbf{R}$ or $\mathbf{C}$ such that $\alpha x \in M$, whenever $x \in M$ and $|\alpha| \leq 1$.

Banach algebra A vector space B, over the complex numbers, with a multiplication defined and satisfying (for $x, y, z \in B$)
(i.) $x \cdot y = y \cdot x$;
(ii.) $x \cdot (y \cdot z) = (x \cdot y) \cdot z$;
(iii.) $x \cdot (y + z) = x \cdot y + x \cdot z$;
and, in addition, with a norm $\|\cdot\|$ making B into a Banach space and satisfying
(iv.) $\|x \cdot y\| \leq \|x\|\|y\|$, for $x, y \in B$.

Banach analytic space A Banach space of analytic functions. (*See* Banach space.) Examples are the Hardy spaces. *See* Hardy space.

Banach area Let $T : A \to \mathbf{R}^3$ be a continuous mapping defining a surface in $\mathbf{R}^3$ and let K be a polygonal domain in A. Let P_0 be the projection of $\mathbf{R}^3$ onto a plane E and let m denote Lebesgue measure on $PT(K)$. The *Banach area* of $T(A)$ is

$$\sup_S \sum_{K \in S} [m^2(A_1) + m^2(A_2) + m^2(A_3)]$$

where A_j are the projections of K onto coordinate planes in $\mathbf{R}^3$ and S is a finite collection of non-overlapping polygonal domains in A.

Banach manifold A topological space M such that every point has a neighborhood which is homeomorphic to the open unit ball in a Banach space.

Banach space A complete normed vector space. That is, a vector space X, over a scalar field ($\mathbf{R}$ or $\mathbf{C}$) with a nonnegative real valued function $\|\cdot\|$ defined on X, satisfying (i.) $\|cx\| = |c|\|x\|$, for c a scalar and $x \in X$; (ii.) $\|x\| = 0$ only if $x = 0$, for $x \in X$; and (iii.) $\|x + y\| \leq \|x\| + \|y\|$, for $x, y \in X$. In addition, with the metric $d(x, y) = \|x - y\|$, X is assumed to be complete.

Banach-Steinhaus Theorem Let X be a Banach space, Y a normed linear space and $\{\Lambda_\alpha : X \to Y\}$, a family of bounded linear mappings, for $\alpha \in A$. Then, either there is a constant $M < \infty$ such that $\|\Lambda_\alpha\| \leq M$, for all $\alpha \in A$, or $\sup_{\alpha \in A} \|\Lambda_\alpha x\| = \infty$, for all x in some subset $S \subset X$, which is a dense G_δ.

Barnes's extended hypergeometric function Let $G(a, b; c; z)$ denote the sum of the hypergeometric series, convergent for

0-8493-0320-6/00/$0.00+$.50

$|z| < 1$:

$$\sum_{j=0}^{\infty} \frac{\Gamma(a+j)\Gamma(b+j)}{\Gamma(c+j)j!} z^j,$$

which is the usual hypergeometric function $F(a,b;c;z)$ divided by the constant $\Gamma(c)/[\Gamma(a)\Gamma(b)]$. Barnes showed that, if $|\arg(-z)| < \pi$ and the path of integration is curved so as to lie on the right of the poles of $\Gamma(a+\zeta)\Gamma(b+\zeta)$ and on the left of the poles of $\Gamma(-\zeta)$, then

$$G(a,b;c;z) =$$

$$\frac{1}{2\pi i} \int_{-\pi i}^{\pi i} \frac{\Gamma(a+\zeta)\Gamma(b+\zeta)\Gamma(-\zeta)}{\Gamma(c+\zeta)} (-z)^{\zeta} d\zeta,$$

thus permitting an analytic continuation of $F(a,b;c;z)$ into $|z| > 1$, $\arg(-z) < \pi$.

barrel A convex, balanced, absorbing subset of a locally convex topological vector space. *See* balanced set, absorbing.

barrel space A locally convex topological vector space, in which every barrel is a neighborhood of 0. *See* barrel.

barrier *See* branch.

barycentric coordinates Let $p_0, p_1, \ldots, p_n$ denote points in $\mathbf{R}^n$, such that $\{p_j - p_0\}$ are linearly independent. Express a point $P = (a_1, a_2, \ldots, a_n)$ in $\mathbf{R}^n$ as

$$P = \sum_{j=0}^{n} \mu_j p_j$$

where $\sum_0^n \mu_j = 1$ (this can be done by expressing P as a linear combination of $p_1 - p_0, p_2 - p_0, \ldots, p_n - p_0$). The numbers $\mu_0, \mu_1, \ldots, \mu_n$ are called the *barycentric coordinates* of the point P. The point of the terminology is that, if $\{\mu_0, \ldots, \mu_n\}$ are nonnegative weights of total mass 1, assigned to the points $\{p_0, \ldots, p_n\}$, then the point $P = \sum_0^n \mu_j p_j$ is the center of mass or *barycenter* of the $\{p_j\}$.

basic vector field Let M, N be Riemannian manifolds and $\pi : M \to N$ a Riemannian submersion. A horizontal vector field X on M is called *basic* if there exists a vector field $\hat{X}$ on N such that $D\pi(p)X_p = \hat{X}_{\pi(p)}$, for $p \in M$.

basis A finite set $\{x_1, \ldots, x_n\}$, in a vector space V such that (i.) $\{x_j\}$ is linearly independent, that is, $\sum_{j=1}^n c_j x_j = 0$ only if $c_1 = c_2 = \ldots = c_n = 0$, and (ii.) every vector $v \in V$ can be written as a linear combination $v = \sum_{j=1}^n c_j x_j$.

An *infinite* set $\{x_j\}$ satisfying (i.) (for every n) and (ii.) (for some n) is called a *Hamel basis.*

BDF *See* Brown-Douglas-Fillmore Theorem.

Bell numbers The number of ways a set of n elements can be partitioned into nonempty subsets, denoted B_n. For example, there are five ways the numbers $\{1, 2, 3\}$ can be partitioned: $\{\{1\}, \{2\}, \{3\}\}$, $\{\{1, 2\}, \{3\}\}$, $\{\{1, 3\}, \{2\}\}$, $\{\{1\}, \{2, 3\}\}$, and $\{\{1, 2, 3\}\}$, so $B_3 = 5$. $B_0 = 1$ and the first few Bell numbers for $n = 1, 2, \ldots$ are 1, 2, 5, 15, 52, 203, 877, 4140, 21147, 115975,.... *Bell numbers* are closely related to Catalan numbers.

The integers B_n can be defined by the sum

$$B_n = \sum_{k=1}^{n} S(n,k),$$

where $S(n,k)$ is a Stirling number of the second kind, or by the generating function

$$e^{e^x - 1} = \sum_{n=0}^{\infty} \frac{B_n}{n!} x^n.$$

Beltrami equation The equation $D_\ell f = 0$. *See* Beltrami operator.

Beltrami operator Given by

$$D_\ell = \sum_{i=1}^{\ell} x_i^2 \frac{\partial^2}{\partial x_i^2} + \sum_{i \neq j} \frac{x_j^2}{x_i - x_j} \frac{\partial}{\partial x_j}.$$

The Beltrami operator appears in the expansions in many distributions of statistics based on normal populations.

Bergman metric The distance function, on a domain $\Omega \subset \mathbf{C}^n$, defined by

$$\delta(z_1, z_2) = \inf_{\{\gamma:\gamma(0)=z_1,\gamma(1)=z_2\}} |\gamma|_{B(\Omega)},$$

where, for a $\mathbf{C}^1$ curve $\gamma : [0, 1] \to \Omega$,

$$|\gamma|_{B(\Omega)} = \int_0^1 \Big(\sum_{i,j} g_{i,j}(\gamma(t))\gamma_i'(t)\overline{\gamma_j'(t)}\Big)^{1/2} dt,$$

and where

$$g_{ij}(z) = \frac{\partial^2}{\partial z_i \partial \bar{z}_j} \log K(z, z),$$

for K the Bergman kernel. *See* Bergman's kernel function.

Bergman space For a domain $\Omega \subset \mathbf{C}^n$, the class of functions

$$A^2(\Omega) = \{f(z) : f \text{ holomorphic in } \Omega \text{ and } \int_\Omega |f(z)|^2 dV(z) < \infty\},$$

where dV is volume on $\mathbf{C}^n$.

Bergman's kernel function For a domain $\Omega \subset \mathbf{C}^n$, the function $K(z, \zeta)$ satisfying

$$f(z) = \int_\Omega K(z, \zeta) f(\zeta) dV(\zeta),$$

for all f belonging to the Bergman space $A^2(\Omega)$. *See* Bergman space.

Bernoulli numbers The numbers $\{B_j, j = 1, 2, \ldots\}$ in the Taylor expansion in $|z| < 2\pi$:

$$\frac{1}{2} z \cot \frac{1}{2} z = 1 - \sum_{j=1}^\infty B_j \frac{z^{2j}}{(2j)!}.$$

The first few values are: $B_1 = \frac{1}{6}, B_2 = \frac{1}{30}, B_3 = \frac{1}{42}, B_4 = \frac{1}{30}, B_5 = \frac{5}{66}$, etc.

Bernoulli polynomials The polynomials $\{\phi_j(x)\}$ appearing as coefficients in the Taylor expansion in $|z| < 2\pi$:

$$z\frac{e^{xz} - 1}{e^z - 1} = \sum_{j=1}^\infty \frac{\phi_j(x) z^j}{j!}.$$

Bernstein's Theorem Let $f \in \mathrm{Lip}_\alpha(\mathbf{T})$ ($\mathbf{T}$ the unit circle) for some $\alpha > \frac{1}{2}$. Then $f \in A(\mathbf{T})$, the space of functions having absolutely convergent Fourier series on $\mathbf{T}$ and

$$\|f\|_{A(\mathbf{T})} = \sum_{-\infty}^{\infty} |\hat{f}| \le c_\alpha \|f\|_{Lip_\alpha}.$$

Bertrand's curves The family of skew curves whose principal normals are the principal normals of a given skew curve.

Besov space The classes B^s_{pq} ($1 \le p, q \le \infty, s \in \mathbf{R}$) of functions $f(z) = f(re^{it})$, analytic in the unit disk, such that

$$\int_0^1 (1-r)^{nq-sq-1} \Big[\int_0^{2\pi} |f^{(n)}(re^{it})|^p dt\Big]^{q/p} dr < \infty$$

if $q < \infty$ and

$$\sup_{0<r<1} (1-r)^{n-1} \Big[\int_0^{2\pi} |f^{(n)}(re^{it})|^p dt\Big]^{1/p} < \infty$$

if $q = \infty$, where n is an integer such that $n > s$.

Bessel function For n a nonnegative integer, the function

$$J_n(x) = \sum_{j=0}^\infty \frac{(-1)^j x^{n+2j}}{2^{n+2j} j!(n+j)!}.$$

Bessel's inequality (1.) The inequality

$$\frac{1}{2} a_0 + \sum_{j=1}^\infty (a_j^2 + b_j^2) \le \frac{1}{\pi} \int_0^{2\pi} f(x)^2 dx,$$

where $\{a_j, b_j\}$ are the Fourier coefficients of the square-integrable function f:

$$a_j = \frac{1}{\pi} \int_0^{2\pi} f(x) \cos jx dx (j = 0, 1, \ldots),$$

and

$$b_k = \frac{1}{\pi} \int_0^{2\pi} f(x) \sin kx dx (k = 1, 2, \ldots).$$

(2.) The inequality

$$\sum_{\alpha\in A} |(x, u_\alpha)|^2 \le \|x\|^2,$$

where $\{u_\alpha, \alpha \in A\}$ is an orthonormal set in a Hilbert space H and x is an element of H.

beta function The function

$$B(m,n) = \int_0^1 x^{m-1}(1-x)^{n-1}dx.$$

The *beta function* satisfies

$$B(m,n) = B(n,m) = \frac{\Gamma(m)\Gamma(n)}{\Gamma(m+n)},$$

for m and n positive real numbers. *See* gamma function.

Beurling's Theorem Every closed subspace M of the Hardy class $H^2(\mathbf{T})$ of the unit disk $\mathbf{T}$ which is *invariant* ($zf \in M$, for all $f \in M$) and *nontrivial* ($M \neq \{0\}$ and $M \neq H^2$) has the form $M = BH^2 = \{Bf : f \in H^2\}$, where B is an *inner function* ($B(z)$ is analytic for $|z| < 1$ and $|B(z)| = 1$, almost everywhere on $\mathbf{T}$).

Generalizations of the theorem include characterizations of the invariant subspaces of $L^2(\mathbf{T})$ and $H^2_{\mathcal{C}}(\mathbf{T})$, based upon a separable Hilbert space $\mathcal{C}$. In both these cases, it is necessary to take into consideration *reducing* subspaces of the operator of multiplication by z. Both generalizations are referred to as the *Lax-Halmos Theorem* or *Beurling-Lax-Halmos Theorem.*

Bianchi's identities For vector fields X, Y, Z on a Riemannian manifold M, let $R(X, Y)Z$ denote the vector field

$$R(X,Y)Z = \nabla_X\nabla_Y Z - \nabla_Y\nabla_X Z - \nabla_{[X,Y]}Z$$

where ∇_X is the covariant derivative. The *Bianchi identities* are

(i.) $R(X, Y)Z+R(Y, Z)X+R(Z, X)Y = 0$, and

(ii.) $(\nabla_X R)(Y,, Z)U + (\nabla_Y R)(Z, X)U + (\nabla_Z R)(X, Y)U = 0$.

bidual The dual of the dual space. *See* dual space.

bifurcation equation Given an equation $G(\lambda, u) = 0$, where $G : \Lambda \times \mathcal{E} \to \mathcal{F}$, with $\mathcal{E}$ and $\mathcal{F}$ Banach spaces and Λ a parameter space with a bifurcation point (λ_0, u_0) (*see* bifurcation point); an associated finite dimensional equation, having the same solutions as $G(\lambda, u) = 0$, near the point (λ_0, u_0).

For example, suppose $\lambda_0 = 0$ and $u_0(0) = 0$. Let $L_0 = G_u(0, 0)$ (Fréchet derivative) have a kernel $\mathcal{K}$ of dimension n. Let P be the projection on $\mathcal{K}$ commuting with L_0 and let $Q = I - P$. From the equation $G(\lambda, u) = 0$ we obtain the system

$$QG(\lambda, v+w) = 0, PG(\lambda, v+w) = 0$$

where $v = Pu$ and $w = Qu$. If $w = w(\lambda, v)$ is a solution of the first of these equations (existing by the Implicit Function Theorem for Banach spaces), then, close to the bifurcation point (λ_0, u_0), solutions of the original equation $G(\lambda, u) = 0$ are in one-to-one correspondence with those of the *bifurcation equation*

$$PG(\lambda, v + w(\lambda, v)) = 0.$$

bifurcation point Let an equation $G(\lambda, u) = 0$ be given, where

$G : \Lambda \times \mathcal{E} \to \mathcal{F}$, with $\mathcal{E}$ and $\mathcal{F}$ Banach spaces and Λ a parameter space. Then a point $(\lambda_0, u_0) \in \Lambda\times\mathcal{E}$ is a *bifurcation point* or *branch point* if, for some smooth curve γ lying in the solution set of the equation and passing through (λ_0, u_0), there is a neighborhood U of (λ_0, u_0) in $\Lambda \times \mathcal{E}$ with $U\backslash\gamma$ intersecting the solution set.

bifurcation theory The study of branch points of nonlinear equations; that is, the study of singular points of equations, where more than one solution comes together. *See* bifurcation point.

biharmonic function A function $u(x, y)$ of class $\mathbf{C}^4$ in a domain $D \subseteq \mathbf{R}^2$ satisfying

$$\nabla^4 u = \frac{\partial^4 u}{\partial x^4} + 2\frac{\partial^4 u}{\partial x^2 \partial y^2} + \frac{\partial^4 u}{\partial y^4} = 0.$$

bilateral Laplace transform Suppose $f \in L^1(-\infty, \infty)$. Then the function g defined by $g(\lambda) = \int_{-\infty}^{\infty} e^{-\lambda t} f(t)dt$ is the *bilateral Laplace transform* of f.

bilateral shift The operator S defined on the Hilbert space l^2 of all square-summable, bi-infinite sequences

$$l^2 = \left\{\{x_j\}_{j=-\infty}^{\infty} : \sum_{-\infty}^{\infty} |x_j|^2 < \infty\right\},$$

by $S\{x_j\} = \{y_j\}$, where $y_j = x_{j-1}$.

See also shift operator.

bilinear form (1.) A function $a : H \times H \to F$, where H is a vector space over a field F, such that $a(c_1x + c_2y, z) = c_1a(x, z) + c_2a(y, z)$ and $a(x, c_1y + c_2z) = c_1a(x, y) + c_2a(x, z)$. Also called *bilinear functional. See also* sesquilinear form.
(2.) A *linear fractional transformation. See* linear fractional transformation.

bilinear mapping (1.) A function $L : V \times V \to W$, where V and W are vector spaces, satisfying

$$L(c_1v_1 + c_2v_2, w) = c_1L(v_1, w) + c_2L(v_2, w)$$

and

$$L(v, c_1w_1 + c_2w_2) = c_1L(v, w_1) + c_2L(v, w_2).$$

(2.) A linear fractional transformation $w = (az + b)/(cz + d)$, of the complex plane to itself, where a, b, c, d are complex numbers (usually with $ad - bc \neq 0$).

Binet's formula Either of the two relations, valid when $\Re z > 0$,

$$\log\Gamma(z) = \left(z - \tfrac{1}{2}\right)\log z - z + \frac{1}{2}\log(2\pi) + \int_0^{\infty}\left(\frac{1}{2} - \frac{1}{t} + \frac{1}{e^t - 1}\right)\frac{e^{-tz}}{t}dt$$

or

$$\log\Gamma(z) = \left(z - \tfrac{1}{2}\right)\log z - z + \frac{1}{2}\log(2\pi) + 2\int_0^{\infty}\frac{\arctan(t/z)}{e^{2\pi t} - 1}dt.$$

binomial coefficient series Given by $\sum_{n=0}^{\infty}\binom{a}{n}$, where for any real number a, $\binom{a}{0} = 1$ and

$$\binom{a}{n} = \frac{a(a-1)(a-2)\cdots(a-n+1)}{n!}$$

$(n \geq 1)$.

binomial series The power series

$$(1 + x)^n = 1 + \sum_{j=1}^{\infty}\frac{n(n-1)\cdots(n-j+1)}{j!}x^j,$$

which converges for $|x| < 1$, unless n is a positive integer, in which case it reduces to a finite sum.

binormal The normal component of the derivative of the principal normal. Let $f : \mathbf{R} \to \mathbf{R}^3$ be a smooth function, with tangent vector $\mathbf{T} = \frac{df}{ds}$. Write $\frac{d\mathbf{T}}{ds} = \kappa\mathbf{N}$ and $\frac{d\mathbf{N}}{ds} = \kappa_2\mathbf{N}_1 + \beta\mathbf{T}$. The unit vector $\mathbf{N}_1$ is called the *binormal*.

bipolar coordinates A two-dimensional system of coordinates, two types of which are commonly defined. The first type is defined by

$$x = \frac{a\sinh v}{\cosh v - \cos u} \qquad y = \frac{a\sin u}{\cosh v - \cos u},$$

where $u \in [0, 2\pi)$, $v \in (-\infty, \infty)$. The following identities show that curves of constant u and v are circles in xy-space.

$$x^2 + (y - a\cot u)^2 = a^2\csc^2 u$$
$$(x - a\coth v)^2 + y^2 = a^2\operatorname{csch}^2 v.$$

The scale factors are

$$h_u = \frac{a}{\cosh v - \cos u}$$

$$h_v = \frac{a}{\cosh v - \cos u}.$$

The Laplacian in bipolar coordinates is

$$\nabla^2 = \frac{(\cosh v - \cos u)^2}{a^2}\left(\frac{\partial^2}{\partial u^2} + \frac{\partial^2}{\partial v^2}\right).$$

Laplace's equation is separable.

Two-center bipolar coordinates are two coordinates giving the distances from two fixed centers r_1 and r_2, sometimes denoted r and r'. For two-center bipolar coordinates with centers at $(\pm c, 0)$,

$$r_1^2 = (x+c)^2 + y^2 \quad r_2^2 = (x-c)^2 + y^2.$$

Combining, we get

$$r_1^2 - r_2^2 = 4cx.$$

Solving for Cartesian coordinates x and y gives

$$x = \frac{r_1^2 - r_2^2}{4c}$$

$$y = \pm\frac{1}{4c}\sqrt{16c^2r_1^2 - (r_1^2 - r_2^2 + 4c^2)}.$$

Solving for polar coordinates gives

$$r = \sqrt{\frac{r_1^2 + r_2^2 - 2c^2}{2}}$$

$$\theta = \tan^{-1}\left[\sqrt{\frac{8c^2(r_1^2 + r_2^2 - 2c^2)}{r_1^2 - r_2^2}} - 1\right].$$

birational transformation Two curves are *birationally equivalent* if their rational function fields are equal. So if $\mathbf{C}(\xi_1, \ldots, \xi_r) = \mathbf{C}(\eta_1, \ldots, \eta_s)$ are the rational function fields for two curves with generic points $(\xi_1, \ldots, \xi_r)$ and $(\eta_1, \ldots, \eta_s)$, respectively, then a birational map between the two curves is $y_j = f_j(x_1, \ldots, x_r)$, $(j = 1, \ldots, s)$ and $x_k = g_k(y_1, \ldots, y_s)$, $(k = 1, \ldots, r)$, where $\eta_j = f_j(\xi_1, \ldots, \xi_r)$, $(j = 1, \ldots, s)$ and $\xi_k = g_k(\eta_1, \ldots, \eta_s)$, $(k = 1, \ldots, r)$, as elements of the rational function field.

BKW method Brillouin-Kramers-Wentzel method. *See* WKB method.

Blaschke product An infinite product of the form

$$B(z) = \prod_{j=1}^{\infty} \frac{\bar{a}_j}{|a_j|}\frac{a_j - z}{1 - \bar{a}_j z},$$

where $\{a_j\}$ is a *Blaschke sequence*, that is,

$$\sum_{j=1}^{\infty}(1 - |a_j|) < \infty.$$

The product converges uniformly in compact subsets of the unit disk $D = \{z : |z| < 1\}$ and therefore shows that, for any Blaschke sequence, there is a function, bounded and analytic in D, having its zeros exactly at those points. The Blaschke product also has the property that $\lim_{r\to 1-} B(re^{it})$ exists and has modulus 1, for almost every $t \in [0, 2\pi]$.

Often the term *Blaschke product* also includes finite products of the above form.

Blaschke sequence An infinite sequence $\{a_j\}$ in the unit disk $D = \{z : |z| < 1\}$ in the complex plane, satisfying

$$\sum_{j=1}^{\infty}(1 - |a_j|) < \infty.$$

See also Blaschke product.

blowing up Let N be an n-dimensional compact, complex manifold ($n \geq 2$), and $p \in N$. Let $\{z = (z_i)\}$ be a local coordinate system, in a neighborhood U, centered at p and define

$$\tilde{U} = \{(z, l) \in U \times P^{n-1} : z \in l\},$$

where P^{n-1} is regarded as a set of lines l in $\mathbf{C}^n$. Let $\pi : \tilde{U} \to U$ denote the projection $\pi(z, l) = z$. Identify $\pi^{-1}(p)$ with P^{n-1} and $\tilde{U}\backslash\pi^{-1}(p)$ with $U\backslash\{p\}$, via the map π and set

$$\tilde{N} = (N \backslash \{p\}) \cup \tilde{U},\ B_p(N) = \tilde{N}/\sim,$$

where $z \sim w$ if $z \in N\backslash\{p\}$ and $w = (z, l) \in \tilde{U}$. The *blowing up* of N at p is $\pi : B_p(N) \to N$. *See also* monoidal transformation.

BMO *See* bounded mean oscillation.

BMOA *See* bounded mean oscillation.

Bochner's Theorem A complex-valued function $f(t)$ on **R** has a representation

$$f(t) = \int_{-\infty}^{\infty} e^{it\lambda} dv(\lambda),$$

with $v(\lambda)$ non-decreasing, continuous from the right and bounded, if and only if $f(t)$ is positive-definite in the sense that

$$\int_{-\infty}^{\infty}\int_{-\infty}^{\infty} f(t-s)u(t)\overline{u(s)}dtds \geq 0$$

for every continuous function u with compact support.

Bohr compactification For a locally compact, Abelian group G, the dual group of G_d, the group G with the discrete topology.

Bonnet's Fundamental Theorem *See* Second Mean Value Theorem.

Borel function A function that is measurable with respect to the σ-algebra of Borel sets. Also, *Borel measurable function.*

Borel isomorphism Suppose $(X, \mathcal{B}(X))$, $(Y, \mathcal{B}(Y))$ are measurable, polish spaces, where $\mathcal{B}(X), \mathcal{B}(Y)$ are the σ-algebras of Borel functions on X and Y, respectively, and let $f : X \to Y$ be a bijection. If f and f^{-1} both map Borel sets to Borel sets, then f is called a Borel isomorphism. *See* Borel set.

Borel measurable function *See* Borel function.

Borel measure A measure on a topological space X, for which all the Borel subsets of X are measurable and which assigns a finite value to every compact set.

Borel set Beginning with the open sets in a topological space, generate a σ-field (closed under complementation and countable unions). The elements of this σ-field are called the *Borel sets.*

Borel's Theorem For each n-tuple $\alpha = (\alpha_1, \ldots, \alpha_n\}$ of nonnegative integers, let there be given a real number C_α. Then there is a $\mathbf{C}^\infty$-function $f : \mathbf{R}^n \to \mathbf{R}$, with the $\{C_\alpha\}$ as its Taylor coefficients about the origin; i.e.,

$$C_\alpha = \frac{1}{\alpha!}\frac{\partial^{\alpha_1}}{\partial x_1^{\alpha_1}} \cdots \frac{\partial^{\alpha_n}}{\partial x_n^{\alpha_n}} f(0).$$

bornologic space A locally convex, topological vector space X such that a balanced, convex set $M \subset X$ which absorbs every bounded set of X is necessarily a neighborhod of 0 in X. *See* bounded set (for a topological vector space).

bound *See* bounded set, bounded function, greatest lower bound, least upper bound.

bound of function *See* bounded function.

boundary The set $\text{cl}(E)\backslash\text{int}(E)$, for a set E in a topological space. Here $\text{cl}(E)$ denotes the closure and $\text{int}(E)$ denotes the interior of E. *See also* Shilov boundary.

boundary function *See* boundary value.

boundary value **(1.)** The value, on the boundary $\partial\Omega$ of a region $\Omega \subset \mathbf{R}^n$, of a function $f(x_1, \ldots, x_n)$ defined and continuous in the closure of Ω. *See* boundary value problem.

(2.) A value on the boundary $\partial\Omega$ of a region $\Omega \subset \mathbf{R}^n$, which can be assigned, through some limit process, to a function, which is originally defined only in the interior or Ω. For example, a function $f(z)$, analytic and bounded in $\Omega = \{z : |z| < 1\}$, has boundary values

$$f(e^{it}) = \lim_{r \to 1-} f(re^{it}),$$

which exist almost everywhere on $\partial\Omega$.

Occasionally, the limit on the boundary is something more general than a function (a distribution, for example). The term *boundary function* may be used for emphasis, when applicable.

boundary value problem A problem in which an unknown function u is desired, satisfying a partial differential equation $P(D)u = 0$ on the interior of some set $\Omega \subset \mathbf{R}^n$ and taking a given value (the *boundary value*) $u(t) = f(t)$ for $t \in \partial\Omega$.

bounded (set) from above A subset S of a partially ordered set X such that $s \leq x$, for some $x \in X$ and for all $s \in S$.

bounded (set) from below A subset S of a partially ordered set X such that $s \geq x$, for some $x \in X$ and for all $s \in S$.

bounded domain A connected, open set D in $\mathbf{R}^n$ or $\mathbf{C}^n$ which is bounded: $|x| \leq C$, for all $x \in D$ and for some real number C.

bounded function (**1.**) A function $f : S \to \mathbf{R}$ or $f : S \to \mathbf{C}$, for some set S, such that $|f(x)| \leq C$, for all $x \in S$ and for some real number C (called a *bound* of f).
(**2.**) More generally, a function with its image contained in a bounded set. *See* bounded set.

bounded linear operator A mapping $T : X \to Y$, where X and Y are normed linear spaces with norms $\|\cdot\|_X$ and $\|\cdot\|_Y$, respectively, which is linear:

$$T(c_1x_1 + c_2x_2) = c_1T(x_1) + c_2T(x_2)$$

for c_1, c_2 scalars and $x_1, x_2 \in X$, and also satisfies $\|Tx\|_Y \leq C\|x\|_X$, for all $x \in X$ and for some real number C.

The *norm* of T is the smallest such constant C. For a linear operator T, the norm condition is equivalent to continuity.

bounded mean oscillation A locally integrable function $f(t) : [0, 2\pi] \to \mathbf{C}$ is *of bounded mean oscillation*, if

$$\frac{1}{|I|}\int_I |f(t) - \frac{1}{|I|}\int_I f(s)ds|dt$$

is bounded, for all intervals I.

The notation BMO is used for the class of such functions. BMOA refers to the functions of class BMO that are boundary values of functions analytic in $|z| < 1$.

bounded set (**1.**) In $\mathbf{R}^n$ or $\mathbf{C}^n$ a set E such that $|x| \leq C$, for all $x \in E$ and for some constant C (called a *bound* of E).
(**2.**) In a metric space S, a set E such that $d(x, x_0) \leq C$ for all $x \in E$, for some $x_0 \in S$, and for some constant C.
(**3.**) In a locally compact space, a subset of a compact set.
(**4.**) In a topological vector space, a set E which is absorbed by any neighborhood U of 0, i.e., $E \subset \alpha U$, for some positive constant α.
(**5.**) In a topological group, a set E such that, for every neighborhood U of the identity, there is a finite set $\{x_1, \ldots, x_n\}$, such that $E \subset \cup_{j=1}^n x_jU$.

brachistochrone A wire is bent into a planar curve from point A to point B, below. The *brachistochrone* problem is to determine the shape of the curve that will result in the shortest time of descent. The problem leads to a differential equation of the form $y[1+(y')^2] = c$, where $y = y(x)$ is the curve sought. Solving by separation of variables, yields the solution $x = a(\theta - \sin\theta)$, $y = a(1 - \cos\theta)$, which are the parametric equations of a cycloid. The word *brachistochrone* derives from the Greek (*brachistos* = shortest + *chronos* = time).

branch Let D_1 and D_2 be disjoint, open, connected sets in the complex plane $\mathbf{C}$. Suppose a function $f(z)$, analytic in D_1, can be continued analytically along a curve, terminating in D_2, and that the continuation leads to a function $f_1(z)$, analytic in D_2. Then f_1 is called a *branch* of f. The terminology is used when different curves lead to more than one analytic function in D_2, so that f may have several distinct branches in D_2.

For example, the function $f(z) = \log(z) = \sum_{j=1}^{\infty}(-1)^j(z-1)^j/j$ is analytic in $D_1 = \{z : |z-1| < 1\}$ and continuation around the two arcs of the circle $z = \pm e^{it}, 0 \leq t \leq \pi$, lead to two branches $f_1(z)$ and $f_2(z)$, analytic in $D_2 = \{z : |z+1| < 1\}$, one satisfying $f_1(-1) = \pi$ and the other satisfying $f_2(-1) = -\pi$.

When two curves around a singularity lead to different branches, the singularity is called a *branch point* or *algebraic singularity.* In order to remove the multi-valued character of an analytic function caused by a branch point, a curve terminating in that point may be removed from the plane. For example, for the logarithm function, the positive imaginary axis might be removed. Such a removed curve is called a *cut* or *barrier.*

branch point (1.) A branch point of an analytic function. *See* branch.
(2.) A branch point of an equation. *See* bifurcation point.

Branges' Theorem [deBranges' Theorem] Let $f(z)$ be univalent in the unit disk and have power series

$$f(z) = z + a_2z^2 + a_3z^3 + \ldots$$

then $|a_n| \leq n$. Equality occurs for the Koebe function $f(z) = z(1-z)^{-2}$. This had been a famous conjecture known as the Bieberbach conjecture until proved by Louis deBranges.

Brianchon's Theorem The dual of Pascal's Theorem, stating that if the sides of a hexagon are tangent to a conic, then the lines joining opposite vertices are concurrent.

Bromwich integral Any contour integral along the the vertical line $\Re(z) = c > 0$ in the complex plane, oriented upward. A particular example is

$$f(w) = \frac{1}{2\pi i}\int_B g(z)e^{zw}dz,$$

where B is the above contour, which is the inversion formula for the Laplace transform.

Brown-Douglas-Fillmore Theorem Let T_1 and T_2 be bounded linear operators on a Hilbert space H. Suppose that $T_1T_2 - T_2T_1$ is compact, T_1 and T_2 have the same essential spectrum $A \subset \mathbf{C}$ and $T_1 - \lambda I$ and $T_2 - \lambda I$ have the same index, for $\lambda \notin A$. Then T_1 is unitarily equivalent to a compact perturbation of T_2.

Actually, an additional hypothesis on the set A is required (A must be homeomorphic to a subset of a 1-complex).

bundle *See* tangent bundle.

bundle of *i*-forms *See* algebra of differential forms.

Busemann function Let M be a complete, open Riemannian manifold of dimension ≥ 2. A geodesic $\gamma : [0, \infty) \to M$, emanating from p and parameterized by arc length, is called a *ray eminating from p* if $d(\gamma(t), \gamma(s)) = |t - s|$, for $t, s \in [0, \infty)$. The *Busemann function b*, corresponding to the ray γ, is

$$b_\gamma(q) = \lim_{t\to\infty}[t - d(q, \gamma(t))].$$

The limit exists because the expression in brackets is monotonically increasing and bounded.

C

calculus (1.) The study of properties of functions of one or several variables, using derivatives and integrals. *Differential calculus* usually refers to the one variable study of the derivative and its applications and *integral calculus* to the study of the Riemann integral of a function of one variable. Classically, it has been referred to as *the* calculus. (2.) Any system of computations based upon some unifying idea, such as the *calculus of residues* or *calculus of variations.*

calculus of differential forms The calculus of Grassmann bundles. *See also* tangent bundle, algebra of differential forms, exterior algebra, tensor field, formal vector field on a manifold.

calculus of residues The evaluation of definite integrals using the Residue Theorem. *See* Residue Theorem.

calculus of variations The study of certain problems of minimization of integrals. Typical is the *fundamental problem of the calculus of variations:* Given a function of three variables, $F(x, y, z)$, and two real numbers a, b, to find a differentiable function $y = y(x)$ which minimizes the integral

$$I = \int_a^b F(x, y, \frac{dy}{dx})dx.$$

Calderón-Zygmund kernel A function $K(x)$, defined on $\mathbf{R}^n$, satisfying
(i.) $|K(x)| \leq C/|x|^n$;
(ii.) $\int_{\alpha<|x|<\beta} K(x)dx = 0$, whenever $0 < \alpha < \beta$;
(iii.) $\int_{|x|>2|h|} |K(x+h) - K(x)|dx \leq C$ for all $h \neq 0$.
The convolution operator $Hf = f * K$ generalizes the Hilbert transform (the case $n = 1$, $K(x) = 1/x$) and Calderón and Zygmund proved that H is bounded on $L^p(\mathbf{R}^n)$, for $1 < p < \infty$.

Calkin algebra The quotient of the algebra of all operators on a Hilbert space by the ideal of compact operators.

canonical 1-form A left invariant $\mathcal{G}$-valued 1-form α, uniquely defined by $\alpha(X) = X$, for $X \in G$, where G is a complex Lie group and $\mathcal{G}$ its complex Lie algebra.

canonical affine connection A system of n^3 smooth real functions on an n-dimensional manifold which may be chosen arbitrarily in one canonical coordinate frame.

canonical bilinear mapping The mapping from the Cartesian product $E \times F$ of two vector spaces to their tensor product, $E \otimes F$, sending (x, y) to $x \otimes y$. *See also* canonical transformation.

canonical coordinate system Let $n \geq 1$. Consider the vector space K^n, over a field K, and let $E_1 = (1, 0, \ldots, 0)$, $E_2 = (0, 1, \ldots, 0)$, ..., $E_n = (0, 0, \ldots, 1)$. The set $\{E_1, \ldots, E_n\}$ is called the *canonical basis* of K^n. Let V be a K-vector space, and A an affine space over V. An affine system of coordinates in the space A is given by a point $O \in A$ and a basis $\{e_1, \ldots, e_n\}$ of V. This coordinate system is denoted $Oe_1 \cdots e_n$. If $A = A^n$, then the affine coordinate system $OE_1 \cdots E_n$ in which $O = (0, 0, \ldots, 0)$ and $\{E_1, \ldots, E_n\}$ is the canonical basis of K^n and is called the *canonical coordinate system.* Note that in this coordinate system, every point $(x_1, \ldots, x_n) \in A^n$ has itself as coordinate n-tuple.

canonical decomposition (1.) Of a polynomial: Let F be a field and let $f(X) \in F[X]$, the polynomial ring in X over F. Then we can write

$$f(X) = f_1(X)^{n_1} \cdots f_r(X)^{n_r} \quad (n_i \geq 1),$$

0-8493-0320-6/00/$0.00+$.50

where $f_1(X), \ldots, f_r(X)$ are distinct irreducible monic polynomials uniquely determined by $f(X)$. We then refer to the above expression as the *canonical decomposition* of $f(X)$.

(2.) Of a vector: Any point x in a Hilbert space E can be written uniquely as

$$x = \sum x_k,$$

where $x_k \in E_k$, for orthogonal subspaces $E_1, E_2, \ldots$, of E, and the sum is convergent (the partial sums converge in norm to x) in E. The above relation is called the *canonical decomposition* of a vector x in a Hilbert space E. Classically, the subspaces $E_1, E_2, \ldots$ are taken to be the eigenspaces for distinct eigenvalues of a self-adjoint operator on E.

canonical divisor A divisor D on a compact Riemann surface M having the form $D = (\omega)$, where ω is a non-zero meromorphic 1-form on M. Here, D is defined so that, if U_a is an open set from a finite open cover if M, and if, locally, $\omega|_{U_a} = f_a(z)dz$, then $D|_{U_a} = (f_a) = \sum a_i p_i - \sum b_j q_j$, where each p_i is a zero of f_a of order a_i and each q_j is a pole of f_a of order b_j.

canonical divisor class The module $\Omega^n(X)$ of n-dimensional regular differential forms on an n-dimensional, smooth variety X has dimension 1 over the algebra of polynomials $k(X)$. This implies that the divisors of all the forms in $\Omega^n(X)$ are equivalent. The divisor class of these equivalent divisors is the *canonical divisor class* of X.

canonical model In the study of a class of mathematical objects, it is sometimes possible to select a special subclass, constructed in a natural way, such that each member x in the original class is isomorphic to some unique member of the subclass, called the *canonical model* of x.

For example, for the class of contraction operators on Hilbert space (operators T with $\|T\| \leq 1$), every such T is similar to the restriction of the adjoint of the unilateral shift (on a vector valued H^2 space) to one of its invariant subspaces. One can call such a restriction the canonical model of T.

canonical parameters A set of parameters $\{p_i, q_i\}$, all of which satisfy the Poisson bracket relations

$$[p_i, p_j]_{PB} = 0, \quad [q_i, q_j]_{PB} = 0,$$
$$[q_i, p_j]_{PB} = \delta_{ij},$$

where the Poisson bracket is defined by

$$[A, B]_{PB} = \sum_{i=1}^{3} \left(\frac{\partial A}{\partial q_i} \frac{\partial B}{\partial p_i} - \frac{\partial A}{\partial p_i} \frac{\partial B}{\partial q_i} \right).$$

canonical transformation The quotient map, from the set of all formal linear combinations $\Lambda(E, F)$ of elements in the Cartesian product $E \times F$ of two vector spaces into their tensor product $E \otimes F$, viewed as the quotient of $\Lambda(E, F)$ by the subspace generated by the elements

$$(\sum \alpha_j x_j, \sum \beta_k y_k) - \sum \sum \alpha_j \beta_k (x_j, y_k).$$

Cantor function Define a function $f(x)$ on $[0, 1]$ by first setting $f(x) = \frac{1}{2}$ on the interval $(\frac{1}{3}, \frac{2}{3})$. Then define $f(x) = \frac{1}{4}$ on the interval $(\frac{1}{9}, \frac{2}{9})$ and $f(x) = \frac{3}{4}$ on $(\frac{7}{9}, \frac{8}{9})$. Proceeding inductively, one can define f on all open intervals of the form $(\frac{j}{3^n}, \frac{j+1}{3^n})$ and extend the definition to all of $[0, 1]$ by continuity. The resulting monotonic function is continuous and nonconstant but has derivative 0 on a subset of $[0, 1]$ of Lebesgue measure 1 (the complement of the *Cantor set*). *See* Cantor set.

Cantor set Generally, any perfect subset of $\mathbf{R}$ which contains no segments.

The usual construction begins with the interval $[0, 1]$. One removes the open middle third $(\frac{1}{3}, \frac{2}{3})$, with $E_1 = [0, \frac{1}{3}] \cup [\frac{2}{3}, 1]$ remaining. Next, one removes the open middle thirds from both the intervals in E_1, with E_2 remaining, etc. Finally, the *Cantor middle thirds set* is $C = \cap_{j=1}^{\infty} E_j$. C has the additional property that it is uncountable, but has Lebesgue measure 0.

Instead of middle thirds, one can remove open middle segments of length δ_n, where $1 > \delta_1 > \delta_2 > \ldots$, from [0, 1]. One obtains a perfect set containing no intervals, but it may have positive measure, depending upon the sequence $\{\delta_n\}$.

See also Cantor function.

Cantor-Lebesgue Theorem Let T denote the unit circle and assume $E \subset T$ is a measurable set having positive measure. Moreover assume that the series

$$\frac{1}{2}a_0 + \sum_{n=1}^{\infty}(a_n \cos nx + b_n \sin nx)$$

converges for all $x \in E$. Then

$$\lim_{n\to\infty} a_n = \lim_{n\to\infty} b_n = 0.$$

CAR algebra The C^* tensor product of countably many copies of the algebra of 2×2 matrices.

Carathéodory construction For each metric space X, each family F of subsets of X, and each function ζ such that $0 \le \zeta(S) \le \infty$ whenever $S \in F$, one can construct measures ϕ_δ on X, corresponding to $0 < \delta \le \infty$, and then a final measure ψ, in the following manner: Whenever $A \subset X$, $\phi_\delta(A) = \inf_G \sum_{S\in G} \zeta(S)$ where G ranges over all countable families such that $G \subset F \cap \{S : \text{diam}(S) \le \delta\}$ and $A \subset \cup G$. Since $\phi_\delta \ge \phi_\gamma$ for $0 < \delta < \gamma \le \infty$ we have the existence of a measure $\psi(A) = \lim_{\delta\to 0+} \phi_\delta(A) = \sup_{\delta>0}(A)$ whenever $A \subset X$. ψ is a measure on X, and it is the result of *Carathéodory's construction* for ζ on F.

Carathéodory outer measure *See* outer measure.

Carathéodory pseudodistance Let X be a locally convex space. Let $\Omega \subset X$ be open and connected and let $\mathcal{A}$ be the set of (Frechet-) analytic maps from X to $\mathbf{C}$. The *Carathéodory pseudodistance* $C_\Omega(a, b)$, for $a, b \in \Omega$ is

$$C_\Omega(a, b) = \sup\left\{ \arg\tanh \left| \frac{\phi(a) - \phi(b)}{1 - \overline{\phi(a)}\phi(b)} \right| \right\}$$

where the sup is over $\phi \in \mathcal{A}$ with $|\phi| < 1$ and where tanh is the hyperbolic tangent.

cardioid The locus of a point on a circle which rolls around a fixed circle of the same radius. This is a special case of an epicycloid. With appropriate choice of coordinates, a cardioid has polar equation $r = a(1 \pm \cos\theta)$ or $r = a(1 \pm \sin\theta)$.

Carleman kernel A complex-valued function k defined on the Cartesian product $X \times Y$ of two measure spaces X and Y, with the property that the function $k(x, \cdot)$ is in $L^2(Y)$.

Carleman's Theorem Carleman's Uniqueness Theorem gives conditions under which a measure $d\mu$ is uniquely determined by its sequence of moments $\int_0^\infty t^n d\mu(t)$.

carrier *See* support.

Cartan atlas A collection of Cartan gauges on a smooth manifold M, with a fixed model (g, h), satisfying a compatibility condition, and such that the open sets in the atlas cover M. *See* Cartan gauge.

Cartan connection Let M be a smooth manifold on which has been defined a Cartan geometry with model (g, h), and let H be the Lie group realizing h. The *Cartan connection* is a certain g-valued 1-form on P, the principal H bundle associated with the Cartan geometry on M. *See* Cartan geometry.

Cartan gauge A model geometry for a Cartan geometry is an effective infinitesimal Klein geometry (g, h), where g and h are Lie algebras and h is a subalgebra of g, and a Lie group H realizing h. If M is a smooth manifold, then a *Cartan gauge* with model (g, h)

is a pair (U, Θ_U), where U is an open set in M and Θ_U is a g-valued 1-form on U that satisfies the condition that for any $u \in U$, the composition of Θ_U on the tangent space to U at u and the canonical projection from g to g/h is a linear isomorphism. *See* Cartan geometry.

Cartan geometry A smooth manifold, on which has been defined a Cartan structure using a model (g, h). *See* Cartan structure.

Cartan pseudoconvex domain A subset of $\mathbf{C}^n$ that is locally Cartan pseudoconvex at every point of its boundary. *See* locally Cartan pseudoconvex domain.

Cartan structure A *Cartan structure with model* (g, h), on a smooth manifold M, is an equivalence class of Cartan atlases on M that use this model. Two atlases are considered equivalent if their union is also an atlas. *See* Cartan atlas.

Cartesian coordinates (1.) A point in a plane has Cartesian coordinates (x, y) that locate the point relative to two lines (usually labelled the x-axis and the y-axis) that intersect at a point referred to as the origin. The axes are usually (but not necessarily) perpendicular. The first coordinate (the *x-coordinate*, or *abscissa*) is the signed distance of the point from the y-axis, measured parallel to the x-axis. The second coordinate (the *y-coordinate*, or *ordinate*) is the signed distance of the point from the x-axis, measured parallel to the y-axis.
(2.) A point in space has Cartesian coordinates (x, y, z) that locate the point relative to three planes (the coordinate planes) that intersect at a common point (the origin). The coordinate planes are usually (but not necessarily) perpendicular. The lines of intersection of pairs of coordinate planes are the coordinate axes, which are usually referred to as the x-axis, the y-axis, and the z-axis. The coordinate planes can be described by the axes they contain. For example, the y, z-plane contains the y- and z-axes. Each coordinate of a point is the signed distance of the point from a coordinate plane, measured parallel to a coordinate axis. For example, the x-coordinate is the signed distance of the point from the y, z-plane measured parallel to the x-axis.

Cartesian plane A plane upon which is imposed a system of Cartesian coordinates. Also called Euclidean plane.

Cartesian product For two sets X and Y, the set of ordered pairs $X \times Y = \{(x, y) : x \in X \text{ and } y \in Y\}$.

Cartesian three-space A three-dimensional space upon which is imposed a system of Cartesian coordinates. Also called Euclidean space.

Cassini's oval The locus obtained as the vertex of a triangle which satisfies the following condition: the side opposite that vertex is fixed with length $2a$, and the product of the lengths of the sides adjacent to the vertex is a constant b. If $b > a^2$ then a single oval is obtained. If $b < a^2$ then two ovals are actually obtained. If $b = a^2$ then the curve obtained is a lemniscate.

catenary If an idealized weightless chain is hung from two points that lie on a horizontal line, the resulting plane curve is referred to as a *catenary.* With appropriate choice of coordinates, a catenary can be described by the equation $y = \cosh x$.

catenoid The surface obtained if a catenary is rotated about its axis.

Cauchy condition A sequence $\{a_n\}$ in a metric space such that $d(a_m, a_n) \to 0$ as $m \to \infty$ and $n \to \infty$. In other words, for every $\epsilon > 0$ there exists an N such that $d(a_m, a_n) < \epsilon$ for all $m > N$ and all $n > N$.

Cauchy criterion The *Cauchy criterion* for convergence of a sequence of real or complex numbers is that a sequence converges if and only if it is a Cauchy sequence. The Cauchy criterion can be applied to convergence of a series by checking for the convergence of the sequence of partial sums of the series. The Cauchy criterion is used to define a complete metric space. By definition, a metric space is complete if and only if all Cauchy sequences converge.

Cauchy integral representation (1.) Suppose f is holomorphic on a simply connected region Ω of the complex plane, and let γ be a simple closed rectifiable curve in Ω. Then, for all z in the interior of γ,

$$f(z) = \frac{1}{2\pi i}\int_\gamma \frac{f(\zeta)}{\zeta - z}d\zeta.$$

(2.) Suppose γ is a closed rectifiable curve in a region Ω of the complex plane, and suppose f is holomorphic on Ω. Then, for all $z \in \Omega$, with z not on the curve γ,

$$f(z)\cdot \mathrm{Ind}_\gamma(z) = \frac{1}{2\pi i}\int_\gamma \frac{f(\zeta)}{\zeta - z}d\zeta$$

where $\mathrm{Ind}_\gamma(z)$ is the index (or winding number) of z with respect to γ.

(3.) Suppose γ is a simple closed rectifiable curve in a region Ω of the complex plane, and suppose f is holomorphic on Ω. Then, for all z in the interior of γ,

$$f^{(n)}(z) = \frac{n!}{2\pi i}\int_\gamma \frac{f(\zeta)}{(\zeta - z)^{n+1}}d\zeta.$$

Cauchy integral test A necessary and sufficient condition for the convergence of a series $\sum a_n$. Suppose f is a positive, monotonically decreasing function with $f(n) = a_n$ for all sufficiently large n. Then $\sum a_n$ is convergent if and only if the improper integral $\int_N^\infty f(x)dx$ converges for some number N.

Cauchy net Suppose that $(D, \geq)$ is a directed set and that $\{S_n : n \in D\}$ is a net in a uniform topological space $(X, \mathcal{U})$. The net is called a *Cauchy net* if for each member U of $\mathcal{U}$ there exists an $N \in D$ such that $(S_m, S_n) \in U$, whenever $m \geq N$ and $n \geq N$.

Cauchy principal value Let f be a function defined almost everywhere on an interval $[a, b]$ and suppose that $\int_a^b f(x)dx$ does not exist. Suppose, also, that there exists a number $c \in (a, b)$ such that, for every $\epsilon > 0$, f is integrable on the intervals $[a, c-\epsilon]$ and $[c+\epsilon, b]$. The *Cauchy principal value* of the integral $\int_a^b f(x)dx$ is the value

$$PV\int_a^b f(x)dx$$

$$= \lim_{\epsilon\to 0^+}\left(\int_a^{c-\epsilon} f(x)dx + \int_{c+\epsilon}^b f(x)dx\right)$$

provided the limit exists.

Cauchy sequence A sequence in a metric space which satisfies the Cauchy condition. *See* Cauchy condition.

Cauchy's integral theorem Suppose f is holomorphic on a simply connected region Ω of the complex plane, and let γ be a simple closed rectifiable curve in Ω. Then $\int_\gamma f(z)dz = 0$.

Cauchy-Riemann equations The equations

$$\frac{\partial u}{\partial x} = \frac{\partial v}{\partial y}, \quad \frac{\partial u}{\partial y} = -\frac{\partial v}{\partial x},$$

where $z = x + iy$ and $f(z) = u + iv$.

A complex-valued function f of a complex variable z is analytic if and only if the Cauchy-Riemann equations are satisfied.

Cauchy-Schwarz inequality In a Hilbert space with inner product (x, y) and norm $\|x\| = \sqrt{(x, x)}$, the inequality

$$|(x, y)| \leq \|x\| \cdot \|y\|$$

with equality if and only if x and y are linearly dependent. A special case of the Cauchy-Schwarz inequality is obtained when the Hilbert space is taken to be ℓ^2:

$$\left|\sum_{i=1}^\infty x_i\bar{y}_i\right| \leq \sqrt{\sum_{i=1}^\infty |x_i|^2}\sqrt{\sum_{i=1}^\infty |y_i|^2}$$

for any ℓ^2 sequences of complex numbers $x_1, x_2, \ldots$ and $y_1, y_2, \ldots$. Another special case is obtained when the Hilbert space is taken to be L^2:

$$\left|\int f\bar{g}\right| \leq \sqrt{\int |f|^2}\sqrt{\int |g|^2}$$

for any functions f and g in L^2.

Cayley transform (**1.**) The mapping of the extended complex plane onto itself defined by $f(\lambda) = (\lambda - i)(\lambda + i)^{-1}$. This transform maps the real line to the unit circle, with $f(\infty) = 1$. The upper half plane transforms to the interior of the unit circle, and the lower half plane transforms to the exterior of the unit circle, with $f(-i) = \infty$.
(**2.**) For a closed operator A on a Hilbert space, the operator defined by $f(A) = (A - iI)(A + iI)^{-1}$, where I is the identity operator. If A is self-adjoint, $f(A)$ is unitary. If A is symmetric, $f(A)$ is isometric.

center If a curve has symmetry about a point, then that point is known as the *center* of the curve.

center of curvature The *center of curvature* of a curve at a point on the curve is the center of the osculating circle at that point.

central conic An ellipse or a hyperbola.

Cesàro means For a sequence $s_0, s_1, \ldots$, the numbers

$$\sigma_n = \frac{1}{n+1}(s_0 + s_1 + \cdots + s_n).$$

A series $\sum a_j$ is *Cesàro summable* (to s) if the Cesàro means of its partial sums converge (to s).

chain of simplices Let K be a simplicial complex and let G be an Abelian group, with the group operation being denoted by addition. An r-dimensional chain (or r-chain) x is defined by $x = g_1S_1 + g_2S_2 + \ldots + g_nS_n = \sum_{i=1}^{n} g_iS_i$, where each S_i, for $i = 1, \ldots, n$, is an oriented r-dimensional simplex of K. Replacing g_iS_i by $-g_iS_i$ in a chain has the same effect as reversing the orientation of the simplex S_i. The r-chains form a group under the addition induced from the group G. G is usually taken to be the integers or the integers modulo n, for some $n \geq 2$; in that case the boundary operator δ can be defined. If S is an oriented r-dimensional simplex, and if $B_1, B_2, \ldots, B_n$ are the oriented $(r-1)$-dimensional faces of S, then the boundary of S is the chain $\delta(S) = \sum_{i=1}^{n} \epsilon_i B_i$, where $\epsilon_i = 1$ if B_i and S are coherently oriented, and $\epsilon_i = -1$ otherwise. The boundary of the chain x is $\delta x = \sum_{i=1}^{n} g_i\delta(S_i)$. If a chain has a boundary equal to zero, it is called a cycle; in particular, the boundary of any chain is a cycle: $\delta(\delta x) = 0$ for any chain x.

chain rule (**1.**) If f and g are differentiable functions, then the composition function h defined by $h(x) = f(g(x))$ is differentiable with derivative given by $h'(x) = f'(g(x))g'(x)$. If we write $y = f(u)$ and $u = g(x)$, then the chain rule can be written

$$\frac{dy}{dx} = \frac{dy}{du}\frac{du}{dx}.$$

More precisely, the chain rule is valid if g is differentiable at $x = x_0$, if f is differentiable at $u = g(x_0)$, and if each deleted neighborhood of x_0 contains points in the domain of the composition function h.
(**2.**) If f is a differentiable function of several variables $u_1, u_2, \ldots, u_m$ and if the variables $u_1, u_2, \ldots, u_m$ are each differentiable functions of several variables $x_1, x_2, \ldots, x_n$, then the composition function h defined by

$$h(x_1, x_2, \ldots, x_n) = f(u_1(x_1, x_2, \ldots, x_n), u_2(x_1, x_2, \ldots, x_n), \ldots, u_m(x_1, x_2, \ldots, x_n))$$

is a differentiable function with

$$\frac{\partial h}{\partial x_k} = \sum_{i=1}^{m} \frac{\partial f}{\partial u_i}\frac{\partial u_i}{\partial x_k}$$

for all $k = 1, 2, \ldots, n$.

character Of a locally compact Abelian group G, a map ϕ from G to the group of complex numbers of modulus 1, having the properties $\phi(a+b) = \phi(a)\phi(b)$ and $\phi(0) = 1$.

For a more general group, the trace of a representation.

characteristic (1.) If a positive (rational) number is written in scientific notation $k \cdot 10^n$, where $1 \leq k < 10$ and n is an integer, then the common logarithm of the number is $\log_{10} k + n$. The *characteristic* of this logarithm is the integer n.
(2.) If a ring has the property that there exists a positive integer n such that $na = 0$ for all elements a in the ring, then the smallest such integer is known as the *characteristic* of the ring. If no such integer exists, the ring is said to have characteristic zero. If a ring is an integral domain with non-zero characteristic, then the characteristic must be prime.

characteristic curves The conjugate curves on a surface S with the property that at every point P of S the tangents to the curves are the characteristic directions on S at P.

characteristic directions At any point P on a surface S (with the exception of umbilical points) there is a unique pair of conjugate directions that are symmetric with respect to the lines of curvature on S through P. These are the *characteristic directions* on S at P. The angle between the characteristic directions is the minimum of the angles between conjugate directions on S at P.

characteristic equation Let A be a square matrix and I the identity matrix with the same order. If det denotes the determinant of a matrix, then the *characteristic equation* of A is $\det(\lambda I - A) = 0$. The roots of the characteristic equation are the characteristic roots, or eigenvalues, of the matrix A.

characteristic exponents Consider the system of differential equations $\frac{dx}{dt} = A(t)x$, where $A(t)$ is an $n \times n$ matrix-valued function which is periodic with period ω. If $X(t)$ is an $n \times n$ matrix-valued function whose columns form a basis for solutions for the system, then $X(t)$ is invertible for each t and the matrix $C = X^{-1}(t)X(t + \omega)$ is a constant matrix, which can be written in the form $C = e^{\omega R}$ for an $n \times n$ matrix R. The eigenvalues of R are called the *characteristic exponents* of the system.

characteristic function (1.) The *characteristic function* of a square matrix A is its characteristic polynomial $\det(\lambda I - A)$, where det denotes determinant and I is the identity matrix with the same order as A.
(2.) If f is the probability density function of a random variable, then the characteristic function of the random variable is the function $g(t) = \int_{-\infty}^{\infty} e^{itx} f(x)dx$.
(3.) The characteristic function of a set A is the function

$$\chi_A(x) = \begin{cases} 1 & \text{if} x \in A \\ 0 & \text{if} x \notin A. \end{cases}$$

characteristic multiplier Consider the system of differential equations $\frac{dx}{dt} = A(t)x$, where $A(t)$ is an $n \times n$ matrix-valued function which is periodic with period ω. If $X(t)$ is an $n \times n$ matrix-valued function whose columns form a basis for solutions for the system, then $X(t)$ is invertible for each t and the matrix $C = X^{-1}(t)X(t + \omega)$ is a constant matrix, which can be used to reduce the system to a system with constant coefficients. The eigenvalues of the matrix C are referred to as the *characteristic multipliers* of the system. If the characteristic multipliers are distinct, then X can be chosen so that C is a diagonal matrix with the characteristic multipliers on the diagonal, and the solutions to the system can be written in terms of the characteristic multipliers.

characteristic operator function (1.) If T is a contraction operator on a Hilbert space, define $Q_T = (I - T^*T)^{1/2}$ and $Q_{T^*} = (I - TT^*)^{1/2}$. Let $\mathcal{D}_T$ be the closure of the range of Q_T, and $\mathcal{D}_{T^*}$ be the closure of the range of Q_{T^*}. The characteristic function for T is the operator valued function

$$\Theta_T(\lambda) = -T + \lambda Q_{T^*}(I - \lambda T^*)^{-1} Q_T|_{\mathcal{D}_T}.$$

For each complex number λ for which it is defined, the operator $\Theta_T(\lambda)$ is a bounded operator between the Hilbert spaces $\mathcal{D}_T$ and $\mathcal{D}_{T^*}$. If T is completely non-unitary, then the characteristic operator function uniquely determines T and can be used to construct a functional model for T.

(2.) If T is a closed operator on a Hilbert space, define $Q_T = |I - T^*T|^{1/2}$, $Q_{T^*} = |I - TT^*|^{1/2}$, $J_T = \text{sgn}(I - T^*T)$, and $J_{T^*} = \text{sgn}(I - TT^*)$. $\mathcal{D}_T$ and $\mathcal{D}_{T^*}$ are defined as above, but these are considered as Kreĭn spaces with indefinite inner products determined by J_T and J_{T^*}. The characteristic function for T is the operator valued function

$$\Theta_T(\lambda) = -T + \lambda Q_{T^*}(I - \lambda T^*)^{-1} Q_T|_{\mathcal{D}_T}.$$

For each complex number λ for which it is defined, the operator $\Theta_T(\lambda)$ is a bounded operator between the Kreĭn spaces $\mathcal{D}_T$ and $\mathcal{D}_{T^*}$. Functional models based on the characteristic function have been developed only in the case where Θ_T is uniformly bounded on the open unit disk.

characteristic polynomial For a square matrix A, the polynomial $\det(\lambda I - A)$, where det denotes determinant and I is the identity matrix with the same order as A.

characteristic root A root of the characteristic polynomial (of a given square matrix).

characteristic vector If λ is a characteristic root (or eigenvalue) of a square matrix A, then a *characteristic vector* (or eigenvector) associated with λ is any non-zero vector x satisfying $Ax = \lambda x$.

Charpit's subsidiary equations Consider a partial differential equation

$$F(x, y, z, \frac{\partial z}{\partial x}, \frac{\partial z}{\partial y}) = 0.$$

In terms of $F = F(x, y, z, p, q)$, *Charpit's subsidiary equations* are

$$\frac{dx}{\partial F/\partial p} = \frac{dy}{\partial F/\partial q} = \frac{dz}{p\partial F/\partial p + q\partial F/\partial q}$$
$$= \frac{-dp}{\partial F/\partial x + p\partial F/\partial z} = \frac{-dq}{\partial F/\partial y + q\partial F/\partial z}.$$

chart *See* atlas.

Christoffel symbols *Christoffel symbols* of the first and second kinds are symbols $\begin{bmatrix} ij \\ k \end{bmatrix}$ and $\begin{Bmatrix} ij \\ k \end{Bmatrix}$, respectively, that represent functions of the coefficients and the derivatives of coefficients of a quadratic differential form. In Euclidean space, with rectangular Cartesian coordinates, the Christoffel symbols of the second kind are identically zero.

circle The set of all points in a plane that are a constant distance (the radius) from a fixed point (the center) in the plane.

circular cone The solid bounded by a circular region (the base) and the line segments joining points on the circular boundary of the region to a fixed point (the vertex) that is not in the plane of the base. The axis of the cone is the line through the vertex and the center of the base. If the axis is perpendicular to the plane of the base, the solid is a *right circular cone;* otherwise it is an *oblique circular cone.*

circular cylinder Consider two circles of the same radius lying in two parallel planes, and the line ℓ joining the centers of the circles. A circular cylinder is the surface consisting of the plane regions bounded by these circles (the bases) and the line segments parallel to ℓ that join points on one circle to points on the other. The line ℓ is the axis of the cylinder. If the axis is perpendicular to the planes of the bases, the cylinder is a *right circular cylinder;* otherwise it is an *oblique circular cylinder.*

circular functions *See* trigonometric function.

cissoid of Diocles The cissoid of a circle and a tangent line, with respect to the point on the circle opposite to the point of tangency. *See* cissoidal curve.

cissoidal curve Let C_1, C_2 be curves and P a fixed point. Let Q_1, Q_2 be the points of intersection of C_1, C_2 with a line L through P. The *cissoid* of C_1, C_2 *with respect to* P is the locus of a point p on such line L satisfying

$$Pp = PQ_2 - PQ_1 = Q_2Q_1.$$

classical theory of the calculus of variations A branch of mathematics which is a sort of generalization of calculus. Calculus of variations seeks to find the path, curve, surface, etc., for which a given function has a stationary value (which, in physical problems, is usually a minimum or maximum). Mathematically, this involves finding stationary values of integrals of the form

$$I = \int_b^a f(y, \dot{y}, x)\,dx.$$

I has an extremum only if the Euler–Lagrange differential equation is satisfied, i.e., only if

$$\frac{\partial f}{\partial y} - \frac{d}{dx}\left(\frac{\partial f}{\partial \dot{y}}\right) = 0.$$

The fundamental lemma of the calculus of variations states that, if

$$\int_a^b M(x)h(x)\,dx = 0$$

for all $h(x)$ with continuous second partial derivatives, then

$$M(x) = 0$$

on (a, b).

closable operator Let T be a linear operator with domain and range in normed linear spaces. T is called *closable* if there is a linear extension of T which is a closed operator.

closed (1.) In a topological space, a *closed set* is the complement of an open set. A set is closed if and only if it contains all of its accumulation points. A set is closed if and only if it contains its boundary.
(2.) Within a topological vector space, a *closed subspace* is a subspace that is also a closed set in the topology of the space. The adjective *closed* may be similarly used to describe an object contained within other structures that include a topology, such as a closed submanifold of a topological manifold, a closed subgroup of a topological group, etc.
(3.) A set is said to be *closed with respect to a binary operation* if applying the binary operation to any pair of members of the set always gives a member of the set.

closed convex curve A closed curve γ such that the set of points interior to γ (having a nonzero winding number with respect to γ) is a convex set. *See* closed curve, convex set.

closed convex surface A closed surface that is convex, in the sense that any plane section of the surface is a convex curve.

closed curve A curve that has no end points. If the curve is viewed as the image of an interval $[a, b]$ under a continuous mapping f, then it is closed if and only if $f(a) = f(b)$. Equivalently, a *closed curve* may be viewed as the image of a circle under a continuous mapping.

closed differential form A differential form ω such that $d\omega = 0$.

Closed Graph Theorem If T is a closed linear operator whose domain is a Banach space and whose range is contained in a Banach space, then T is continuous. The name of the theorem derives from the fact that T closed is equivalent to the graph $G(T) = \{(x, Tx)\}$ being closed in the Cartesian product of the two Banach spaces.

closed interval The *closed interval* $[a, b]$, with $-\infty < a$ and $b < \infty$, is the set of all points x on the real number line with $a \leq x \leq b$. If $a = -\infty$ [resp. $b = \infty$], the interval $(-\infty, b] = \{x : -\infty < x \leq b\}$ [resp. $[a, \infty) = \{x : a \leq x < \infty\}$] is also a closed interval.

closed linear subspace A closed subset of a topological vector space **V**, which is closed under the operations of vector addition and scalar multiplication of **V**. *See* topological vector space.

closed operator Suppose A and B are Banach spaces. A linear operator $T : A \to B$ is called closed if it has a closed graph. This means that, if $\mathcal{D}$ is the domain of T in A, then the set $\{(x, Tx) : x \in \mathcal{D}\}$ is a closed subspace of the Banach space $A \times B$. Equivalently, T is closed if and only if the following condition holds: Suppose x_n is in $\mathcal{D}$ for each n, x_n converges to x, and Tx_n converges to y; then $x \in \mathcal{D}$ and $Tx = y$.

Closed Range Theorem For a densely defined operator T on a Banach space, the range $R(T)$ is closed if and only if $R(T^*)$ is closed.

closed surface A surface that has no boundary curves. A *closed surface* can be characterized as a connected compact metric space in which every point has a neighborhood that is homeomorphic with the interior of a circle.

closure (1.) The *closure of a set* S is the intersection of all closed sets containing S, and is therefore the smallest closed set containing S. The closure of S can also be described as the union of S and the set of accumulation points of S, or the union of S and the boundary of S.
(2.) The *closure of a closable operator* is the closed extension of that operator, with minimal domain.

cluster point A point x is a *cluster point* (or *accumulation point*) of a subset A of a topological space if every neighborhood of x contains points of A other than x.

cluster set A *cluster value* of a function $f(z)$ at a point z_0 is a point α such that $f(z_j) \to \alpha$, as $j \to \infty$, for some sequence $z_j \to z_0$. The *cluster set* of f at z_0 is the set of all cluster values.

cluster value *See* cluster set.

co-analytic set The complement of an analytic set. *See* analytic set.

co-latitude For a point P on the surface of the earth, the complement of the latitude of P.

coboundary Let K be a simplicial complex on which is defined a boundary operator δ. Then δ maps the group of r-dimensional chains into the group of $(r-1)$-dimensional chains. If the r-dimensional oriented simplices are denoted by $S_1^r, S_2^r, \ldots, S_m^r$, and the $(r-1)$-dimensional oriented simplices by $S_1^{r-1}, S_2^{r-1}, \ldots, S_n^{r-1}$, then there are group elements $\{g_{ij}\}$ such that $\delta(S_i^r) = \sum_{j=1}^n g_{ij} S_j^{r-1}$ for $i = 1, \ldots, m$. For $j = 1, \ldots, n$, the coboundary of S_j^{r-1} is defined by $\nabla(S_j^{r-1}) = \sum_{i=1}^m g_{ij} S_i^r$. If x is a chain $x = \sum g_i \sigma_i$, for some oriented simplices σ_i and group elements g_i, then the *coboundary* of x is the chain $\nabla(x) = \sum g_i \nabla(\sigma_i)$.

cocycle A chain of simplices in a simplicial complex having coboundary zero.

codimension The *codimension* of a subspace A of a vector space X is the dimension of the quotient space X/A.

Codazzi equation The *Codazzi equations* are certain equations involving the fundamental coefficients of the first and second order of a surface. Together with the Gauss equation, the Codazzi equations uniquely determine a surface, up to its position in space.

coechelon space *See* echelon space.

coercive boundary condition Consider a partial differential equation of the form $u_t = Au = \sum_{i=1}^{n} A_i u_{x_i}$, where $u(x,t)$ is defined for $x = (x_1, x_2, \dots, x_n)$ in a domain of $\mathbf{R}^n$ and for $t \geq 0$; u takes values in k- dimensional complex Euclidean space $\mathbf{C}^k$; and each A_i is a $k \times k$ matrix-valued function of x. Suppose D is a domain in $\mathbf{R}^n$ with boundary ∂D. Let f be an initial condition for the equation, i.e., u is required to satisfy $u(x,0) = f(x)$, where f is defined on D and is required to satisfy certain conditions on the boundary ∂D. A boundary condition is said to be *coercive* for A if there are constants K_1 and K_2 such that, for every smooth function f defined on D and satisfying the boundary condition,

$$(\sum_{i=1}^{n} |\frac{\partial f}{\partial x_i}|^2)^{1/2} \leq K_1 \|Af\| + K_2 \|f\|$$

where $\|\cdot\|$ is the L^2 norm over D.

cofactor The *cofactor* of an element a_{ij} of a square $n \times n$ matrix A is $A_{ij} = (-1)^{i+j} M_{ij}$, where M_{ij} (the minor of a_{ij}) is the determinant of the matrix obtained by deleting row i and column j from A. The determinant of A can be expanded by the elements of row i and their cofactors by $\det(A) = \sum_{j=1}^{n} A_{ij} a_{ij}$. Similarly, the determinant can be expanded by column j: $\det(A) = \sum_{i=1}^{n} A_{ij} a_{ij}$.

coherently oriented Let S be an oriented n-dimensional simplex with an ordered set of vertices $(p_0 p_1 \dots p_n)$, and let B be an oriented $(n-1)$-dimensional face of S. The vertices of B will then consist of the vertices of S with one vertex p_i discarded. S and B are said to be *coherently oriented* if the orientation of B is given by the order of its vertices (up to an even permutation) $(p_{i+1} p_{i+2} \dots p_n p_0 \dots p_{i-1})$.

Cohn-Vossen Theorem Two isometric ovaloids differ by an orthogonal, linear transformation of $\mathbf{R}^3$.

cohomology Let K be a simplicial complex with coboundary operator ∇. Let T^r be the group of r-dimensional cocycles of K and let H^r be the group whose elements are cycles that are either zero or coboundaries of $(r-1)$-dimensional chains in K. Since ∇ maps $(r-1)$-dimensional chains to r-dimensional chains, and since $\nabla h = 0$ for all $h \in H^r$, it follows that H^r is a subgroup of T^r. The r-dimensional *cohomology* group is the quotient group T^r/H^r.

coincident Any two geometric figures that have exactly the same points are said to be *coincident.* For example, two curves that have equivalent equations are coincident.

colinear Line segments that lie on the same line are *colinear* line segments. Planes that intersect in a common line are *colinear* planes.

combination A subset of a given set, that has been selected without any consideration of the order of the elements selected. For a set of n elements, the number of combinations of k elements (or the number of combinations of n objects taken k at a time) is given by the binomial coefficient

$$\binom{n}{k} = {}_nC_k = \frac{n!}{k!(n-k)!}.$$

common denominator A *common denominator* for a set of fractions is a denominator that is a common multiple of the denominators of all the fractions in the set.

common difference In an arithmetic sequence $a, a+d, a+2d, a+3d, \dots$, the value d.

common logarithm A logarithm with base 10, abbreviated $\log_{10}$. The common logarithm is defined by the requirement that $\log_{10} x = y$ if and only if $x = 10^y$. On most calculators, and in some undergraduate texts, the common logarithm is abbreviated to log; in other contexts this abbreviation usually refers to the natural logarithm.

commutative (1.) A binary operation $\circ$ on a set A such that $a \circ b = b \circ a$ for every pair of elements a and b in A.
(2.) An algebraic structure with a single binary operation (such as a group), such that the operation is commutative.

commutative diagram A *diagram* is a configuration of three or more sets with maps between the sets. The diagram is *commutative* if, whenever there are two compositions mapping one of the sets A to another set B, those two compositions are equal. Thus, we may have sets A, B, C and maps $\alpha : A \to B, \beta : B \to C, \gamma : A \to C$. This produces two maps, $\beta \circ \alpha$ and γ, mapping A to C. Commutativity requires that they be equal.

commutator (1.) The *commutator* of two elements a and b of a group is $a^{-1}b^{-1}ab$.
(2.) The *commutator* subgroup of a group is the subgroup of all products of commutators of elements of the group.
(3.) Given two operators A and B that act on a vector space, the *commutator* of A and B is the operator $[A, B] = AB - BA$. *See also* self-commutator.

commute Let A be a set on which is defined a binary operation $\circ$. Two elements a and b of A *commute* if $a \circ b = b \circ a$.

compact (1.) A topological space X is said to be *compact* if every open cover of X (i.e., any collection of open sets whose union is X) has a finite subcover.
(2.) A subset A of a topological space is said to be *compact* if, with the relative topology, it is a compact topological space. Equivalently, A is compact if and only if every open cover of A (i.e., any collection of open sets whose union contains A) has a finite subcover.

In a compact space, all closed subsets are compact. The Heine-Borel theorem characterizes the compact subsets of a finite-dimensional Euclidean space $\mathbf{R}^n$: a subset of $\mathbf{R}^n$ is compact if and only if it is closed and bounded.

compact degeneracy *See* degeneracy.

compact kernel A kernel is *compact* if the integral operator that it induces is a compact operator. *See* compact operator.

compact operator Let T be a bounded operator mapping a Banach space X to a Banach space Y, and let B be the closed unit ball in X. T is said to be a *compact operator* if, using the strong topology on Y, the closure of TB is a compact subset of Y. T is said to be a *weakly compact* operator if, using the weak topology on Y, the closure of TB is a compact subset of Y.

compactification If X is a topological space, then a *compactification* of X is a compact topological space Y that contains X as a subspace, or which has a subspace that is homeomorphic with X. *See* Alexandrov compactification, Stone-Cech compactification.

comparison test Suppose $\sum_{n=1}^{\infty} a_n$ and $\sum_{n=1}^{\infty} b_n$ are series such that, for some $N \geq 1$, $0 \leq a_n \leq b_n$ for all $n \geq N$. If the series $\sum_{n=1}^{\infty} b_n$ converges, then the series $\sum_{n=1}^{\infty} a_n$ also converges.

compatible charts *See* atlas.

complement (1.) The *complement* of an acute angle θ is an angle ϕ such that the measures of θ and of ϕ sum to the measure of a right angle. If degree measure is used, the complement of θ is $90° - \theta$; if radian measure is used, the complement of θ is $\pi/2 - \theta$.
(2.) The *complement* of a subset A of a set B is the set of all elements of B that are not contained in A, denoted $B \backslash A$ or $B - A$.

complementary angles Two angles that are complements are referred to as *complementary angles*. *See* complement.

complementary modulus A complex number denoted by k, appearing in the definition of elliptic Legendre integrals of the first, second, and third kind. In particular we define by

$$F(k,\varphi) \equiv \int_0^{\varphi} \frac{d\varphi}{\sqrt{1-k^2\sin^2\varphi}},$$

$$E(k,\varphi) \equiv \int_0^{\varphi} \sqrt{1-k^2\sin^2\varphi}\,d\varphi,$$

and

$$\pi(c,k,\varphi) \equiv \int_0^{\varphi} \frac{d\varphi}{(\sin^2\varphi - c)\sqrt{1-k^2\sin^2\varphi}}$$

(c a complex number), normal Legendre elliptic integrals of first, second, and third kind respectively.

complementary subspace Suppose X is a vector space and that A is a subspace of X. A *complementary subspace* of A is a subspace B such that $A \cap B = \{0\}$ and $A + B = X$.

complete (1.) A field is algebraically *complete* if every polynomial with coefficients in the field has zeros in the field. For example, the complex numbers are algebraically complete, whereas the real numbers are not.
(2.) An ordered field is *complete* if every nonempty subset of the field with an upper bound has a least upper bound. For example, the real numbers are a complete ordered field.
(3.) A uniform topological space is called *complete* if every Cauchy net in the space converges to a point in the space. A metric space is called *complete* if every Cauchy sequence in the space converges to a point in the space.
(4.) A measure is called *complete* if all subsets of sets of measure zero are measurable.

complete additivity of integral Let μ be a finite measure on a measure space (X, Ω, μ) and $f \in L^1(\mu)$. Define a set function ν by

$$\nu(E) = \int_E f(x)d\mu(x),$$

for $E \in \Omega$. Then ν is *completely additive* (= countably additive), that is, if $\{E_j\}$ is a disjoint family of sets in Ω, we have

$$\nu(\cup_{j=1}^{\infty} E_j) = \sum_{j=1}^{\infty} \nu(E_j).$$

complete orthonormal set An orthonormal set of vectors in a Hilbert space $\mathcal{H}$ such that its closed linear span is $\mathcal{H}$. *See* orthonormal set.

completely continuous operator *See* compact operator.

completely nonunitary contraction An operator T on a Hilbert space H which is a contraction ($\|T\| \leq 1$) and has the property that, if $H = H_1 \oplus H_2$, where H_1 and H_2 are invariant under T and $T|_{H_1}$ is unitary, then $H_1 = \{0\}$. Abbreviated *c.n.u.* contraction.

completeness (1.) The property of a metric space that every Cauchy sequence in the space converges to a point in the space. In an ordered field, the completeness property is that every nonempty subset with an upper bound has a least upper bound. For the real numbers, the two concepts of completeness are equivalent.
(2.) In a measure space, the property that every subset of a zero set is a zero set. That is, a measure space (X, Ω, μ) is complete if, for every $E \in \Omega$ with $\mu(E) = 0$, $E' \subset E$ implies $E' \in \Omega$ and $\mu(E') = 0$.

completing the square *Completing the square* for a quadratic polynomial ax^2+bx+c is the process of rewriting the quadratic in the form $a(x-h)^2 + k$.

completion (1.) If a field has a minimal extension that is algebraically complete, then that extension is called an *algebraic completion* of the field. For example, the complex numbers are the algebraic completion of the real numbers.
(2.) Let X be a uniform topological space. Then there is a complete uniform space X^*

and a uniform isomorphism f from X to a dense subset of X^*. Such a pair (f, X^*) is known as a completion of X. If X is a Hausdorff uniform space, then it has a Hausdorff completion that is unique up to uniform isomorphism. In particular, a metric space X has a unique completion to a metric space X^*. The *completion* can be identified with a space of equivalence classes of Cauchy sequences, where two Cauchy sequences $\{x_n\}$ and $\{y_n\}$ are equivalent if $x_n - y_n \to 0$ as $n \to \infty$.
(3.) Suppose μ is a measure defined on a σ-algebra Ω of subsets of a set X. The μ-completion of Ω is the σ-algebra Ω^* consisting of all subsets E of X such that there are sets $A \in \Omega$ and $B \in \Omega$ with $A \subset E \subset B$ and $\mu(B \backslash A) = 0$. The measure μ is extended to μ^* on Ω^* by defining $\mu^*(E) = \mu(A)$ in this case. This extension $(X, \Omega,^* \mu^*)$ is a complete measure space and is known as the *completion* of (X, Ω, μ).

complex analytic space Suppose D is an open set in the complex plane. The *complex analytic space* $A(D)$ is the space of all complex functions that are bounded and continuous on the closure of D and analytic on D. The space is equipped with the supremum norm: $\|f\| = \sup\{|f(z)| : z \in D\}$.

complex form A *complex form* is a differential form $\omega = \omega_1 + i\omega_2$, where ω_1 and ω_2 are real differential forms.

complex function A function taking values in the complex numbers $\mathbf{C}$.

complex Hilbert space A vector space over the complex field, equipped with a positive definite inner product (x, y), and which is complete in the metric induced by the norm $\|x\| = \sqrt{(x, x)}$. Sometimes it is required that a Hilbert space be infinite dimension.

complex linear space A vector space over the field of complex numbers.

complex manifold A topological manifold which is locally homeomorphic to $\mathbf{C}^n$. *See* topological manifold.

complex number A number of the form $a+bi$ where a and b are real numbers and i is a number (the imaginary unit) satisfying $i^2 = -1$. The real numbers a and b are referred to as the real part and the imaginary part, respectively, of the complex number. Addition and multiplication of complex numbers are defined by $(a+bi)+(c+di) = (a+c)+(b+d)i$ and $(a+bi)(c+di) = (ac-bd)+(ad+bc)i$. The absolute value of a complex number is $|a+bi| = \sqrt{a^2+b^2}$. The set of all complex numbers is often denoted $\mathbf{C}$.

complex of lines A complex of lines is a three-parameter submanifold of straight lines in three-dimensional projective space.

complex plane A representation of the complex numbers in the Cartesian plane, with the complex number $a+bi$ represented by the point (a, b). The absolute value of a complex number is the distance from the origin to the point representing the complex number. *See* complex number.

complex space *See* analytic space.

complex structure (1.) An atlas making a topological space satisfy the definition of a complex manifold. *See* atlas, complex manifold, topological manifold.
(2.) If X is a complex Banach space, then it is also a Banach space over the real numbers. This Banach space X_r is the real Banach space associated to X. If Y is a real Banach space, then it is said to admit a complex structure if $Y = X_r$ for some complex Banach space X.

complex variable *See* function of a complex variable.

complex vector space A vector space over the field of complex numbers.

complex-valued function A function taking values in the complex numbers **C**.

composite function of one variable Suppose f and g are functions of a single variable. The *composite function* h of f and g is the function defined by $h(x) = f(g(x))$. The domain of h is the set of all x such that $g(x)$ is contained in the domain of f. This composite function is usually written $h = f \circ g$ or $h = fg$.

composite function of several variables Suppose f is a function of m variables, and $g_1, g_2, \dots, g_m$ are each functions of n variables. The composite function h is a function of n variables defined by

$$h(x_1, x_2, \dots, x_n) = f(g_1(x_1, x_2, \dots, x_n),$$

$$g_2(x_1, x_2, \dots, x_n), \dots, g_m(x_1, x_2, \dots, x_n)).$$

composite relation Given two relations R and S, the *composite relation* $R \circ S$ is the relation consisting of pairs (x, z) such that there is an element y with $(x, y) \in R$ and $(y, z) \in S$.

composition of functions The forming of the composite function. *See* composite function of one variable, composite function of several variables.

composition of relations The forming of the composite relation. *See* composite relation.

concave A curve in the plane is *concave* toward a point or line if, whenever a segment of the curve is cut off by a chord, that segment lies on the chord or on the opposite side of the chord from the point or line.

concave downward A curve in the plane is *concave downward* if, whenever a segment of the curve is cut off by a chord, that segment lies on or above the chord.

concave function The negative of a convex function.

concave to the left A curve in the plane is *concave to the left* if, whenever a segment of the curve is cut off by a chord, that segment lies on or to the right of the chord.

concave to the right A curve in the plane is *concave to the right* if, whenever a segment of the curve is cut off by a chord, that segment lies on or to the left of the chord.

concave upward A curve in the plane is *concave upward* if, whenever a segment of the curve is cut off by a chord, that segment lies on or below the chord.

concentric Two circles in the same plane and with a common center. More generally, two figures that have centers are called *concentric* if they have a common center.

conchoid of Nicomedes The conchoid of a straight line with respect to a point not on the line. *See* conchoidal curve.

conchoidal curve Let C be a curve and P a fixed point not on C. Let L be a straight line passing through P and intersecting C at the point Q. The *conchoid of C with respect to P* is the locus of points p_1, p_2 such that

$$p_1Q - Qp_2 = c,$$

where c is a constant.

concurrent Geometric figures that share a common point. For example, the altitudes of a triangle are concurrent lines.

conditional convergence A series $\sum a_n$ which is convergent but not absolutely convergent, i.e., the series $\sum |a_n|$ is divergent.

conditon of transversality In the calculus of variations, a generalization of the principle that the shortest line segment from a point to a curve is orthogonal to the curve at the point where it meets the curve. For a curve C given by parametric equations $x = f(t)$ and $y = g(t)$, and a function $h = h(x, y, y')$, the *condition of transversality* is $(h - y'h_{y'})f_t + h_{y'}g_t = 0$.

cone The solid bounded by a plane region (the base) and the line segments joining points on the boundary of the base to a fixed point (the vertex) that is not in the plane of the base. The cone is called a circular cone if the boundary of the base is a circle. In this case, the line joining the center of the circular base and the vertex is called the axis of the cone. A right circular cone is a circular cone whose axis is perpendicular to its base.

confluent differential equation The general *confluent differential equation* is of the form

$$y'' + \left[\frac{2a}{x} + 2f' + \frac{bh'}{h} - h' - \frac{h''}{h'}\right]y' + \left[\left(\frac{bh'}{h} - h' - \frac{h''}{h'}\right)\left(\frac{a}{x} + f'\right) + \frac{a(a-1)}{x^2} + \frac{2af'}{x} + f'' + (f')^2 - \frac{a(h')^2}{h}\right]y = 0$$

confluent hypergeometric differential equation The equation $xy'' + (\gamma - x)y' - \alpha y = 0$.

See also Whittaker's differential equation.

confocal conic sections Two conic sections with coincident foci.

conformal correspondence *See* conformal equivalence.

conformal equivalence Two regions Ω_1 and Ω_2 in the complex plane are *conformally equivalent* if there is a function f, that is holomorphic and injective on Ω_1, for which $f(\Omega_1) = \Omega_2$.

conformal mapping A *conformal mapping* (or *conformal transformation*) is a mapping $z \to f(z)$ that preserves angles; that is,

$$\lim_{r \to 0} e^{-i\theta} \frac{f(z_0 + re^{i\theta}) - f(z_0)}{|f(z_0 + re^{i\theta}) - f(z_0)|}$$

exists and is independent of θ, at every z_0. If f is a holomorphic function defined on a region Ω of the complex plane, and if $f'(z) \neq 0$ for all $z \in \Omega$, then f is a conformal mapping.

conformal parameters Let $x = x(u, v)$, $y = y(u, v)$, and $z = z(u, v)$ be parametric equations of a surface S. If this transformation from Cartesian (u, v)-space to S is conformal, then the coordinates (u, v) are known as *conformal parameters*.

conformal transformation *See* conformal mapping.

congruence A statement that two integers are congruent modulo n, i.e., a statement of the form $a \equiv b \pmod{n}$.

congruence of lines (1.) If A and B are distinct points in the Euclidean plane, the line segment $\overline{AB}$ consists of A and B with all the points which lie between A and B. We say that $\overline{uv}$ and $\overline{xw}$ are *congruent lines* in $\mathbf{R}^2$, where $\mathbf{R}$ is the real line, equipped with the usual Euclidean inner product $(\cdot, \cdot)$, if

$$d(u, v) = d(x, w),$$

where

$$d(u, v) = (v - u, v - u)^{1/2}.$$

(2.) A two-parameter submanifold of straight lines in three-dimensional projective space.

congruent (1.) Two geometric figures in the plane or in three-space such that one can be made coincident with the other by translations and rotations in three-space. For geometric figures in three-space, this relation is often referred to by saying that the figures are *directly congruent*. If one geometric figure in three-space is directly congruent to the reflection in a plane of another, then the figures are called *oppositely congruent*.

(2.) Two matrices A and B such that there exists an invertible matrix P such that $B = P^T AP$. Congruent matrices can be interpreted as describing the same linear transformation with respect to different bases, and the columns of the matrix P describe one basis in terms of the other.

(3.) Two integers a and b are called *congruent modulo n*, written $a \equiv b \pmod{n}$, if $a - b$ is a multiple of n.

conic *See* conic section.

conic section A curve obtained as the intersection of a plane with a right circular conical surface. Depending on the orientation of this plane, the *conic section* obtained is a circle, ellipse, hyperbola, parabola, or (if the vertex of the conical surface lies in the plane) a degenerate conic section—a point, a line, or a pair of intersecting lines.

Any *conic section* can be described in Cartesian coordinates by an algebraic equation of degree two in two variables. Conversely, any such equation which has a real solution has as its graph a conic section (possibly degenerate).

A *conic section* can also be described as the locus of a point such that the ratio of its distance to a fixed point (the focus) and its distance from a fixed line (the directrix) is a constant (the eccentricity e). The curve obtained is an ellipse if $0 < e < 1$, a parabola if $e = 1$, or a hyperbola if $e > 1$. By convention, the eccentricity of a circle is taken to be $e = 0$. In polar coordinates, with the focus at the origin and with a horizontal or vertical directrix at a distance d from the focus, the equation of a conic takes one of the forms

$$r = \frac{de}{1 \pm e\cos\theta}, r = \frac{de}{1 \pm e\sin\theta}.$$

conical surface The surface formed by the union of the lines passing through a fixed point (the vertex) and intersecting a fixed curve (the directrix). Each of these lines is called a *generator* of the surface. A *circular conical surface* is one that has a directrix which is a circle. The *axis* of a circular conical surface is the line through the center of this circle and the vertex. A circular conical surface is called *right* if the axis is perpendicular to the plane containing the directrix.

conjugate For a complex number $z = a + bi$, the number $\bar{z} = a - bi$.

conjugate axis For a hyperbola, the axis of symmetry that does not intersect the hyperbola. The *conjugate axis* is perpendicular to the transverse axis, which is the axis of symmetry on which lie the foci and the vertices of the hyperbola.

conjugate curves Two curves such that the principal normals of one curve are the principal normals of the other.

conjugate diameters The locus of the midpoints of a family of parallel chords. Two diameters of a conic section are *conjugate* if each diameter is one of the chords of the family that defines the other.

conjugate directions *Conjugate directions* at an elliptic or hyperbolic point on a surface are the directions of a pair of conjugate diameters to the Dupin indicatrix at that point.

conjugate exponent A real number p paired with a real number q such that $1 < p < \infty$, $1 < q < \infty$, and

$$\frac{1}{p} + \frac{1}{q} = 1.$$

By extension, 1 and ∞ are also a pair of *conjugate exponents*.

conjugate Fourier integral The integral $\frac{1}{\pi}\int_0^\infty \int_{-\infty}^\infty f(u) \sin t(u-x)du\,dt$, which is said to be conjugate to $\frac{1}{\pi}\int_0^\infty \int_{-\infty}^\infty f(u)$ $\cos t(u-x)du\,dt$.

conjugate function *See* conjugate harmonic function.

conjugate harmonic function A real harmonic function v arising from a given real harmonic function u so that the function $u + iv$ is analytic. When the domain of the functions is the open unit disk, v is usually chosen so that $v(0) = 0$.

conjugate hyperbola A hyperbola paired with another hyperbola so that both have the same asymptotes and the conjugate axis of each is the transverse axis of the other. If, with a suitable choice of coordinate axes, a hyperbola has the equation $\frac{x^2}{a^2} - \frac{y^2}{b^2} = 1$, then its conjugate hyperbola has the equation $-\frac{x^2}{a^2} + \frac{y^2}{b^2} = 1$.

conjugate hyperboloid A hyperboloid paired with another hyperboloid so that under a suitable choice of coordinate axes, the equation of the hyperboloid is $\frac{x^2}{a^2} + \frac{y^2}{b^2} - \frac{z^2}{c^2} = 1$, and the equation of its conjugate hyperboloid is $-\frac{x^2}{a^2} - \frac{y^2}{b^2} + \frac{z^2}{c^2} = 1$. A pair of *conjugate hyperboloids* have the property that one is an elliptic hyperboloid of one sheet if and only if the other is an elliptic hyperboloid of two sheets.

conjugate point (1.) A complex number z is conjugate to a complex number w with respect to a circle with center z_0 and radius r if $(z - z_0)\overline{(w - z_0)} = r^2$. That is, z and w are *conjugate points* with respect to the circle if they lie on the same radius (and its continuation) and the distance from z to z_0 times the distance from w to z_0 equals r^2. By extension, z and w are conjugate points with respect to a line if the line is the perpendicular bisector of the line segment joining z and w.

(2.) A complex number z is conjugate to a complex number w with respect to a conic if each lies on the polar of the other.

(3.) If a linear homogeneous ordinary differential equation has a solution on an interval $[a, b]$ that vanishes for $x = a$ and $x = b$ but is not identically zero, then the point $(b, 0)$ is said to be conjugate to the point $(a, 0)$.

(4.) If $\tilde{a} > a$ and the Euler equation $-\frac{d}{dx}(Ph') + Qh = 0$ corresponding to the quadratic functional $\int_a^b (Ph'^2 + Qh^2)\,dx$ has a solution that vanishes for $x = a$ and $x = \tilde{a}$ but is not identically zero, then $\tilde{a}$ is said to be a *conjugate point* to a.

(5.) Let M be a complete Riemannian manifold and $\gamma: [0, a] \to M$ be a geodesic. A point $\gamma(t_0)$, $t_0 \in (0, a]$, is said to be conjugate to $\gamma(0)$ along γ if there exists a Jacobi field J along γ such that $J(t_0) = J(0) = 0$ but J is not identically zero.

conjugate series A formal trigonometric series $\sum_{n=1}^{\infty}(-b_n \cos n\theta + a_n \sin n\theta)$ arising from the Fourier series $\frac{a_0}{2} + \sum_{n=1}^{\infty}(a_n \cos n\theta + b_n \sin n\theta)$. Under certain restrictions, conjugate series converge to the boundary functions of conjugate harmonic functions in the unit disk.

connection Let M be a $\mathbf{C}^\infty$ manifold with tangent bundle TM, and let $\Gamma^1(TM)$ be the collection of all $\mathbf{C}^1$ vector fields on M. A connection on M is a mapping $\nabla: TM \times \Gamma^1(TM) \to TM$ that for each $p \in M$ satisfies the following conditions for all ξ, η in the tangent space M_p, all Y, Y_1, Y_2 in $\Gamma^1(TM)$, all real numbers α, β, and all $\mathbf{C}^1$ functions f on M : $\nabla(\xi, Y) \in M_p$,

$\nabla(\alpha\xi + \beta\eta, Y) = \alpha\nabla(\xi, Y) + \beta\nabla(\eta, Y)$,

$\nabla(\xi, Y_1 + Y_2) = \nabla(\xi, Y_1) + \nabla(\xi, Y_2)$,

$\nabla(\xi, fY) = (\xi f)Y + f\nabla(\xi, Y)$,

and additionally satisfies the condition that $\nabla(X, Y)$ is a $\mathbf{C}^\infty$ vector field on M whenever X and Y are $\mathbf{C}^\infty$ vector fields on M.

connection form A 1-form ω_{ij} of a frame field E_1, E_2, E_3 on $\mathbf{R}^3$ such that for each tangent vector $\mathbf{v}$ to $\mathbf{R}^3$ at the point $\mathbf{p}$, $\omega_{ij}(\mathbf{v})$ is the dot product of the covariant derivative of $\mathbf{v}$ relative to E_i with E_j at $\mathbf{p}$.

connection formula A formula that relates locally valid solutions of a linear homogeneous differential equation or system of such equations so as to form a global solution.

connection problem The problem of determining connection formulas.

constant A quantity that does not vary. A symbol that represents the same quantity throughout a discussion.

constant of integration The symbol used to indicate that an indefinite integral represents *all* antiderivatives of a given function, not just a single one. If F is an antiderivative of f on an interval I, then the collection of all antiderivatives on I is represented by $F+C$, where C is a *constant of integration.* This can be written as $\int f(x)\,dx = F(x)+C$.

constant variational formula A formula for the solution of the system of linear equations $\dot{x}(t) = f(t, x(t+s)) + r(t)$ through the point (τ, ξ) of the form $x(t) = [L(t,\tau)\xi](0) + \int_\tau^t [L(t,s)\Gamma](0) r(s)\,ds$.

constituent One of uncountably many Borel sets in the decomposition of an analytic set.

contact element An equivalence class of submanifolds of a given dimension of a manifold M with the same order contact at a common point.

contact form A 1-form ω on a $(2n+1)$-dimensional differentiable manifold such that the exterior product $\omega \wedge (d\omega)^n \neq 0$, where $d\omega$ denotes the exterior derivative of ω.

contact manifold A $(2n+1)$-dimensional differentiable manifold with a contact form.

contact metric structure A triple (ϕ, ξ, ω) for a contact manifold consisting of a suitably chosen tensor field ϕ of type $(1, 1)$, vector field ξ, and contact form ω.

contact structure An infinitesimal structure of order one on a $(2n+1)$-dimensional differentiable manifold M^{2n+1} that is determined by defining on M^{2n+1} a contact form.

contact transformation A diffeomorphism f of a contact manifold such that $f^*\omega = \sigma\omega$, where σ is a non-zero function on the manifold.

continuity The property of being continuous (for a function).

continuity principle A theorem of Hartogs stating that a function of several complex variables that is holomorphic in each complex variable separately, must be continuous (in all variables simultaneously).

continuous at a point A function f is continuous at a point p_0 if for each neighborhood N of $f(p_0)$, $f^{-1}(N)$ is a neighborhood of p_0. Thus, f is continuous at p_0 if f has a limit at p_0 and that limit equals $f(p_0)$. This can be written as $\lim_{p\to p_0} f(p) = f(p_0)$.

continuous from the left A function f of a real variable is *continuous from the left* at a point x_0 if f has a limit from the left at x_0 and that limit from the left equals $f(x_0)$. This can be written as $\lim_{x\to x_0^-} f(x) = f(x_0)$.

continuous from the right A function f of a real variable is *continuous from the right* at a point x_0 if f has a limit from the right at x_0 and that limit from the right equals $f(x_0)$. This can be written as $\lim_{x\to x_0^+} f(x) = f(x_0)$.

continuous function A function that is continuous at each point in its domain.

continuous plane curve A continuous function from a closed bounded interval into the Euclidean plane. Also the image of such a function.

continuous spectrum (1.) If A is a linear operator whose domain and range are subspaces of a complex normed linear space X, the *continuous spectrum* of A is the set of all complex numbers λ for which the resolvent operator $(\lambda I - A)^{-1}$ exists and has domain dense in X, but is not bounded.
(2.) The part of the spectrum of the

Schroedinger equation that consists of intervals.

continuous with respect to a parameter A family of distributions $\{\Lambda_\alpha\}$ is said to be *continuous with respect to the parameter* α if for each complex valued function ϕ of class $\mathbf{C}^\infty$ on a subset of Euclidean space $\mathbf{R}^n$ with compact support, $\Lambda_\alpha(\phi)$ is a continuous function of α.

continuum A compact connected set with at least two points.

continuum of real numbers The set of all real numbers.

contour A piecewise smooth curve in the complex plane.

contour integral *See* line integral.

contracted tensor A tensor resulting from the contraction of a tensor.

contraction (1.) A mapping T of a metric space into itself for which there is a number $K < 1$ such that $\text{dist}(T(x), T(y)) \leq K\,\text{dist}(x, y)$ for all (x, y) in the space.
(2.) A linear Hilbert space operator of norm ≤ 1.
(3.) The process of forming a tensor from a given one by setting one covariant index and one contravariant index equal to common values and then adding over all such common values.

contraction principle The principle that any contraction T defined on a Banach space has a unique fixed point x_0: $Tx_0 = x_0$; and that, furthermore, Tx_n converges to x_0 for any sequence of points x_n in the Banach space given by $x_{k+1} = Tx_k$, $(k = 1, 2, \ldots)$ and x_1 arbitrary.

contractive analytic operator function A holomorphic function whose values are contractions.

contragradient The change in a vector, thought of as being in a dual space, when it and another vector are acted on by linear transformations in such a way that the inner product remains unchanged.

contravariant index The index i when the coordinates of a contravariant vector y are written as y^i, $i = 1, 2, \ldots, n$. The *contravariant index* is usually written as a superscript. *See* covariant index.

contravariant tensor A tensor of type $(r, 0)$. A *contravariant tensor* of rank (or order) r is an element of the tensor product $E \otimes \cdots \otimes E$ of r copies of a vector space E.

contravariant tensor field A mapping of a subset of a $\mathbf{C}^\infty$-manifold of dimension n whose values are contravariant tensors of fixed rank arising from a vector space of dimension n.

contravariant vector (1.) A row vector y appearing in an expression yAx, where x is a column vector, A is a square matrix, and y is thought of as being in a dual space.
(2.) A contravariant tensor of rank 1.

contravariant vector field A mapping of a subset of a $\mathbf{C}^\infty$-manifold of dimension n whose values are n-dimensional contravariant vectors.

convergence The state of being convergent.

convergence domain The set of points at which a sequence or series of functions is convergent.

convergence in mean order Convergence in the L^p norm. A sequence of functions $\{f_n\}$ converges in the mean of order p, $1 \leq p < \infty$, to f on a set X if $\int_X |f_n(x) - f(x)|^p\,dx \to 0$ as $n \to \infty$. If the order p is not stated explicitly, it is assumed to be 2.

convergence in measure If $f, f_1, f_2, \ldots$ are measurable functions whose values are finite almost everywhere, then the sequence $\{f_n\}$ converges to f in measure if for each $\epsilon > 0$, the measure of the set $\{t: |f_n(t) - f(t)| \geq \epsilon\}$ converges to 0 as $n \to \infty$.

convergent Tending toward a limit, as a convergent sequence or series.

convergent filter A filter F on a topological space X for which there exists a point $x \in X$ such that F is finer that the neighborhood filter of x. F converges to $x \in X$ if each neighborhood of x is a member of F.

convergent integral (1.) The improper integral $\int_a^\infty f(x)\,dx$ converges if the function f is Riemann integrable on $[a, b]$ for all $b > a$ and $\lim_{b\to\infty} \int_a^b f(x)\,dx$ exists (and is finite). In this case, this limit is denoted by $\int_a^\infty f(x)\,dx$.
(2.) The improper integral $\int_{-\infty}^b f(x)\,dx$ converges if the function f is Riemann integrable on $[a, b]$ for all $a < b$ and $\lim_{a\to-\infty} \int_a^b f(x)\,dx$ exists (and is finite). In this case, this limit is denoted by $\int_{-\infty}^b f(x)\,dx$.
(3.) The improper integral $\int_{-\infty}^\infty f(x)\,dx$ converges if for any (and consequently for all) c, both $\int_c^\infty f(x)\,dx$ and $\int_{-\infty}^c f(x)\,dx$ converge. In this case, the sum of $\int_c^\infty f(x)\,dx$ and $\int_{-\infty}^c f(x)\,dx$ is denoted by $\int_{-\infty}^\infty f(x)\,dx$.
(4.) If f is Riemann integrable on $[a, c]$ for all $c < b$ but not for $c = b$, then the improper integral $\int_a^b f(x)\,dx$ converges if $\lim_{c\to b^-} \int_a^c f(x)\,dx$ exists (and is finite). In this case, this limit is denoted by $\int_a^b f(x)\,dx$.
(5.) If f is Riemann integrable on $[c, b]$ for all $c > a$ but not for $c = a$, then the improper integral $\int_a^b f(x)\,dx$ converges if $\lim_{c\to a^+} \int_c^b f(x)\,dx$ exists (and is finite). In this case, this limit is denoted by $\int_a^b f(x)\,dx$.
(6.) If $a < c < b$ and f is Riemann integrable on every subinterval of $[a, b]$ missing c but not on $[a, b]$ and if both $\int_a^c f(x)\,dx$ and $\int_c^b f(x)\,dx$ converge, then the improper integral $\int_a^b f(x)\,dx$ converges and equals the sum of the first two integrals.

convergent net A net ϕ from a directed set D to a topological space X is said to converge to a point $p \in X$ if ϕ is eventually in every neighborhood of p. That is, for each neighborhood N of p, there exists a residual subset U of D such that $\phi(U) \subset N$.

convergent sequence A sequence of points having a limit. If $\{p_n\}$ is a sequence of points in a metric space X, then $\{p_n\}$ converges if there is a (unique) point p in X such that for each $\epsilon > 0$, there exits an integer N such that $\text{dist}(p_n, p) < \epsilon$ for all $n \geq N$.

convergent series A series whose sequence of partial sums converges. The series $\sum_{k=1}^\infty a_k$ is said to converge (to the sum S) if the sequence of partial sums $\{s_n\}$ converges (to S), where $s_n = \sum_{k=1}^n a_k$ for $n = 1, 2, \ldots$.

convex analysis The study of convex sets and convex functions.

convex downward Bulging downward. A curve in the plane is *convex downward* if whenever a segment of the curve is cut off by a chord, that segment lies on or below the chord. *See also* concave upward.

convex function A function ϕ defined on a convex set S that satisfies the inequality $\phi(\lambda x + (1-\lambda)y) \leq \lambda\phi(x) + (1-\lambda)\phi(y)$ whenever $x \in S$, $y \in S$, and $0 < \lambda < 1$. If ϕ is a function of one real variable and is differentiable on an interval (a, b), then ϕ is convex on (a, b) if and only if $\phi'(s) \leq \phi'(t)$ whenever $a < s < t < b$.

convex hull The *convex hull* of a set A is the intersection of all convex sets that contain A. It consists of all points expressible in the form $\lambda_1 x_1 + \cdots + \lambda_k x_k$, where $x_1, \ldots, x_k$ are points in A, $\lambda_j \geq 0$ for all j, and $\sum_j \lambda_j = 1$.

convex set A set S in a real linear space with the property that for every pair of points x and y in S, the point $\lambda x + (1 - \lambda)y$ is in S for all $0 < \lambda < 1$.

convex to the left Bulging to the left. A curve in the plane is *convex to the left* if whenever a segment of the curve is cut off by a chord, that segment lies on or to the left of the chord.

convex to the right Bulging to the right. A curve in the plane is *convex to the right* if whenever a segment of the curve is cut off by a chord, that segment lies on or to the right of the chord.

convex upward Bulging upward. A curve in the plane is *convex upward* if whenever a segment of the curve is cut off by a chord, that segment lies on or above the chord. *See also* concave downward.

convolution The function $f * g$ formed from two functions f and g, both integrable on $(-\infty, \infty)$, by $f*g(x) = \int_{-\infty}^{\infty} f(t)g(x - t)\,dt$. *Convolution* can also be defined by a similar formula for two functions, integrable with respect to Haar measure, on a locally compact Abelian group.

coordinate A scalar in an ordered set of scalars that determine the position of a point relative to a frame of reference. If $B = \{\mathbf{v}_1, \ldots, \mathbf{v}_n\}$ is an ordered basis of a vector space V and if $\mathbf{v} \in V$, then the uniquely determined scalars $x_1 \ldots x_n$ such that $\mathbf{v} = x_1\mathbf{v}_1 + \cdots + x_n\mathbf{v}_n$ are *coordinates* of $\mathbf{v}$ with respect to B. In particular, if $\mathbf{p} = (x_1, \ldots x_n)$ is a point in n-dimensional Euclidean space $\mathbf{R}^n$, then $x_1, \ldots x_n$ are *coordinates* of $\mathbf{p}$ (with respect to the standard basis for $\mathbf{R}^n$).

coordinate axis A line along which directed distance is measured from a specified point.

coordinate curve A curve in a hypersurface resulting from setting one curvilinear coordinate (different from the curvilinear coordinate that determines the hypersurface) constant.

coordinate geometry *See* analytic geometry.

coordinate hypersurface The hypersurface resulting from setting one curvilinear coordinate constant.

coordinate neighborhood If a family of pairs (U_i, ϕ_i) where each U_i is an open subset of a topological space M with $\bigcup_i U_i = M$ and each ϕ_i is a homeomorphism defined on U_i, forms an atlas of M, then each U_i, supplied with a local coordinate system, is a *coordinate neighborhood.*

coordinate plane **(1.)** A plane along with two coordinate axes that are used to locate points in the plane.
(2.) A plane in 3-dimensional Euclidean space determined by one of the coordinates being zero. For example, one coordinate plane is the yz–plane, which is determined by $x = 0$.

coordinate space *See* sequence space.

coordinate system A convention by which an ordered n-tuple of numbers locates the position of a point in n-dimensional space. Some common examples are Cartesian coordinates and polar coordinates in 2-dimensional Euclidean space and Cartesian coordinates, cylindrical coordinates, and spherical coordinates in 3-dimensional Euclidean space.

coordinates *See* coordinate.

Cornu spiral A curve in the plane satisfying the equation $r = k/s$, where r is the radius of curvature, k is a constant, and s is the arc length. It can be parameterized by the Fresnel integrals $x = \int_0^t \cos(\frac{1}{2}k\tau^2)\,d\tau$, $y = \int_0^t \sin(\frac{1}{2}k\tau^2)\,d\tau$. Also known as Euler's spiral or clothoid.

corona problem If M is the maximal ideal space of H^∞, the set of all bounded analytic functions on the open unit disc D with the supremum norm, then the corona problem is to determine whether (the natural embedding in M of) D is dense in M under the Gelfand topology. The corona problem can also be posed for domains other than D and Banach algebras other than H^∞.

Corona Theorem The theorem of Lennart Carleson that states that if M is the maximal ideal space of H^∞, then (the natural embedding in M of) the open unit disc D is dense in M under the Gelfand topology.

correspondence of Combescure A correspondence of two curves that has the property that corresponding points on the curves have parallel tangents.

correspondence A triple (R, X, Y), where X and Y are distinct sets and R is a subset of the Cartesian product $X \times Y$. *See also* relation.

cosigma function One of three specific elliptic functions of the third kind defined in terms of the Weierstrass sigma function.

cosine integral The function on $(0, \infty)$ given by $Ci(x) = -\int_x^\infty \frac{\cos t}{t}\, dt$.

covariance Given n sets of random variables denoted $\{x_1\}, \ldots, \{x_n\}$, a quantity called the covariance matrix is defined by

$$\begin{aligned} V_{ij} &= \mathrm{cov}(x_i, x_j) \\ &\equiv \langle (x_i - \mu_i)(x_j - \mu_j) \rangle \\ &= \langle x_i x_j \rangle - \langle x_i \rangle \langle x_j \rangle, \end{aligned}$$

where $\mu_i = \langle x_i \rangle$ and $\mu_j = \langle x_j \rangle$ are the *means* of x_i and x_j, respectively. An individual element V_{ij} of the *covariance matrix* is called the covariance of the two random variables x_i and x_j, and provides a measure of how strongly correlated these variables are. In fact, the derived quantity

$$\mathrm{cor}(x_i, x_j) \equiv \frac{\mathrm{cov}(x_i, x_j)}{\sigma_i \sigma_j},$$

where σ_i, σ_j are the standard derivations, is called the correlation of x_i and x_j. Note that if x_i and x_j are taken from the same set of random variables (say, x), then

$$\mathrm{cov}(x, x) = \langle x^2 \rangle - \langle x \rangle^2 = \mathrm{var}(x),$$

giving the usual variance $\mathrm{var}(x)$. The covariance is also symmetric since

$$\mathrm{cov}(x, y) = \mathrm{cov}(y, x).$$

For two variables, the covariance is related to the variance by

$$\mathrm{var}(x + y) = \mathrm{var}(x) + \mathrm{var}(y) + 2\,\mathrm{cov}(x, y).$$

For two independent random variables $x = x_i$ and $y = x_j$,

$$\begin{aligned} \mathrm{cov}(x, y) &= \langle xy \rangle - \mu_x \mu_y \\ &= \langle x \rangle \langle y \rangle - \mu_x \mu_y = 0, \end{aligned}$$

so the covariance is zero. However, if the variables are correlated in some way, then their covariance will be nonzero. In fact, if $\mathrm{cov}(x, y) > 0$, then y tends to increase as x increases. If $\mathrm{cov}(x, y) < 0$, then y tends to decrease as x increases.

The covariance obeys the identity

$$\begin{aligned} \mathrm{cov}(x + z, y) &= \langle (x + z)y - \langle x + z \rangle \langle y \rangle \rangle \\ &= \langle xy \rangle + \langle zy \rangle - (\langle x \rangle + \langle z \rangle)\langle y \rangle \\ &= \langle xy \rangle - \langle x \rangle \langle y \rangle + \langle zy \rangle - \langle z \rangle \langle y \rangle \\ &= \mathrm{cov}(x, y) + \mathrm{cov}(z, y). \end{aligned}$$

By induction, it therefore follows that

$$\mathrm{cov}\left(\sum_{i=1}^n x_i, y \right) = \sum_{i=1}^n \mathrm{cov}(x_i, y)$$

$$\begin{aligned} &\mathrm{cov}\left(\sum_{i=1}^n x_i, \sum_{j=1}^m y_j \right) \\ &= \sum_{i=1}^n \mathrm{cov}\left(x_i, \sum_{j=1}^m y_j \right) \end{aligned}$$

$$= \sum_{i=1}^{n} \text{cov}\left(\sum_{j=1}^{m} y_j, x_i\right)$$

$$= \sum_{i=1}^{n}\sum_{j=1}^{m} \text{cov}(y_j, x_i)$$

$$= \sum_{i=1}^{n}\sum_{j=1}^{m} \text{cov}(x_i, y_i).$$

covariant derivative The analog for surfaces of the usual differentiation of vectors in the plane. Let w be a differentiable vector field in an open subset U of a surface S, and let $p \in U$. Let v be a tangent vector to S at p corresponding to a curve γ in S so that $\gamma(0) = p$ and $\gamma'(0) = v$, and let $w(t)$ be the restriction of the vector field w to the curve γ. The *covariant derivative* at p of the vector field w relative to the vector v is the normal projection vector of $(dw/dt)(0)$ onto the tangent plane to S at p.

covariant differential If α is a k-form on a C^∞ manifold with values in a vector space V, then the covariant differential of α is the $(k+1)$-form $D\alpha$ given by $D\alpha(Y_1, \dots, Y_{k+1}) = (d\alpha)(hY_1, \dots, hY_{k+1})$, where d is the ordinary differential and hY is the horizontal component of Y.

covariant index The index i when the coordinates of a covariant vector x are written as $x_i, i = 1, 2, \dots, n$. The *covariant index* is usually written as a subscript. *See* contravariant index.

covariant tensor A tensor of type $(0, s)$. A covariant tensor of rank s is an element of the tensor product $E^* \otimes \cdots \otimes E^*$ of s copies of E^*, the dual space of a vector space E.

covariant tensor field A mapping of a subset of a C^∞-manifold of dimension n whose values are covariant tensors of fixed rank arising from a vector space of dimension n.

covariant vector (1.) A column vector x appearing in an expression yAx, where y is a row vector, A is a square matrix, and y is thought of as being in a dual space.
(2.) A covariant tensor of rank 1.

covariant vector field A mapping of a subset of a C^∞-manifold of dimension n whose values are n-dimensional covariant vectors.

cramped module *See* Hilbert module.

critical point (1.) *See* critical value.
(2.) A point at which the gradient of a function f of several variables either vanishes or does not exist. The *critical points* of f are the points at which f might possibly, but not necessarily, have extrema.
(3.) For a system of differential equations

$$x_1'(t) = F_1(x_1, \dots, x_n)$$
$$\vdots$$
$$x_n'(t) = F_n(x_1, \dots, x_n),$$

$\mathbf{c}$ is a critical point if $F_i(\mathbf{c}) = 0$ for $i = 1, \dots, n$. In this situation, $\mathbf{c}$ is sometimes referred to as an equilibrium point.

critical value A point at which the first derivative of a function f of one real variable either vanishes or does not exist. The critical values of f are the points at which f might possibly, but not necessarily, have extrema. Also, often referred to as critical point.

cross product *See* vector product of vectors.

curl A vector field

$$\nabla \times \mathbf{F} = \left(\frac{\partial P}{\partial y} - \frac{\partial N}{\partial z}\right)\mathbf{i} + \left(\frac{\partial M}{\partial z} - \frac{\partial P}{\partial x}\right)\mathbf{j} + \left(\frac{\partial N}{\partial x} - \frac{\partial M}{\partial y}\right)\mathbf{k}$$

corresponding to a given vector field $\mathbf{F} = M\mathbf{i} + N\mathbf{j} + P\mathbf{k}$. In matrix notation,

$$\nabla \times \mathbf{F} = \begin{vmatrix} \mathbf{i} & \mathbf{j} & \mathbf{k} \\ \frac{\partial}{\partial x} & \frac{\partial}{\partial y} & \frac{\partial}{\partial z} \\ M & N & P \end{vmatrix}.$$

If **F** is a velocity field of a fluid, then $\nabla \times \mathbf{F}$ gives the direction of the axis about which the fluid rotates most rapidly and $\|\nabla \times \mathbf{F}\|$ is a measure of the speed of this rotation.

current An exterior differential form whose coefficients are distributions.

curvature (1.) A measure of how sharply a curve bends. Consider a smooth curve C in two- or three-dimensional Euclidean space that is traced out by a vector function $\mathbf{r}(t)$. Let $s(t)$ represent arc length and let $\mathbf{T}(t) = r'(t)/\|r'(t)\|$ be the unit tangent vector for the curve. The *curvature* κ of C is defined as $\kappa = \|d\mathbf{T}/ds\|$. (The vector $d\mathbf{T}/ds$ is said to be the curvature vector.) In particular, if the curve is the graph in two-dimensional space of $y = g(x)$, then $\kappa = \frac{|y''|}{[1+y'^2]^{3/2}}$.
(2.) A measure of the manner in which a surface bends. Consider a point **p** on a smooth surface S in three-dimensional Euclidean space, and let **n** be a normal vector to the surface at **p**. For each plane that is through **p** and parallel to **n**, the intersection of the plane with S is a curve, called a normal section. For each normal section, put $k = \pm\kappa$, where κ is the curvature of the normal section and the plus is chosen if the curvature vector points in the same direction as **n** while the minus is chosen if the curvature vector points in the opposite direction. The principal curvatures, k_1 and k_2, at **p** are by definition the extreme values of the set of all values of k. The Gaussian curvature is $G = k_1k_2$; the mean curvature is said by some to be $H = k_1 + k_2$, while others say $H = (k_1 + k_2)/2$.

curvature form The covariant differential of a connection form. Measures the deviation of a connection from a locally flat connection. A 2-form $\Omega_{ij} = d\omega^i_j + \omega^i_k \wedge \omega^k_j$, where the 1-forms ω^i_j are connection forms for a local basis.

curvature tensor A tensor of type $(1, 3)$ obtained from the decomposition of a curvature form Ω_{ij}. The curvature tensor R satisfies $(R(e_i, e_j)e_k, e_l) = \Omega_{ij}(e_k, e_l)$, where $\{e_i\}$ is an orthonormal basis for the tangent space.

curve (1.) A continuous function defined on an interval, into some higher dimensional space. Often the interval is required to be a closed, bounded interval $[a, b]$.
(2.) The image of such a continuous function.

curve of constant width A simple, closed curve in the plane whose interior is a convex set with constant width. For a smooth curve, all parallel tangents are the same distance from each other. Also known as an orbiform curve.

curve of pursuit The path taken by a point in the plane moving with constant velocity if it is always heading toward another point that is moving with constant velocity along a given curve.

curve of quickest descent The curve among all curves between two points that minimizes the time it takes a particle to move from the higher point to the lower point, under the assumption that the particle is subject to a constant gravitional force but no friction. The problem of finding this curve is known as the brachistochrone problem. Its answer is the portion of a suitably chosen cycloid joining the two points.

curve of the second class A curve in the Cartesian plane enveloped by the family of lines $ux+vy+w = 0$, where the coefficients u, v, w satisfy an equation of the form $Au^2+Buv+Cv^2+Duw+Evw+Fw^2 = 0$, with $A^2 + B^2 + C^2 \neq 0$.

curve of the second order A curve in the Cartesian plane given by a second degree polynomial equation in the variables x and y, $Ax^2 + Bxy + Cy^2 + Dx + Ey + F = 0$, where $A^2 + B^2 + C^2 \neq 0$.

curve tracing The process of finding and sketching the shape of a curve, usually in the plane, from the equation that determines the curve.

curvilinear coordinates Local coordinates in n-dimensional Euclidean space $\mathbf{R}^n$. If $x_1, \ldots, x_n$ are rectangular coordinates in $\mathbf{R}^n$ and if for all $i = 1, \ldots, n$, x_i is a continuously differentiable function of $u_1, \ldots, u_n$ such that the Jacobian determinant $|\partial(x_1, \ldots, x_n)/\partial(u_1, \ldots, u_n)|$ is non-zero throughout some open set, then $u_1, \ldots, u_n$ are said to be *curvilinear coordinates* of $\mathbf{R}^n$.

curvilinear integral *See* line integral.

curvilinear integral with respect to line element A line integral $\int_C f(\mathbf{r})\, ds = \int_a^b f(\mathbf{r}(t)) \|\mathbf{r}'(t)\|\, dt$, where C is a piecewise smooth curve traced out by $\mathbf{r}(t)$ for $a \leq t \leq b$ and f is a continuous scalar field on C. Also known as a line integral with respect to arc length.

cusp (1.) A point p on a plane curve at which there is no tangent and yet tangents to the two branches of the curve meeting at p have the same limit as p is approached. If the curve is thought of as being traced out by a moving particle, then a cusp is a point at which the particle reverses direction. For example, the cycloid $x = R(t - \sin t)$, $y = R(1 - \cos t)$ has a cusp at each point for which $y = 0$.
(2.) If $\bar{H}$ is the union of the upper half plane $\{z = x + iy : y > 0\}$ and the point at infinity and the set of rational numbers on the real axis, then the point at infinity and the rational numbers on the real axis are said to be cusps of $\bar{H}$.

cusp form A cusp form f of weight k on $\Gamma_0(N)$, the group of all 2×2 integer matrices of determinant 1 which are upper triangular mod N, is an analytic function on the upper half plane $\{z = x+iy : y > 0\}$ that vanishes at infinity and satisfies the relation $f\left(\frac{az+b}{cz+d}\right) = (cz + d)^k f(z)$ for all $\begin{bmatrix} a & b \\ c & d \end{bmatrix} \in \Gamma_0(N)$.

cut *See* branch.

cut locus The set of all cut points for geodesics starting from a given point p on a manifold.

cut point A point for which a geodesic starting from a given point in a manifold stops globally minimizing arc length. If p is a point on a complete Riemannian manifold and $\gamma: [0, \infty) \to M$ is a normalized geodesic with $\gamma(0) = p$, $\gamma(t_0)$ is the *cut point* of p along γ if $\text{dist}(\gamma(0), \gamma(t_0)) = t_0$ but $\text{dist}(\gamma(0), \gamma(t)) > t$ for all $t > t_0$.

cyclic element A point x in a topological linear space X is a *cyclic element* corresponding to a linear operator T on X if the linear span of $\{T^n x : n = 0, 1, 2, \ldots\}$ is dense in X.

cyclic vector *See* cyclic element.

cyclide A quartic surface in $\mathbf{R}^n$ that envelopes the family of hyperspheres tangent to n fixed hyperspheres.

cyclide of Dupin A surface for which all lines of curvature are circles. A *cyclide of Dupin* can be obtained by inverting a torus in a sphere.

cycloid A curve traced out by a point on a circle that rolls without slipping along a line. If R is the radius of the circle and t is the angle of rotation, the parametric equations $x = R(t - \sin t)$, $y = R(1 - \cos t)$ determine a cycloid.

cylindrical coordinates A coordinate system for 3-dimensional Euclidean space consisting of a plane with polar coordinates and an axis through the pole (or origin) of the plane that is perpendicular to the plane. The *cylindrical coordinates* of a point are commonly written as (r, θ, z), where r and θ are polar coordinates and z is a Cartesian coordinate for the perpendicular axis.

cylindrical surface A surface swept out by a line, called the generator, moving parallel to itself and intersecting a given curve, called the directrix.

C* algebra An *abstract* **C***-*algebra* is a Banach algebra $\mathcal{A}$ over the complex numbers, equipped with an involution * satisfying $\|x^*x\| = \|x\|^2$ for each $x \in \mathcal{A}$.

A *concrete* **C***-*algebra* is a self-adjoint algebra of Hilbert space operators which is closed in the norm topology.

D

D'Alembert's method of reduction of order A method for finding a second solution ϕ_2 of a second order linear homogeneous ordinary differential equation given a first solution ϕ_1 by assuming that $\phi_2 = u\phi_1$ for some unknown function u, whose derivative u' is found by solving a first order equation.

More generally, suppose that ϕ_1 is a solution of a linear homogeneous equation of order n on an interval I and ϕ_1 is never zero on I. Then $\{\phi_1, u_2\phi_2, \ldots, u_n\phi_n\}$ is a basis for the solution space of the equation on I for appropriately chosen functions $u_2, \ldots, u_n$, whose first derivatives are solutions of a certain equation of order $n - 1$.

Daniell integral Consider an arbitrary set X and let $\mathcal{L}$ denote a linear lattice of real-valued functions on X. This means that $\mathcal{L}$ is a vector space and for every $f, g \in \mathcal{L}$, the functions $(f \vee g)(x) = \max\{f(x), g(x)\}$ and $(f \wedge g)(x) = \min\{f(x), g(x)\}$ are in $\mathcal{L}$. A linear functional $I : \mathcal{L} \to \mathbf{R}$ on the linear lattice $\mathcal{L}$ is said to be *positive* if $I(f) \geq 0$ for every nonnegative function $f : X \to \mathbf{R}$ in $\mathcal{L}$.

On $\mathcal{L}$, a *Daniell integral* is a positive functional $I : \mathcal{L} \to \mathbf{R}$ such that for every sequence $\{f_n : n \geq 1\}$ of nonnegative functions in $\mathcal{L}$,

$$I(f) \leq \sum_{n=1}^{\infty} I(f_n)$$

for every function $f : X \to \mathbf{R}$ in $\mathcal{L}$ satisfying

$$f(x) \leq \sum_{n=1}^{\infty} f_n(x),$$

$(x \in X)$.

Darboux sum An *upper Darboux sum* of a function f on an interval $[a, b]$ with respect to a partition $P = \{x_0, \ldots, x_n\}$ of $[a, b]$ is a sum $S^+(f, P) = \sum_{i=1}^{n} M_i \Delta x_i$, where $M_i = \sup\{f(x): x \in (x_{i-1}, x_i)\}$ and $\Delta x_i = x_i - x_{i-1}$. A *lower Darboux sum* of f on $[a, b]$ with respect to $P = \{x_0, \ldots, x_n\}$ of $[a, b]$ is a sum $S_-(f, P) = \sum_{i=1}^{n} m_i \Delta x_i$, where $m_i = \inf\{f(x): x \in (x_{i-1}, x_i)\}$ and $\Delta x_i = x_i - x_{i-1}$.

Darboux's curve The curve on a surface in three-dimensional projective space having third order contact with Darboux's quadric.

Darboux's frame Corresponding to a point $\mathbf{p}$ on a surface S, *Darboux's frame* is a triple of vectors, one being a unit normal vector $\mathbf{n}$ to S at $\mathbf{p}$, and the other two being mutually orthogonal principal unit vectors $\mathbf{r}_1$ and $\mathbf{r}_2$ to S at $\mathbf{p}$, such that $\mathbf{n} = \mathbf{r}_1 \times \mathbf{r}_2$.

Darboux's quadric For a surface S in three-dimensional projective space, *Darboux's* quadratic is a surface of the second order having second order contact to S at a point p in such a way that the line of intersection has certain specified properties.

Darboux's tangent The tangent to Darboux's curve.

Darboux's Theorem (1.) If f is a differentiable real valued function on an interval $[a, b]$ such that $f'(b) \neq f'(a)$, and V is any number between $f'(a)$ and $f'(b)$, then there exists at least one $c \in (a, b)$ such that $f'(c) = V$.

(2.) For any bounded function f on an interval $[a, b]$, let l_- be the least upper bound of the set of all Darboux lower sums and l^+ be the greatest lower bound of the set of all Darboux upper sums of f. A theorem of Darboux says that for any $\epsilon > 0$, there exists $\delta > 0$ such that $|S_-(f, P) - l_-| < \epsilon$ and $|S^+(f, P) - l^+| < \epsilon$ for every Darboux lower sum $S_-(f, P)$ and upper sum $S^+(f, P)$ corresponding to any partition P of $[a, b]$ of norm $< \delta$.

0-8493-0320-6/00/$0.00+$.50

decreasing function (1.) A real-valued function f on an interval I such that $f(x_2) < f(x_1)$ for all x_1 and x_2 in I with $x_2 > x_1$. Correspondingly, f is said to be nonincreasing if $f(x_2) \leq f(x_1)$ for all x_1 and x_2 in I with $x_2 > x_1$.
(2.) A real-valued function f on an interval I such that $f(x_2) \leq f(x_1)$ for all x_1 and x_2 in I with $x_2 > x_1$. Correspondingly, f is said to be strictly decreasing if $f(x_2) < f(x_1)$ for all x_1 and x_2 in I with $x_2 > x_1$.

decreasing sequence (1.) A sequence $\{a_n\}$ such that $a_{n+1} < a_n$ for all n. Correspondingly, $\{a_n\}$ is said to be nonincreasing if $a_{n+1} \leq a_n$ for all n.
(2.) A sequence $\{a_n\}$ such that $a_{n+1} \leq a_n$ for all n. Correspondingly, $\{a_n\}$ is said to be strictly decreasing if $a_{n+1} < a_n$ for all n.

deficiency For a linear transformation $T: X \to Y$, the *deficiency* of T is the dimension of the vector space complement of the range of T.

deficiency index For an unbounded, symmetric operator T on a Hilbert space H, the numbers

$$\nu_{\pm i}(T) = \text{dimker}(T^* \pm iI).$$

The *deficiency indices* measure how far the operator T comes to being self-adjoint; they are both 0 if and only if the closure of T is self-adjoint and they are equal if and only if T has a self-adjoint extension in the space H.

defining function For a domain $\Omega \subset \mathbf{R}^n$ [resp., $\mathbf{C}^n$] a function $\rho : \mathbf{R}^n \to \mathbf{R}$ [resp., $\rho : \mathbf{C}^n \to \mathbf{R}$] such that

$$\Omega = \{z : \rho(z) < 0\}.$$

One can then formulate properties of Ω in terms of ρ; for example, by definition, Ω has $\mathbf{C}^k$ boundary if ρ is a $\mathbf{C}^k$ function with $\text{grad}\rho(z) \neq 0$ for $z \in \partial\Omega$.

definite integral The Riemann integral of a function over a set in $\mathbf{R}$, as opposed to the *indefinite* integral of a function f, which refers to an integral over an interval $[a, x]$ with variable endpoint, or to an antiderivative of f.

degeneracy (1.) Given a pseudodistance d on a topological space X (i.e., d satisfies the axioms of a metric, except that $d(p, q)$ may $= 0$ for certain $p, q \in X$ with $p \neq q$) the *degeneracy set* $\Delta(p)$ of d at $p \in X$ is

$$\Delta(p) = \{q \in X : d(p, q) = 0\}.$$

The pseudodistance d is said to have *compact degeneracy* if $\Delta(p)$ is compact, for every $p \in X$.
(2.) Given a pseudolength F on a vector space X (a nonnegative real valued function F on X satisfying $F(av) = |a|F(v)$, for $a \in \mathbf{C}$ and $v \in X$) a *degeneracy point* of F is a point $x \in X$ such that $F(x) = 0$. The *degeneracy set* of F is the set of all degeneracy points of F.

degenerate conic Either a *degenerate ellipse:* an ellipse with both axes 0, that is, a point; a *degenerate hyperbola:* a hyperbola whose axes are 0, that is, a pair of intersecting lines; or a *degenerate parabola:* the locus of a quadratic equation that is a pair of straight lines.

degenerate function *See* degenerate mapping.

degenerate mapping A mapping f, into a projective space P^n, such that the image of f is contained in a subspace of P^n of lower dimension.

degree of divisor class On a curve, a divisor (class) is a linear combination of points: $D = \sum \lambda_j x_j$ and the degree is $\sum_j \lambda_j$.

For an irreducible variety V, a divisor is a sum of the form $D = \sum \lambda_j C_j$, where $C_1, \ldots, C_r$ are irreducible subvarieties of codimension 1 of V, with preassigned integral multiplicities $\lambda_1, \ldots, \lambda_r$. The *degree* of D is defined to be $\text{deg}D = \sum \lambda_j \text{deg}C_j$.

deMoivre's Theorem The relation

$$(\cos\theta + i\sin\theta)(\cos\phi + i\sin\phi)$$

$$= \cos(\theta+\phi) + i\sin(\theta+\phi).$$

Denjoy integrable function *See* Denjoy integral.

Denjoy integral A function $f : [a,b] \to \mathbf{R}$ is *Denjoy integrable* provided there exists a function $F : [a,b] \to \mathbf{R}$, which is of generalized absolute continuity in the restricted sense and such that $F' = f$, almost everywhere on $[a,b]$. *See* generalized absolute continuity in the restricted sense.

The Denjoy integral of f over $[a,b]$ is $F(b) - F(a)$. *See* Denjoy-Lusin Theorem.

Denjoy integral in the restricted sense Construct a transfinite sequence (of the type $\Omega + 1$) of more and more general integrals which we call D_ξ-integrals, and which we denote by

$$(D_\xi)\int_a^b f(x)\,dx \qquad (1)$$

for a function $f(x)$ defined on the closed interval $[a,b]$.

The integrals (1) are defined inductively. Namely,

$$(D_0)\int_a^b f(x)\,dx$$

is to be understood as simply the Lebesgue integral

$$(L)\int_a^b f(x)\,dx.$$

We shall now assume that we have defined integrals (1) for all $\xi < \eta$, where $\eta \leq \Omega$. If η is a number of the first kind and $\eta - 1$ is its immediate predecessor, then, setting

$$\overset{b}{\underset{a}{\mathbf{T}}}(f) = (D_{\eta-1})\int_a^b f(x)\,dx,$$

we define the integral D_η by means of the formula

$$(D_\eta)\int_a^b f(x)\,dx = \overset{b}{\underset{a}{\mathbf{T}_*}}(f).$$

But if η is a number of the second kind, then we include in the class of functions which are D_η-integrable on the closed interval $[a,b]$ all functions $f(x)$ which are D_ξ-integrable on this segment for at least one $\xi < \eta$, and we set, by definition,

$$(D_\eta)\int_a^b f(x)\,dx = (D_{\xi_0})\int_a^b f(x)\,dx,$$

where ξ_0 is the smallest of these ξ. Thus, if η is a number of the second kind, then the class of functions which are D_η-integrable is the set-theoretic sum (over all $\xi < \eta$) of classes of D_ξ-integrable functions. Therefore the definition of the integrals (1) which we have just given can be written in terms of the following symbolic relations:

$$D_0 = L, \quad D_\eta = (D_{\eta-1})_*, \quad D_\eta = \sum_{\xi<\eta} D_\xi,$$

where the second relation must be used if η is a number of the first kind and the third if η is of the second kind.

In particular, taking $\xi = \Omega$, we arrive at the *Denjoy integral in the restricted sense.* The *Denjoy integral in the restricted sense* of the function $f(x)$, defined on the closed interval $[a,b]$, is the integral

$$(D_\Omega)\int_a^b f(x)\,dx. \qquad (2)$$

This integral is called the *Denjoy–Perron integral.* Instead of the notation (2), it is usually denoted by

$$(D)\int_a^b f(x)\,dx.$$

Using the same notation as above, we can write the formula

$$D = \sum_{\xi<\Omega} D_\xi,$$

so that every function which is Denjoy-integrable is also D_ξ-integrable for some $\xi < \Omega$ and if the smallest of these ξ is ξ_0, then

$$(D)\int_a^b f(x)\,dx = (D_{\xi_0})\int_a^b f(x)\,dx.$$

Denjoy integral in the wide sense Suppose $\overset{b}{\underset{a}{\mathbf{T}}}(f)$ is an arbitrary given integral. We shall construct a generalization, $\overset{b}{\underset{a}{\mathbf{T}^*}}(f)$.

Namely, we include in the class $T^*([a,b])$ every function $f(x)$ which is defined on $[a,b]$ and satisfies the following four conditions:

1) the set $S_T = S_T(f;[a,b])$ is nowhere dense and $f(x)$ is not Lebesgue-summable on S_T;

2) if $\{(a_n,b_n)\}$ is the set of intervals complementary to S_T, then the limit

$$I_n = \lim \overset{\beta}{\underset{\alpha}{\mathbf{T}}}(f)$$

$(a_n < \alpha < \beta < b_n, \alpha \to a_n, \beta \to b_n)$ exists and is finite for every n;

3) the inequality

$$\sum_n |I_n| < +\infty \tag{1}$$

holds;

4) if there are an infinite number of intervals (a_n,b_n) and

$$W_n = \sup|\overset{\beta}{\underset{\alpha}{\mathbf{T}}}(f)| \quad (a_n < \alpha < \beta < b_n),$$

then

$$\lim W_n = 0. \tag{2}$$

We see that the difference between the definitions of the classes $T^*([a,b])$ and $T_*([a,b])$ consists in replacing the single requirement

$$\sum_n W_n < +\infty \tag{3}$$

by the two requirements (1) and (2). Since $|I_n| \ll W_n$, (1) and (2) follow from (3), from which we have

$$T_*([a,b]) \subset T^*([a,b]).$$

It follows from this, in particular, that the classes $T^*([a,b])$ are non-void. We can show that the system of these classes is legitimate.

Having introduced the classes $T^*([a,b])$, we can define a functional $\overset{b}{\underset{a}{\mathbf{T}^*}}$ on each of them by associating the number

$$\overset{b}{\underset{a}{\mathbf{T}^*}}(f) = \sum_n I_n + (L)\int_{S_T} f(x)\,dx$$

with each $f \in T^*([a,b])$.

This integral is constructed with the aid of the integral T^*, in exactly the same way that the integral in the restricted sense was constructed with the aid of T_*.

Namely, we introduce the transfinite (of type $\Omega+1$) sequence of integrals

$$(D_\xi)\int_a^b f(x)\,dx, \tag{4}$$

defined by induction. For $\xi = 0$, integral (4) is simply the Lebesgue integral. If integrals (4) have been defined for all $\xi < \eta$, where $\eta \leq \Omega$, then we set

$$D^\eta = (D^{\eta-1})^*$$

for all η of the first kind and we set

$$D^\eta = \sum_{\xi<\eta} D^\xi$$

for η of the second kind.

The integral

$$D^\Omega = \sum_{\xi<\Omega} D^\xi$$

is the *Denjoy integral in the wide sense.*

It can be proved that the Denjoy integral in the wide sense is more general than the Denjoy integral in the restricted sense. Moreover, we have in general that the integral D^{ξ} is more general than D_{ξ} for arbitrary $\xi \leq \Omega$.

Denjoy-Carleman Theorem Let $\mathbf{C}^{\#}\{M_n\}$ denote the class of infinitely differentiable functions f on the circle $\mathbf{T}$ satisfying

$$\|f^{(n)}\|_{L^2} \leq K^n M_n, n = 1, 2, \ldots$$

for some constant $K = K_f$. Assume that $M_n > 0$ and $\log M_n$ is a convex function of n. Then the following are equivalent
(i.) $\mathbf{C}^{\#}\{M_n\}$ is a quasi-analytic class;
(ii.) the function $\tau(r) = \inf_{n \geq 0} M_n/r^n$ satisfies

$$\int_0^{\infty} \frac{\log \tau(r)}{1+r^2} dr = -\infty;$$

(iii.) $\sum \frac{M_n}{M_{n+1}} = \infty$.

Denjoy-Lusin Theorem Suppose

$$\sum_{j=-\infty}^{\infty} c_j e^{ijx}$$

$$= \frac{1}{2} a_0 + \sum_{j=1}^{\infty} (a_j \cos jx + b_j \sin jx),$$

with a_j, b_j real, and assume that $\sum_{j=0}^{\infty} |c_j e^{ijx} + c_{-j} e^{-ijx}|$

$$= |a_0| + \sum_{j=1}^{\infty} |a_j \cos jx + b_j \sin jx|$$

converges for x in some measurable set E of positive Lebesgue measure. Then $\sum_{j=-\infty}^{\infty} |c_j| < \infty$. Equivalently, $\sum_{j=1}^{\infty} (|a_j| + |b_j|) < \infty$.

denumerable set A set in one-to-one correspondence with the nonnegative integers. *At most denumerable* means finite or denumerable. Sometimes, *denumerable* is used to include finite sets. Also *countable.*

dependent variable The variable in a functional relationship whose value is determined by the other variable(s) (the *dependent variables*). Thus if $y = f(x_1, x_2, \ldots, x_n)$, the variable y is the *dependent* variable and $x_1, x_2, \ldots, x_n$ are the *dependent* variables. Also *ordinate.*

deRham cohomology group Let M be an n-dimensional manifold. A differential i-form ω on M is called *closed* if $d\omega = 0$ and *exact* if $\omega = d\sigma$. The *deRham cohomology group,* $H^i(M)$ is defined to be the group of closed i-forms modulo the exact i-forms.

deRham cohomology ring Let $H^i(M)$ be the deRham group of dimension i of a manifold M of dimension n. (*See* deRham cohomology group.) The *deRham cohomology ring* of M is the direct sum

$$H^*(M) = H^0(M) \oplus \cdots \oplus H^n(M).$$

In fact, $H^*(M)$ is an algebra with the multiplication induced by the exterior product on differential forms.

deRham's Theorem The deRham cohomology ring (*See* deRham cohomology ring) is isomorphic to the usual cohomology ring (the direct sum of the cohomology groups with cup product).

derivate The *upper derivate* at $x = c$ of a function $f : [c, b) \to \mathbf{R}$ is $\lim_{h \to 0+} \frac{f(c+h)-f(c)}{h}$. *Lower derivates* are similarly defined, for a function $f : (a, c] \to \mathbf{R}$. Clearly, f is *differentiable* at $x = c$ if and only if both the upper and lower derivates exist at $x = c$ and they are equal.

derivative For a function $f : \mathbf{R} \to \mathbf{R}$ (or $f : \mathbf{C} \to \mathbf{C}$) the limit

$$\lim_{h \to 0} \frac{f(x+h)}{h},$$

whenever it exists. Any of the notations $\frac{df}{dx}$, $D_x f$ or f' are standard, for the above limit, or, if the variable is time, the notation

$\dot{f}(t)$ may be used. *See also* partial derivative, differential, Radon-Nikodym derivative.

derivative at a point The derivative function (*See* derivative) evaluated at a particular value of the variable. Hence, the limit

$$f'(a) = \lim_{h \to 0} \frac{f(a+h) - f(a)}{h},$$

when it exists. The number $f'(a)$ represents the slope of the tangent to the curve $y = f(x)$ at $x = a$, and a variety of instantaneous rates of change.

derivative of a function *See* derivative, derivative at a point.

developable surface A surface generated by the motion of a straight line such that every generating line intersects the following one.

diffeomorphism A map $F : M_1 \to M_2$, between two differentiable manifolds, such that, whenever (U_p, ϕ_p), (U_q, ϕ_q) are coordinate charts on M_1 and M_2, respectively, the maps $\phi_q \circ F \circ \phi_p^{-1}$ and $\phi_p \circ F \circ \phi_q^{-1}$ are differentiable maps (between open subsets of $\mathbf{R}^n$). *See* atlas.

diffeomorphism of class $\mathbf{C}^r$ A map $F : M_1 \to M_2$, between two $\mathbf{C}^r$-manifolds, such that, whenever (U_p, ϕ_p), (U_q, ϕ_q) are coordinate charts on M_1 and M_2, respectively, the maps $\phi_q \circ F \circ \phi_p^{-1}$ and $\phi_p \circ F \circ \phi_q^{-1}$ are $\mathbf{C}^r$ maps (between open subsets of $\mathbf{R}^n$). *See* atlas.

difference equation An equation in an unknown sequence $\{x_j\}$ and its "differences," $\Delta x_j = x_j - x_{j-1}, \Delta^2 x_j = x_j - 2x_{j-1} + x_{j-2}, \ldots$. Analogues of theorems from differential equations can be obtained by considering the nth difference as analogous to the nth derivative.

difference quotient For a real or complex function $f(x)$, defined on a subset of $\mathbf{R}$ or $\mathbf{C}$, the expression

$$\frac{f(a+h) - f(a)}{h}.$$

For a real-valued function of a real variable, the difference quotient represents the slope of the line *secant* to the curve $y = f(x)$, between the points $(a, f(a))$ and $(a+h, f(a+h))$. The notation $\frac{\Delta f}{\Delta x}$ is sometimes used for the difference quotient.

The derivative of f is the limit of the difference quotient and represents the slope of the *tangent* to $y = f(x)$ at the point $(a, f(a))$.

differentiability The property of having a derivative at a point. *See* differentiable.

differentiable Having a derivative at a point. Hence, a function $f(x)$ (from $\mathbf{R}$ to $\mathbf{R}$ or $\mathbf{C}$ to $\mathbf{C}$), such that the limit

$$\lim_{h \to 0} \frac{f(a+h) - f(a)}{h}$$

exists, is said to be *differentiable* at $x = a$.

differentiable manifold A topological manifold with an atlas, where compatibiity of two charts (U_p, ϕ_p), (U_q, ϕ_q) means differentiability of the compositions $\phi_p \circ \phi_q^{-1}$, whenever $U_p \cap U_q \neq \emptyset$. *See* atlas.

differentiable manifold of class $\mathbf{C}^r$ A topological manifold with an atlas, where compatibility of two charts (U_p, ϕ_p), (U_q, ϕ_q) means that the composition $\phi_p \circ \phi_q^{-1}$ is of class $\mathbf{C}^r$, whenever $U_p \cap U_q \neq \emptyset$. *See* atlas.

differentiable manifold with boundary of class $\mathbf{C}^r$ A topological manifold with boundary, equipped with an atlas, where compatibility of two charts (U_p, ϕ_p), (U_q, ϕ_q) means that the compositions $\phi_p \circ \phi_q^{-1}$ are of class $\mathbf{C}^r$, whenever $U_p \cap U_q \neq \emptyset$. *See* atlas. (For an atlas on a topological manifold with boundary, the maps ϕ_p map the neighborhoods U_p either into $\mathbf{R}^n$ or $\mathbf{R}^n_+ = \{(x_1, \ldots, x_n) : x_j \geq 0, \text{for } j = 1, \ldots, n\}$.)

differentiable mapping of class $\mathbf{C}^r$ (1.) A mapping $F : \mathbf{R}^n \to \mathbf{R}^m$ having continuous partial derivatives of all orders up to and including r.
(2.) A mapping $F : M \to N$ between two differentiable manifolds, M having a $\mathbf{C}^r$ atlas $\{(U_p, \phi_p)\}$ and N having a $\mathbf{C}^r$ atlas $\{(V_q, \psi_q)\}$, such that, for each point $p \in M$, there are neighborhoods U_p and $V_{F(p)}$ such that $F(U_p) \subset V_{F(p)}$ and

$$\psi_{F(p)} \circ F \circ \phi_p^{-1}$$

is of class $\mathbf{C}^r$, as a mapping from $\mathbf{R}^n$ to $\mathbf{R}^m$.

differentiable structure An atlas on a topological manifold M, making M into a differentiable manifold. *See* differentiable manifold.

differentiable structure of class C^r An atlas on a topological manifold M making M into a differentiable manifold of class $\mathbf{C}^r$. *See* differentiable manifold of class $\mathbf{C}^r$.

differentiable with respect to a parameter A function, $f_r(x)$, depending upon a parameter r, such that

$$\lim_{h \to 0} \frac{f_{r+h}(x) - f_r(x)}{h}$$

exists, for x in some domain, independent of r.

differential Let $U \subseteq \mathbf{R}^n$ be an open set and $f : U \to \mathbf{R}^m$ a $\mathbf{C}^r$-map. The *differential* of f at $x \in U$ is the linear map $D_f(x) : \mathbf{R}^n \to \mathbf{R}^m$ with the matrix $(\frac{\partial f_i}{\partial x_j})$, where $f(x) = (f_1(x), \ldots, f_m(x))$. Also called total derivative.

differential analyzer An instrument used for solving differential equations.

differential calculus The study of the properties of a real valued function $f(x)$, of a real variable x, analyzed with the use of the derivative of f. *See* derivative.

differential equation A relation $F(x, y, y', \ldots, y^{(n)}) = 0$, where F is a function of $n+1$ variables, y is an unknown function of x and $y', \ldots, y^{(n)}$ are the derivatives

$$\frac{dy}{dx}, \ldots, \frac{d^n y}{dx^n}.$$

differential form *See* algebra of differential forms.

differential form of order i *See* algebra of differential forms.

differential geometry The study of the geometry of manifolds using the tools of differential forms.

differential i-form *See* algebra of differential forms.

differential invariant A tensor quantity, obtained through outer and/or inner multiplication of other tensor quantities by the invariant vector differential operator

$$\nabla \equiv e^j \frac{D}{\partial x^j}$$

whose *components* $D/\partial x^j$ transform like covariant vector components.

differential laws In differential calculus, the laws of differentiation, adapted to computing $\mathrm{d}f(x) = f'(x)\mathrm{d}x$. For example,

$$\mathrm{d}x^n = nx^{n-1}\mathrm{d}x, \mathrm{d}\log x = \frac{1}{x}\mathrm{d}x,$$

$$\mathrm{d}\sin x = \cos x\mathrm{d}x, \mathrm{d}u \cdot v = u\mathrm{d}v + v\mathrm{d}u, \text{etc.}$$

See differentiation.

differential of differentiable mapping The *differential* of a mapping f from an open set $D \subseteq \mathbf{R}^n$ to $\mathbf{R}^m$ at a point $p \in \mathbf{R}^n$ is a linear transformation $L : \mathbf{R}^n \to \mathbf{R}^m$ (whenever it exists) such that

$$\lim_{u \to 0} \frac{|f(p+u) - f(p) - Lu|}{|u|} = 0.$$

differential of function Given a differentiable ($\mathbf{C}^1$) function $f(p)$ on a manifold M, the *differential* of f, denoted $\mathrm{d}f$, is a covariant vector field on M (i.e., for each $p \in M$, a linear operator $\mathrm{d}f_p$ on $T_p(M)$), defined as follows. For the coordinate functions x^i, $\mathrm{d}x^i$ assigns to a vector $X \in T_p(M)$ its i-th component. For $f : M \to \mathbf{R}$, we define

$$\mathrm{d}f = \frac{\partial f}{\partial x_1}\mathrm{d}x^1 + \cdots + \frac{\partial f}{\partial x^n}\mathrm{d}x^n.$$

differential operator A linear operator, between two vector spaces of functions, whose action consists of taking certain derivatives and multiplying by certain functions. A differential operator may not be bounded and, hence, may be defined only on a proper subspace (the *domain*). For example, a differential operator on an L^2 space usually has only functions with appropriate degrees of differentiability in its domain.

differential-difference equation Any equation which approximates a system of ordinary differential equations by replacing each space derivative by a difference operator, in the manner of (for example) the Runge–Kutta methods.

differentiation The process of taking the derivative of a function. *See* derivative. Often the evaluation of the limit in the definition of derivative can be avoided, with the use of certain *rules of differentiation.* Examples of such rules are

$$\frac{d}{dx}x^n = nx^{n-1}, \ \frac{d}{dx}\log x = \frac{1}{x},$$

$$\frac{d}{dx}\sin x = \cos x, \ \frac{d}{dx}u \cdot v = u' \cdot v + u \cdot v', \text{etc.}$$

digamma function The derivative of $\log\Gamma(x + 1)$. *See* gamma function. Thus the digamma function $F(x)$ is given by

$$F(x) = \lim_{j \to \infty} [\log j - \sum_{k=1}^{j+2} \frac{1}{x+k}],$$

and $F(0) = -\gamma$, where γ is Euler's constant. Sometimes defined as the derivative of $\log\Gamma(x)$.

See also trigamma function, tetragamma function.

dilation For a linear operator $T : H \to H$ on a Hilbert space H, an operator $L : H' \to H'$ on a Hilbert space $H' \supset H$, such that $T^n x = PL^n x$, for $n = 1, 2, \ldots$ and for $x \in H$, where P is the orthogonal projection of H' on H. *See* unitary dilation.

dilation theorem Any theorem asserting the existence of a dilation of a particular type, for certain classes of operators on Hilbert space. *See* dilation. For example, B. Sz.-Nagy proved that any contraction operator (operator of norm ≤ 1) has a unitary dilation, which is unique, if certain minimality conditions are satisfied.

dimension (1.) For a vector space, the number of elements in a basis, when a basis with a finite number of elements exists, is called the *dimension* of the vector space. When no such basis exists, we say the vector space is *infinite dimensional.*

(2.) For a manifold M which, by definition, must be locally homeomorphic to a Euclidian space $\mathbf{R}^n$, the dimension is the number n. A manifold which is locally homeomorphic to an infinite dimensional vector space (such as a *Banach manifold*) is said to be *infinite dimensional.*

dimension of a divisor class For a divisor D on an irreducible, projective variety X, let $\mathcal{L}(D)$ denote the vector space consisting of the function 0 and all nonzero rational functions f on X such that $(f) + D$ is effective. (*See* effective divisor.) The dimension of $\mathcal{L}(D)$ is called the *dimension of the divisor D*. The dimension is invariant under equivalence of divisors and so can be defined to be the dimension of the divisor class of D.

Dini's surface A surface of constant negative curvature, obtained by twisting a pseudosphere and given by the parametric equations

$$x = a \cos u \sin v$$
$$y = a \sin u \sin v$$
$$z = a\{\cos v + \ln[\tan(\tfrac{1}{2}v)]\} + bu.$$

Dini's test A criterion for the convergence of the Fourier series of a function $f(t)$, at a point x. Let

$$s_n(x) = \frac{a_0}{2} + \sum_{j=1}^{n}(a_j \cos jx + b_j \sin jx),$$
$$a_j = \frac{1}{\pi}\int_{-\pi}^{\pi} f(t)\cos jt dt,$$
$$b_j = \frac{1}{\pi}\int_{-\pi}^{\pi} f(t)\sin jt dt.$$

Suppose that the function

$$\phi_x(t)/t = [f(x+t) - f(x-t) - 2f(x)]/t$$

is Lebesgue integrable with respect to t in a neighborhood of $t = 0$, then $s_n(x) \to f(x)$, as $n \to \infty$.

Dini-Lipschitz test *See* Dini's test, a corollary of which is the following Lipschitz criterion: Suppose that, in some neighborhood of $t = 0$,

$$|\phi_x(t)| = |f(x+t) - f(x-t) - 2f(x)| \leq c|t|^\alpha,$$

then the partial sums of the Fourier series of f converge to f, at x.

Dirac's distribution For $\Omega \subseteq \mathbf{R}^n$ an open set and $p \in \Omega$, the generalized function T defined on $u \in \mathbf{C}^\infty(\Omega)$ by

$$T_p u = u(p).$$

direct analytic continuation Given two open sets $D_1, D_2 \subset \mathbf{C}$ (or $\mathbf{C}^n$) and functions $f_i(z)$, analytic in $D_i, i = 1, 2$, we say f_2 is a *direct analytic continuation* of f_1 if $D_1 \cap D_2 \neq \emptyset$ and $f_1(z) = f_2(z)$, for $z \in D_1 \cap D_2$.

direct circle For an ellipse, with axes of length a and b, the circle, centered at the center of the ellipse, and having radius $\sqrt{a^2 + b^2}$.

direct limit Let $\{S_\alpha : \alpha \in \Lambda\}$ be a direct system of sets (*See* direct system of sets) and consider the disjoint union U of the $S_\alpha, \alpha \in \Lambda$. Introduce an equivalence relation $\sim$ on U, by

$$x \sim y (x \in S_\alpha, y \in S_\beta), \text{ if } \rho_{\gamma\alpha}(x) = \rho_{\gamma\beta}(y),$$

for some $\gamma \geq \alpha, \beta$. The disjoint union U, modulo the relation $\sim$ is the *direct limit* of the system, denoted $\lim_{\rightarrow} S_\alpha$.

direct method in the calculus of variations Any method for maximizing or minimizing a given definite integral

$$I = \int_a^b F[y(x), y'(x), x]\, dx$$

which attempts to approximate the desired function $y(x)$ successively by a sequence of functions $u_1(x), u_2(x), \ldots$ selected so as to satisfy the boundary conditions imposed on $y(x)$. Each function $u_r(x)$ is taken to be a differentiable function of x and of r parameters $c_{r1}, c_{r2}, \ldots, c_{rr}$. The latter are then chosen so as to maximize or minimize the function

$$I_r(c_{r1}, c_{r2}, \ldots, c_{rr}) = \int_a^b F[u_r(x), u_r'(x), x]\, dx$$

with the aid of the relations $\frac{\partial I_r}{\partial c_{ri}} = 0$, $i = 1, 2, \ldots, r$, $r = 1, 2, \ldots$. Every tentative solution $y(x)$ obtained in this manner as the limit of a sequence of approximating functions, still needs to be proved to maximize or minimize the definite integral.

Analogous methods of solution apply to variational problems involving more than one unknown function and/or more than one independent variable. If accessory conditions are given, they are made to apply to each approximating function; the approximating integrals may then be maximized or minimized by the Lagrange-multiplier method. Direct methods can yield numerical approximations and/or exact solutions. Note also that every direct method for the solution of a variation problem is also an approximation method for solving differential equations.

The Rayleigh–Ritz is another direct method in the calculus of variations.

direct sum The union of two algebraic structures A and B, inheriting the algebraic properties of A and B, but with no interaction between the two structures. For example, if A and B are vector spaces, $A \oplus B$ is the vector space of pairs (a, b), with $a \in A, b \in B$ and with addition and scalar multiplication defined by $(a_1, b_1) + (a_2, b_2) = (a_1 + a_2, b_1 + b_2)$, $c(a, b) = (ca, cb)$.

If A and B are subsets of X, $A \cup B$ is the direct sum of A and B if and only if $A \cap B = \emptyset$.

direct system of sets A collection $\{S_\alpha : \alpha \in \Lambda\}$ of sets, indexed by a directed set Λ (*See* directed set), together with maps $\rho_{\beta\alpha} : S_\alpha \to S_\beta$, for $\alpha \leq \beta$, such that $\rho_{\alpha\alpha} = Id$, $\rho_{\gamma\alpha} = \rho_{\gamma\beta} \circ \rho_{\beta\alpha}$, for $\alpha \leq \beta \leq \gamma$.

direct variation A relationship between two variables x and y of the form $y = kx$, where k is a constant, known as the constant of variation (or constant of proportionality). In this case, we say that y *varies directly* as x. More generally, y *varies directly* as x^n, where $n > 0$, if $y = kx^n$.

directed angle An angle in which one ray is identified as the initial side of the angle and the other ray as the terminal side.

directed family A nonempty collection F of nonempty sets such that $F_1, F_2 \in F$ implies that $F_1 \cap F_2$ contains at least one element F_3 of F.

directed line A number line, with one point on the line designated as the origin, and represented by the number zero. Directed numbers are used to represent distances from the origin: positive numbers are used for points on one side of the origin; negative numbers are used for points on the other side of the origin.

directed number A number that includes a sign, positive or negative. In a geometric context, a number with a negative sign represents a distance measured in a direction opposite to that of a number with a positive sign.

directed set A partially ordered set Λ with the property that $\alpha, \beta \in \Lambda$ implies that there is $\gamma \in \Lambda$ such that $\alpha \prec \gamma$ and $\beta \prec \gamma$.

direction angles For a nonzero vector $\mathbf{v} = (v_1, v_2, v_3) \in \mathbf{R}^3$, the angles that $\mathbf{v}$ makes with the basis vectors $\mathbf{i}, \mathbf{j}, \mathbf{k}$.

direction cosines For a nonzero vector $\mathbf{v} = (v_1, v_2, v_3) \in \mathbf{R}^3$, the cosines of the angles that $\mathbf{v}$ makes with the basis vectors $\mathbf{i}, \mathbf{j}, \mathbf{k}$.

direction field *See* vector field.

direction numbers On a line in space, three numbers (not all zero) that are proportional to the direction cosines of the line. If α, β, and γ are the angles between the line and the positive directions of the x, y, and z-axes, respectively, then any direction numbers will be of the form $r \cos\alpha, r \cos\beta, r \cos\gamma$, for some $r \neq 0$.

direction of curve The slope of the tangent line to the curve at a point. *See* slope.

direction of line The slope of the line. *See* slope.

directional derivative Of a function $f(x, y, z)$, at a point $P \in \mathbf{R}^3$ in the direction of the unit vector $\mathbf{u}$:

$$D_{\mathbf{u}} f(P) = \nabla f(P) \cdot \mathbf{u},$$

where ∇f is the gradient vector of f.

directrix By definition, a *conic section* is the locus of a point such that the ratio of its distance from a fixed point to its distance from a fixed line is constant. The fixed line is called the *directrix*.

Dirichlet domain An open, connected set $D \subset \mathbf{C}$, in which the Dirichlet problem has a solution. That is, whenever a continuous function $g(z)$ on ∂D is given, there exists a harmonic function $f(z) \in \mathbf{C}^2(D)$, such that $f(z) = g(z)$ on ∂D.

Dirichlet function The function

$$D(x) = \begin{cases} 0 & \text{for } x \text{ irrational} \\ 1/b & \text{for } x = a/b \end{cases}$$

with a, b relatively prime integers.

The function is discontinuous at rational x and continuous on the irrationals.

Dirichlet integral The integral

$$D[u] = \int\int_D \left(\left(\frac{\partial u}{\partial x}\right)^2 + \left(\frac{\partial u}{\partial y}\right)^2\right) dxdy,$$

where $u(z) = u(x+iy)$ is a real, piecewise $\mathbf{C}^2$ function on a domain D in $\mathbf{C}$ or in a Riemann surface. For a complex-valued function the Dirichlet integral is the sum of the Dirichlet integrals of the real and imaginary parts. The *Dirichlet space* of a domain $D \subset \mathbf{C}$ is the vector space of functions, analytic and having finite Dirichlet integral in D, with the norm $\|f\| = D[f]^{\frac{1}{2}}$.

Dirichlet kernel The function $D_n(t) = \sum_{j=-n}^{n} e^{ijt} = \sin(n+\frac{1}{2})t/\sin(t/2)$, which has the property that, for $f(t)$ continuous on $[0, 2\pi]$, the nth partial sum of the Fourier series of f is given by

$$\sum_{j=-n}^{n} a_j e^{ijx} = \frac{1}{2\pi}\int_0^{2\pi} f(t) D_n(x-t)dt.$$

Dirichlet principle Let D be a domain in $\mathbf{C}$ with boundary ∂D a finite union of Jordan curves. Suppose $w(z)$ is continuous on the closure $\bar{D}$ of D and is piecewise $\mathbf{C}^2$ on D, with finite Dirichlet integral $D[w]$. Let $\mathcal{F}$ be the family of functions $u(z)$, continuous on $\bar{D}$, piecewise $\mathbf{C}^2$ on D, with $D[u] < \infty$ and $u(z) = w(z)$ on ∂D. Then there exists a function $h(z) \in \mathcal{F}$, harmonic in D and such that

$$D[h] = \inf_{u \in \mathcal{F}} D[u].$$

See Dirichlet integral.

Dirichlet problem Given an open connected set $D \subset \mathbf{C}$, with boundary ∂D and a continuous function $g(z)$, defined on ∂D, to find a function $f(z) = f(x+iy)$ which is harmonic in D, that is, f is in $\mathbf{C}^2(D)$ and satisfies

$$\Delta f = \left(\frac{\partial^2}{\partial x^2} + \frac{\partial^2}{\partial y^2}\right) f(x+iy) = 0$$

in D, and $f(z) = g(z)$ on ∂D.

When D is the unit disk, the Dirichlet problem is solved by the Poisson integral of g:

$$f(re^{i\theta}) =$$

$$\frac{1}{2\pi}\int_0^{2\pi} \frac{1-r^2}{1-2r\cos(\theta-t)+r^2} g(e^{it})dt.$$

Dirichlet series A formal series of the form

$$\sum_{j=1}^{\infty} a_j e^{-\lambda_j s},$$

where a_j is a complex number, for $j = 1, 2, \ldots$, $0 < \lambda_1 < \lambda_2 \ldots \to \infty$ and s is a complex variable. Also the function represented by the series, where it converges.

The choice $\lambda_j = \log j$ defines an ordinary Dirichlet series. An ordinary Dirichlet series with $a_1 = a_2 = \ldots = 1$ yields the *Riemann zeta function*.

Dirichlet space *See* Dirichlet integral.

Dirichlet's divisor problem To investigate the *divisor function* $\tau(n)$, which, for each positive integer n, is the number of divisors on n. In particular, *Dirichlet's divisor* *problem* is to investigate the *average order* $\sum_{n \leq x} \tau(n)$. *See also* lattice-point problem.

Dirichlet's test A sufficient condition for the convergence of a series of the form $\sum a_j b_j$. Although there are several versions, most are variations on the following one. Suppose the partial sums of the infinite series $\sum a_j$ are bounded, the sequence $\{b_j\}$ is monotonically decreasing ($b_1 \geq b_2 \geq \ldots$) and b_j tends to 0. Then $\sum a_j b_j < \infty$.

There is a corresponding test for infinite integrals.

disconjugate differential equation A homogeneous, linear, second order, ordinary differential equation with real coefficient functions, on an interval J, such that every solution that is not identically 0 has at most one zero on J.

discontinuity A point where a function fails to be continuous. *See* continuous function.

discontinuity point of the first kind For a real-valued function $f(x)$ of a real variable, a point x_0 where $\lim_{x \to x_0+} f(x)$ and $\lim_{x \to x_0-} f(x)$ exist, but are not equal.

discontinuity point of the second kind For a real-valued function $f(x)$ of a real variable, a point x_0 such that either $\lim_{x \to x_0+} f(x)$ or $\lim_{x \to x_0-} f(x)$ does not exist. Thus, a point where $f(x)$ is not continuous, which is not a discontinuity point of the first kind.

discontinuous function A function that is not continuous. *See* continuous function. Typically, a function that has an isolated point of discontinuity, but possibly a function that is discontinuous on a larger set; for example, $f(x) = 1$, for x rational and $= -1$ for x irrational, is discontinuous at every real number.

discrete Fourier transform The mapping $\mathcal{D}_N$ that sends a finite sequence $\{x_0, x_1, \ldots, x_{N-1}\}$ to the sequence $\{X_0, X_1, \ldots, X_{N-1}\}$, given by

$$X_j = \sum_{n=0}^{N-1} x_n \exp\{-nij2\pi/N\}.$$

Also called *fast Fourier transform.*

discriminant of curve of the second order For the general second order equation

$$Ax^2 + Bxy + Cy^2 + Dx + Ey + F = 0,$$

the discriminant is

$$4ACF + BDE - AE^2 - CD^2 - FB^2.$$

Disk Theorem Let f and g be two embeddings of the closed unit disk $D \subset \mathbf{R}^k$ into the interior if a connected manifold M of dimension n. If $k = n$, assume further that f and g are either both orientation preserving or orientation reversing. Then f and g are ambient isotopic, meaning that there is a smooth map $F : M \times \mathbf{R} \to M$ such that $F(p, 0) = p$ and $F(f, 1) = g$.

displacement A term used in the calculus of finite differences to mean the operator that sends a function $f(x)$ to $f(x + h)$.

dissipative operator A linear (possibly unbounded) operator T on a domain D in a Hilbert space, satisfying

$$\Re(Tx, x) \leq 0,$$

for $x \in D$. T is *dissipative* if and only $-T$ is *accretive.*

distance formula Between two points $P_1 = (x_1, y_1)$ and $P_2 = (x_2, y_2)$ in the plane:

$$d(P_1, P_2) = \sqrt{(x_1 - x_2)^2 + (y_1 - y_2)^2}.$$

Between two points $P_1 = (x_1, y_1, z_1)$ and $P_2 = (x_2, y_2, z_2)$ in $\mathbf{R}^3$:
$d(P_1, P_2)$

$$= \sqrt{(x_1 - x_2)^2 + (y_1 - y_2)^2 + (z_1 - z_2)^2}.$$

distance/rate/time formula If an object is moving at a constant velocity v for a time t, then the distance s traveled by the object is given by $s = vt$. If the velocity is not constant, but is given as a function of t, then the distance traveled from time $t = a$ to time $t = b$ is given by $s = \int_a^b v dt$.

distinguished boundary For a polydisk $D^n = D^n(a, \rho) = \{z = (z_1, \ldots, z_n) \in \mathbf{C}^n : |z_j - a_j| < \rho_j, j = 1, \ldots, n\}$, the *distinguished boundary* is the set

$$T^n(a, \rho) = \{z \in \mathbf{C}^n : |z_j - a_j| = \rho_j, j = 1, \ldots, n\}.$$

In several variables, Cauchy's integral formula involves integration over the distinguished boundary, as opposed to the topological boundary.

distribution *See* generalized function.

distribution function A nondecreasing function $F(x)$ satisfying $F(-\infty) = \lim_{x \to -\infty} F(x) = 0$ and $F(\infty) = \lim_{x \to \infty} F(x) = 1$.

distribution of finite order A generalized function u in an open set $\Omega \subset \mathbf{R}^n$ such that there is an integer k for which the inequality

$$|u(\phi)| \leq C \sum_{|\alpha| \leq k} \sup |D^\alpha \phi|$$

holds for all compact sets $K \subset \Omega$.

distribution with compact support *See* support of a distribution.

distributional derivative Let T be a generalized function on an open set $\Omega \subseteq \mathbf{R}^n$. The *distributional derivative* of T is the generalized function S, defined on a function $u \in \mathbf{C}^\infty$ by

$$Su = -T\left(\frac{\partial u}{\partial x_j}\right).$$

The notation $S = \frac{\partial}{\partial x_j} T$ is used for the distributional derivative.

divergence For a $\mathbf{C}^1$ vector field $\mathbf{u} = (u_1, u_2, u_3)$ in a domain in $\mathbf{R}^3$, the function

$$\nabla \cdot \mathbf{u} = \operatorname{div} \mathbf{u} = \frac{\partial u_1}{\partial x} + \frac{\partial u_2}{\partial y} + \frac{\partial u_3}{\partial z}.$$

Divergence Theorem Let D be a closed, bounded region in $\mathbf{R}^3$ such that each point of ∂D has a unique outward normal vector. Let $\mathbf{v}$ be a $\mathbf{C}^1$ vector field on an open set $G \supset D \cup \partial D$. Then

$$\int\int\int_D \operatorname{div} \mathbf{v} \, dV = \int\int_{\partial D} \mathbf{v} \cdot \mathbf{n} \, dS$$

where $\mathbf{n}$ is the unit outward normal function from D.

divergent Referring to a sequence (series or integral) which fails to be convergent. *See* convergent. *See also* divergent integral, divergent series.

divergent integral A formal definite integral

$$I = \int_c^\infty f(x)dx,$$

where the integrals

$$I_r = \int_c^r f(x)dx$$

exist and are finite for all $r \in (c, \infty)$, but I_r does not tend to a limit, as $r \to +\infty$.

An analogous definition can be made for a divergent integral over the interval $(-\infty, c]$.

divergent series A formal infinite series $\sum_{j=1}^\infty a_j$ such that $s_n = \sum_{j=1}^n a_j$ fails to converge to a limit, as $n \to \infty$.

divisor (1.) A section of the sheaf of divisors on a domain $\Omega \subseteq \mathbf{C}^n$. *See* sheaf of divisors.

(2.) On a curve, a *divisor* is a linear combination of points: $D = \sum \lambda_j x_j$.

For an irreducible variety V, a divisor is a sum of the form $D = \sum \lambda_j C_j$, where $C_1, \ldots, C_r$ are irreducible subvarieties of codimension 1 of V, with preassigned integral multiplicities $\lambda_1, \ldots, \lambda_r$.

divisor class The *divisor class group*, $Cl(X)$, of a variety X, is $Div(X)/P(X)$, the group of all divisors, modulo the subgroup of principal divisors. A *divisor class* is a coset in $Div(X)/P(X)$.

divisor class group *See* divisor class.

domain of function *See* function.

domain of holomorphy An open set in $\mathbf{C}^n$ which does *not* have the following property: there is a connected open set $\Omega_1 \supset \Omega$ such that every function $u(z)$, analytic on Ω can be continued analytically into Ω_1.

domain of integration A bounded subset D of $\mathbf{R}^n$ with boundary a set of Jordan content 0. *See* set of Jordan content 0.

domain of operator A subspace of a normed vector space on which a linear operator is defined. Typically, a differential operator cannot be defined on all elements of a natural function space such as an L^2 space, and a dense subspace, such as the continuous functions of bounded variation, is chosen.

domain with regular boundary Many treatments of real and complex analysis use such a term for a subset of $\mathbf{R}^n$ or $\mathbf{C}^n$ sufficiently nice for certain theorems. An example is the following definition, which is sufficient for Green's Theorem in the plane. A region $D \subset \mathbf{R}^n$ is *a domain with regular boundary* if
(i.) D is bounded
(ii.) ∂D is the union of a finite number of piecewise smooth, simple curves, and
(iii.) at each point $p \in \partial D$, there is a Cartesian coordinate system with p as the origin and, for a, b sufficiently small, the rectangle $R = \{(x_1, x_2) : -a \leq x_1 \leq a, -b \leq x_2 \leq b\}$ has the property that $\partial D \cap R = \{(x_1, f(x_1)) : -a \leq x_1 \leq a\}$ for f a piecewise smooth function.

dot product of vectors For two vectors $\alpha = (a_1, a_2, \ldots, a_n)$ and $\beta = (b_1, b_2, \ldots, b_n)$ in $\mathbf{R}^n$,

$$\alpha \cdot \beta = a_1 b_1 + a_2 b_2 + \ldots + a_n b_n.$$

When $n = 2, 3$, the dot product can be interpreted geometrically as the product of the length of α, the length of β, and the cosine of the (smallest) angle between α and β.

Douady space The set of all compact analytic subspaces of a compact, complex manifold.

double integral A definite integral of a function of two variables, over the Cartesian product of two one-variable sets A, B:

$$\int\int_{A \times B} f(x, y) dx dy$$
$$= \int_A \int_B f(x, y) dx dy.$$

There are several different types of *double integrals*. *See* Riemann integral, Lebesgue integral, Lebesgue-Stieltjes integral. *See also* multiple integral.

doubly periodic function A single-valued analytic function $f(z)$ with only isolated singularities on the entire finite plane (i.e., excluding the point at infinity), and such that there are two numbers p_1 and p_2 whose quotient is not a real number with

$$f(z + p_1) = f(z + p_2) = f(z).$$

(If p_1/p_2 is real and rational, then $f(z)$ is a simply-periodic function; if p_1/p_2 is real and irrational, then $f(z)$ is a constant.) Meromorphic doubly periodic functions are called elliptic functions.

du Bois Reymond problem The du Bois Reymond problem asks whether the Fourier series of a continuous function converges almost everywhere. P. du Bois Reymond showed in 1876 that there is a continuous function whose Fourier series diverges at a point. In 1966 L. Carleson showed that the Fourier series of an L^2 function converges almost everywhere, thus answering the problem in the affirmative. R.A. Hunt later (1967) showed that the Fourier series of an L^p function for $1 < p < \infty$ converges almost everywhere.

dual basis Let V be a (finite-dimensional) vector space over the field F with basis $B = \{\alpha_1, \alpha_2, \ldots, \alpha_n\}$. Define

$$f_i(\alpha_j) = \delta_{ij}.$$

Then

$$B^* = \{f_1, f_2, \ldots f_n\}$$

is a basis for the dual space V^* and is called the dual basis of B.

dual cone Let X be a vector space, and X_+ denote a positive cone in X. Then the set

$$X_+^* = L_+^*(X, \mathbf{R}),$$

where $L_+^*(X, \mathbf{R})$ denotes the set of all positive linear maps from X into $\mathbf{R}$ (real number system) is called the dual cone of X_+.

dual group The set of all continuous characters (homomorphisms into the multiplicative group of complex numbers of modulus 1) of a (locally compact, Abelian) group G, with the weakest topology making all the Fourier transforms of functions in $L^1(G)$ continuous.

dual operator Let X and Y be normed linear spaces, and let T be a linear operator with domain $D(T)$ dense in X and with range $R(T) \subset Y$. Suppose there is a pair (f, g) with $f \in Y'$, the dual space of Y, and $g \in X'$, the dual space of X, satisfying the equation $(Tx, f) = (x, g)$ identically for $x \in D(T)$. Since g is uniquely determined by f, define $g = T'f$. Then T' is a linear operator called the *dual operator* (or conjugate operator) of T. This is an extension of the notion of transpose of a matrix in matrix theory.

dual space Let V be a vector space over the field F. The space $L(V, F)$ of all linear transformations from V into F (the so-called *linear functionals of* V) is called the *dual space* of V and is denoted by V^*.

duality mapping Two spaces E and F over K are said to be *in duality* with *duality mapping* B if there is a non-degenerate bilinear form $B : E \times F \to K$, such that

1. if $B(x, y) = 0$ for all $y \in F$, then $x = 0$, and

2. if $B(x, y) = 0$ for all $x \in E$, then $y = 0$.

dummy index A symbol, usually representing discrete values, that is internal to an expression, so that the value of the entire expression is independent of it. For example, in the sum

$$S = \sum_{j=1}^{n} a_j,$$

the symbol j is a dummy index, since S does not depend upon j.

dummy variable A symbol, usually representing a continuous set of values, that is internal to an expression, so that the value of the entire expression is independent of it. For example, in the integral

$$I = \int_a^b f(x)dx,$$

the symbol x is a *dummy variable*, since I does not depend upon x.

Dunford integral Let X be a complex Banach space, T a bounded linear operator on X, and $F(T)$ the set of all functions holomorphic in a neighborhood of the spectrum of T. We define an operator $f(T)$, for $f \in F(T)$ by the *Dunford integral*

$$f(T) = \frac{1}{2\pi i} \int_C f(t)(tI - T)^{-1} dt,$$

where C is a closed curve consisting of a finite number of rectifiable Jordan arcs, which contains the spectrum of T in its interior, and lies with its interior completely in the domain in which f is holomorphic.

Dupin indicatrix Let p_0 be a point on a surface S in Euclidean n-space, and let (X, Y) be the coordinates of a point on the tangent space at p_0 with respect to the Gaussian frame of S at p_0. The *Dupin indicatrix* is the second order curve defined by

$$PX^2 + 2QXY + RY^2 = \epsilon,$$

for ϵ a suitable constant. Here, P, Q, and R are defined by the inner products $P = (-\mathbf{x}_u, \mathbf{N}_u)$, $Q = (-\mathbf{x}_u, \mathbf{N}_v)$, $R = (-\mathbf{x}_v, \mathbf{N}_v)$, ($\mathbf{N}_u = \frac{\partial \mathbf{N}}{\partial u}$, and $\mathbf{N}_v = \frac{\partial \mathbf{N}}{\partial v}$). Furthermore, $\mathbf{x}_u$, $\mathbf{x}_v$ are the tangent vectors at p_0 to the u-curve and the v-curve respectively, when the surface S is expressed by the vector representation $\mathbf{x} = \mathbf{x}(u, v)$.

E

eccentric angle The parameter in the parametric forms for the equations of ellipses and hyperbolas. For the ellipse $x^2/a^2 + y^2/b^2 = 1$ with parametrization $x = a\cos\theta$, $y = b\sin\theta$, θ is called the *eccentric angle* of a point (x, y) on the ellipse. Similarly, for the hyperbola, $x^2/a^2 - y^2/b^2 = 1$ with parametrization $x = a\sec\theta$, $y = b\tan\theta$, θ is called the *eccentric angle* of (x, y).

eccentricity A conic section can be defined as the set of all points P in the plane such that the ratio of the undirected distance of P from a fixed point (the focus) to the undirected distance of P from a fixed line (the directrix) that does not contain the fixed point is a positive constant, denoted e. The constant e is called the *eccentricity* of the conic section. It can be shown that if $e = 1$, the conic is a parabola, if $0 < e < 1$, it is an ellipse, and if $e > 1$, it is a hyperbola. (If $e = 0$, the conic section is actually a circle.) For an ellipse and a hyperbola, the eccentricity can be defined as the ratio of the distances between the foci and the length of the major axis. *See also* conic section.

eccentricity of ellipse *See* eccentricity.

eccentricity of hyperbola *See* eccentricity.

eccentricity of parabola *See* eccentricity.

echelon space Let $(\alpha^{(k)})$ be a sequence, whose elements $\alpha^{(k)} = (a_1^{(k)}, a_2^{(k)}, \ldots)$ are sequences of real [complex] numbers. The space of all real [complex] sequences (x_j) satisfying $\sum |a_j^{(k)} x_j| < \infty$ is denoted $\lambda_{\alpha^{(k)}}$. The $\alpha^{(k)}$ are called *steps*. The linear span $\sum_k \lambda^{\times}_{\alpha^{(k)}}$ is called the *co-echelon space*, where the superscript $\lambda^{\times}$ denotes the dual, that is, the set of sequences (x_j) for which $\sum \lambda_j x_j < \infty$, for all $(\lambda_j) \in \lambda$. The *echelon space* corresponding to $(\alpha^{(k)})$ is the intersection $\cap_k \lambda_{\alpha^{(k)}}$. Also called *Köethe space.*

Edge of the Wedge Theorem A theorem on analytic continuation for functions of several complex variables. Let S be an open cone in $\mathbf{R}^n$ (i.e., an open set such that $y \in S \Rightarrow ty \in S$, for every positive t) and let V be the intersection of S with an open ball, centered at $0 \in \mathbf{R}^n$. For $E \in \mathbf{R}^n$ a nonempty set, let $W^+ = E + iV$, $W^- = E - iV \subset \mathbf{C}^n$. The sets $W^{\pm}$ are the wedges and E is the edge. One version of the *Edge of the Wedge Theorem* states:

There is an open set $\Omega \subset \mathbf{C}^n$, which contains $W^+ \cup E \cup W^-$ with the property that every $f(z)$, continuous on $W^+ \cup E \cup W^-$ and holomorphic on $W^+ \cup W^-$ has an analytic continuation into Ω.

effective divisor A divisor $D = \sum \lambda_j C_j$, with all $\lambda_j > 0$. Here $C_1, \ldots, C_r$ are irreducible subvarieties of codimension 1 of a variety V, with preassigned integral multiplicities $\lambda_1, \ldots, \lambda_r$.

eigenfunction Let T be a linear operator on a vector space V over the field F. If c is an eigenvalue for T, then any vector $v \in V$ such that $Tv = cv$ is called an *eigenfunction* of T associated with the scalar c. Also *eigenvector, characteristic vector.*

eigenfunction expansion This is a technique for the solution of a linear differential equation (either ordinary or partial) with linear boundary conditions for which a complete set of eigenfunctions which satisfy the boundary conditions are known. The solution function, if sufficiently smooth, can be expanded as an infinite series in the eigenfunctions with unknown coefficients. From the given equation and the boundary conditions, equations can then be determined for the unknown coefficients. (*See also* eigenfunction.)

0-8493-0320-6/00/$0.00+$.50

eigenvalue Let T be a linear operator on a vector space V over the field F. A scalar c in F is an *eigenvalue* for T if there is a non-zero vector v in V such that $Tv = cv$. Eigenvalues are also known as *characteristic values, proper values, spectral values,* and *latent roots.*

eigenvalue problem Given a linear operator $L(\cdot)$, with boundary conditions $B(\cdot)$ there will sometimes exist non-trivial solutions to the equation $L[y] = \lambda y$. When such a solution exists, the value of λ is called an eigenvalue. Finding such solutions is called the eigenvalue problem. (Note: The solutions may or may not be required to also satisfy $B[y] = 0$.)

eikonal A first-order partial differential equation of the form

$$\sum_{i=1}^{m}\left(\frac{\partial r}{\partial x^i}\right)^2 = \frac{1}{c^2(x^1, \cdots, x^m)}.$$

Here m is the dimension of the space and c is a smooth function bounded away from zero. In applications, c is the speed of the wave, and the surfaces $r(x^1, \cdots, x^m) = \text{const.}$ are the wave fronts. The rays are the characteristics of the eikonal equation.

The mathematical theory of geometrical optics can be regarded as the theory of the eikonal equation. The solution of the eikonal equation may have singularities. Their theory is part of that of the singularities of differentiable mappings.

Einstein's convention A simplifying notational convention concerning expressions involving symbols with subcripts, under which, whenever a subscript occurs twice, a summation over that subscript is understood. Thus $a_{ij}b_{jk}$ is understood to mean

$$\sum_{j=1}^{n} a_{ij}b_{jk}.$$

When this may cause confusion, the convention is sometimes modified so that summation is implied when a letter appears as a subscript and also a superscript, so that $a_{ij}b^{jk}$ means

$$\sum_{j=1}^{n} aijb^{jk}.$$

Einstein-Kahler metric On a compact Kahler manifold (M, g), with Kahler form ω, the metric g is called an *Einstein-Kahler* metric if $\rho = k\omega$, for k real, where ρ is the Ricci form of g.

Eisensein–Poincaré series This series is defined as

$$E_{S,k} = \sum_{\sigma \in \Gamma_n(S)\backslash\Gamma_n} e^{2\pi i \text{tr}(S\sigma(z))} \det(CZ + D)^{-k}.$$

where S is an $n \times n$ rational symmetric matrix ≥ 0 and

$$\Gamma_n(S) = \left\{ \begin{pmatrix} U & T^tU \\ 0 & {}^tU^{-1} \end{pmatrix} \in \Gamma_n : {}^t USU = S, e^{2\pi i \text{tr}(ST)} = \det(U)^k \right\}.$$

Eisenstein series For $k \geq 3$, the (extended) *Eisenstein series* is defined as

$$G_k(z; c_1, c_2, N) = \sum_{m_i \equiv c_i (mod\ N)} \frac{1}{(m_1 + m_2 z)^k},$$

where c_1, c_2 are integers such that $(c_1, c_2, N) = 1$ and the sum is taken over all pairs excepting the pair $(m_1, m_2) = (0, 0)$.

element of arc length The infinitesimal, or differential, that must be integrated to obtain arc length. Hence, for the length of arc of $y = f(t)$, from $t = a$ to $t = b$, the quantity $\sqrt{1 + f'(t)^2}dt$.

element of integration The infinitesimal, or differential, that must be integrated to obtain a definite integral. Hence, for the Riemann integral of $f(x)$ over the interval $[a, b]$, the quantity $f(x)dx$.

elementary function A function which belongs to the class of functions consisting of the polynomials (*see also* polynomial function), the exponential functions (*see also* exponential function), the logarithmic functions (*see also* logarithmic function), the trigonometric functions (*see also* trigonometric function), the inverse trigonometric functions (*see also* inverse trigonometric function), and the functions obtained from those listed by the four arithmetic operations and/or by composition, applied finitely many times. This class of functions occurs most frequently in mathematics, but so do functions which are not elementary. For example, while the derivative of an elementary function is also an elementary function, the indefinite integral or antiderivative of an elementary function cannot always be expressed as an elementary function. As a specific example, consider

$$\int e^{x^2}dx,$$

which is the antiderivative of an elementary function but is not itself elementary.

elementary function of class *n* An algebraic function of a finite number of complex variables is called an elementary function of class 0. The functions e^z and $\log z$ are by definition elementary functions of class 1. Inductively, assuming elementary functions of class at most $n-1$ have been defined, define elementary functions of class n as follows: let $g(t)$ and $g_j(w_1, \dots, w_n)$ $(1 \leq j \leq m)$ be elementary functions of class at most 1 and $f(z_1, .., z_m)$ be an elementary function of class at most $n-1$. Then the composite functions $g(f(z_1, \dots, z_m))$ and $f(g_1(w_1, \dots w_n), \dots, g_m(w_1, \dots, w_n))$ are called elementary functions of class at most n. An elementary function of class at most n which is not of class at most $n-1$ is called an *elementary function of class n*.

elementary set The Cartesian product of intervals (open, closed or semi-open) in $\mathbf{R}^n$.

ellipse A planar curve, one of the so-called conic sections (*see also* conic section), obtained by the intersection of a double-mapped right circular cone with a plane which is not parallel to any generator of the cone and does not pass through the vertex. (A circle is a special case of the ellipse, in which the cutting plane is also perpendicular to the axis of the cone.) An ellipse may also be defined as the locus of points in the plane, the sum of whose distances from two fixed points, F_1 and F_2, is a constant. Each such point is called a focus. In an ellipse, the distance between the foci is called the *focal distance* and is usually denoted by $2c$. The midpoint of the line segment joining the two foci is called the center of the ellipse. The line on which the foci lie is called the major (or first) axis, and the line through the center of the ellipse perpendicular to the major axis is called the minor (or second) axis. Using this definition it is possible to derive equations for the standard ellipses centered at (h, k). If the principal axis is horizontal, the equation is

$$\frac{(x-h)^2}{a^2} + \frac{(y-k)^2}{b^2} = 1, a > b.$$

If the principal axis is vertical, the equation is

$$\frac{(x-h)^2}{b^2} + \frac{(y-k)^2}{a^2} = 1, a > b.$$

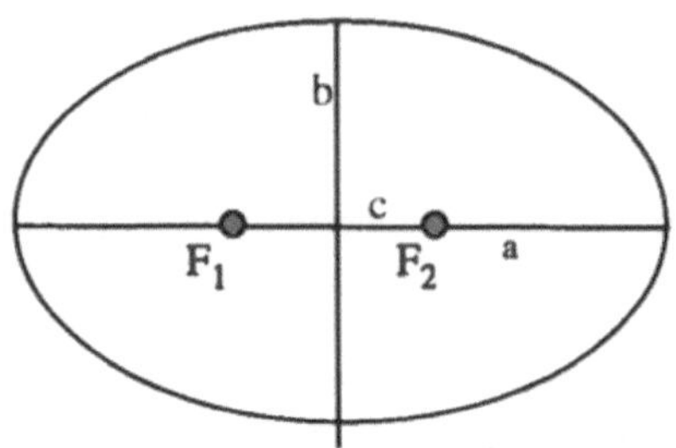

ellipsoid A type of quadric surface (i.e., the graph in three-space of a second degree equation in three variables). The standard form for an ellipsoid centered at $(0, 0, 0)$ is

$$\frac{x^2}{a^2} + \frac{y^2}{b^2} + \frac{z^2}{c^2} = 1,$$

where a, b, and c are positive. Note that if z is zero, the resulting cross section of the ellipsoid in the xy plane is the ellipse

$$\frac{x^2}{a^2} + \frac{y^2}{b^2} = 1.$$

The cross section of the ellipsoid with the plane $z = k$ yields the equation

$$\frac{x^2}{a^2} + \frac{y^2}{b^2} = 1 - \frac{k^2}{c^2}.$$

In this case, if $|k| < c$, the resulting cross section is a planar ellipse and the lengths of its semiaxes decrease to 0 as $|k|$ increases to the value c. If $|k| = c$, the intersection of the plane $z = k$ with the ellipsoid is the single point $(0, 0, k)$. If $|k| > c$, there is no intersection. Similar cross sections result if we consider intersections by planes parallel to either of the other two axes. The numbers a, b, and c are the lengths of the semiaxes of the ellipsoid. If any two of these three numbers are equal, we have an ellipsoid of revolution, also known as a *spheroid*.

ellipsoidal coordinates Coordinates defined to be the three roots ρ, σ, and γ of the cubic equation in s

$$\frac{x^2}{s - e_1} + \frac{y^2}{s - e_2} + \frac{z^2}{s - e_3} = 1,$$

where e_1, e_2, and e_3 are given real numbers. These roots are real for real x, y and z, and if $e_1 > e_2 > e_3$, they can be labeled so as to satisfy the inequalities

$$\rho \geq e_1 \geq \sigma \geq e_2 \geq \gamma \geq e_3.$$

The surfaces ρ =constant, σ =constant, and γ = constant are ellipsoids, hyperboloids of one sheet, and hyperboloids of two sheets respectively. Cartesian coordinates are expressed in terms of ellipsoidal coordinates by the equations

$$x^2 = \frac{(\rho - e_1)(\sigma - e_1)(\gamma - e_1)}{(e_1 - e_2)(e_1 - e_3)},$$

$$y^2 = \frac{(\rho - e_2)(\sigma - e_2)(\gamma - e_2)}{(e_2 - e_3)(e_2 - e_1)},$$

$$z^2 = \frac{(\rho - e_3)(\sigma - e_3)(\gamma - e_3)}{(e_3 - e_1)(e_3 - e_2)}.$$

ellipsoidal harmonic of the first species The first solution to Lamé's differential equation, denoted $E_n^m(x)$ for $m = 1, \ldots, 2n + 1$. They are also called Lamé functions. The product of two ellipsoidal harmonics of the first kind is a spherical harmonic.

See also ellipsoidal harmonic of the second species.

ellipsoidal harmonic of the second species The functions given by

$$F_m^p(x) = (2m + 1)E_m^p(x) \times \int_x^\infty \frac{dx}{(x^2 - b^2)(x^2 - c^2)[E_m^p(x)]^2}.$$

See ellipsoidal harmonic of the first species.

elliptic coordinates Let a family of confocal central quadric surfaces be represented by the equations

$$\frac{x^2}{a + k} + \frac{y^2}{b + k} + \frac{z^2}{c + k} = 1,$$

$a > b > c > 0$ and k a parameter. Given an ellipsoid F and a point $X(x, y, z)$ not in the principal plane, construct three more quadric surfaces, F', F'', and F''' with the properties that they are confocal, pass through X, intersect each other, and are mutually perpendicular. One is an ellipsoid, one is a hyperboloid of one sheet, and one is a hyperboloid of two sheets. Let k_1, k_2, and k_3 be the values of the parameter k in the equation above corresponding to these three surfaces. The parameters k_1, k_2, and k_3 are called the *elliptic coordinates* of the point X. The coordinates (x, y, z) of X are given by

$$x = \sqrt{\frac{(a + k_1)(a + k_2)(a + k_3)}{(b - a)(c - a)}},$$

$$y = \sqrt{\frac{(b + k_1)(b + k_2)(b + k_3)}{(a - b)(c - b)}},$$

$$z = \sqrt{\frac{(c + k_1)(c + k_2)(c + k_3)}{(a - c)(b - a)}}.$$

elliptic differential operator A differential operator $\sum_{i,j=1}^n a_{ij} \frac{\partial^2}{\partial x_i \partial x_j}$ such that the quadratic form $\mathbf{x}^T A\mathbf{x}$, where $A = (a_{ij})$, is positive definite whenever $\mathbf{x} \neq \mathbf{0}$. If the $\{a_{ij}\}$ are functions of some variable, say t, and the operator is elliptic for all values of t of interest, then the operator is called *uniformly elliptic*.

elliptic equation A second order linear partial differential equation in two independent variables

$$A\frac{\partial^2 u}{\partial x^2} + 2B\frac{\partial^2 u}{\partial x \partial y} + C\frac{\partial^2 u}{\partial y^2} + D\frac{\partial u}{\partial x} + E\frac{\partial u}{\partial y} + Fu + G = 0,$$

where $A, B, \ldots, G$ are functions of x and y, satisfying $B^2 - 4AC < 0$.

elliptic function *See* doubly periodic function.

elliptic function of the first kind A doubly periodic function $f(z)$ meromorphic on the complex plane. Also referred to simply as an elliptic function. *See* doubly periodic function.

elliptic function of the second kind A meromorphic function f satisfying the relations

$$f(u + 2\omega_1) = \mu_1 f(u),$$

and

$$f(u + 2\omega_3) = \mu_3 f(u),$$

(μ_1, μ_3 constants) with fundamental periods $2\omega_1, 2\omega_3$.

elliptic function of the third kind A meromorphic function f satisfying

$$f(u + 2\omega_i) = e^{a_i u + b_i} f(u), i = 1, 3,$$

(a_i and b_i are constants) with periods $2\omega_1$, $2\omega_3$. The Weierstrass sigma function $\sigma(u)$ is an example of an elliptic function of the third kind.

elliptic integral of the first kind The (incomplete) elliptic integral of the first kind is defined as

$$u = F(k, \phi) = \int_0^\phi \frac{d\theta}{\sqrt{1 - k^2 \sin^2\theta}}, \; 0 < k < 1.$$

Here ϕ is the amplitude of $F(k, \phi)$ or u, written $\phi = \text{am } u$, and k is its modulus, written $k = \text{mod } u$. The integral is also called Legendre's form for the elliptic integral of the first kind. If $\phi = \frac{\pi}{2}$, the integral is called the *complete* integral of the first kind and is denoted by $K(k)$ or K. This integral arises in connection with calculating the oscillation of a pendulum. *See also* Legendre-Jacobi standard form.

elliptic integral of the second kind The (incomplete) elliptic integral of the second kind is defined as

$$E(k, \phi) = \int_0^\phi \sqrt{1 - k^2 \sin^2\theta} d\theta, \quad 0 < k < 1.$$

The integral is also known as Legendre's form for the elliptic integral of the second kind. If $\theta = \frac{\pi}{2}$, the integral is the complete elliptic integral of the second kind and is denoted $E(k)$ or E. The integral arises in the determination of the arc length of an ellipse, hence the term elliptic integral. *See also* Legendre-Jacobi standard form.

elliptic integral of the third kind The incomplete elliptic integral of the third kind is defined as

$$\Pi(k, n, \phi) = \int_0^\phi \frac{d\theta}{(1 + n\sin^2\theta)\sqrt{1 - k^2 \sin^2\theta}},$$

for $0 < k < 1$. The integral is also known as Legendre's form for the elliptic integral of the third kind. Here n is a non-zero constant. (If $n = 0$, the integral reduces to the elliptic integral of the first kind.) If $\phi = \frac{\pi}{2}$, the integral is called the *complete* elliptic integral of the third kind. *See also* Legendre-Jacobi standard form.

elliptic paraboloid The (noncentral) quadric surface in $\mathbf{R}^3$ whose canonical equation is given by

$$\frac{x^2}{a^2} + \frac{y^2}{b^2} = \frac{z}{c},$$

where a and b are positive and $c \neq 0$. The axis of symmetry of an elliptic paraboloid is called its axis and the point of intersection of the axis with the surface, its vertex.

elliptic point If S is a surface in Euclidean n-space, and $p \in S$, then p is called an *elliptic point* on S if the Dupin indicatrix at p is an ellipse. (*See* Dupin indicatrix.)

elliptic theta function The elliptic theta function (introduced by Jacobi, who based all the theory of elliptic functions on its properties) is defined, for s a complex number and z a quantity in the upper half plane, as

$$\Theta(z, s) = \sum_{-\infty}^{\infty} e^{\pi i t^2 z + 2\pi i t s}.$$

In the eighteenth century, the theta function also appeared in a different connection in the work of Euler, and still earlier, in the work of Johann Bernoulli. Before Jacobi, Gauss recognized the importance of the function for the theory of elliptic functions and developed its essential properties without publishing these important discoveries.

embedding A homeomorphic mapping from one topological space into another. *See also* imbedding.

end Let a continuous plane, curve, or arc C be defined by $x = f(t)$, $y = g(t)$, for $0 \leq t \leq 1$, with continuous functions f, g defined on $[0, 1]$. The points $(f(0), g(0))$ and $(f(1), g(1))$ are called ends of the arc.

energy (1.) For a function $c \in \mathbf{C}([a, b])$, the integral

$$E(c) = \frac{1}{2}\int_a^b \|\dot{c}(t)\|^2 dt$$

where $\dot{c} = dc/dt$.
(2.) Let Ω be a space with a measure μ (≥ 0) and $\Phi(P, Q)$ a real-valued function on the product space $\Omega \times \Omega$. The mutual energy of μ and ν, defined for two measures ≥ 0, is

$$(\mu, \nu) = \int\int \Phi(P, Q) d\mu(Q) d\nu(P).$$

In particular, (μ, μ) is called the *energy* of μ.

energy density Let $U(t, x)$ be a distribution in $\mathbf{R}^{n+1}$ such that $U_{tt} - \Delta_x U = 0$ in $\mathbf{R}^{n+1}$, $U\big|_{t=0} = 0$, $U_t\big|_{t=0} = \delta(x)$, the Dirac measure in $\mathbf{R}^n$. Then the energy density is the function of (x, t) in the region $|x|^2 \neq t^2$ given by

$$\epsilon(x, t) = |U_t|^2 + |\text{grad}_x U|^2.$$

Enskog's method of solving integral equations A method used to solve symmetric integral equations. Consider a positive definite kernel $K(s, t)$, the first eigenvalue of which is greater than 1, i.e., a kernel for which the inequality

$$\int [\phi(s)]^2 ds - \int\int K(s, t)\phi(s)\phi(t) ds dt > 0$$

holds for all ϕ. The integral equation

$$f(s) = \phi(s) - \int K(s, t)\phi(t) dt$$

leads to the integral operator

$$J(\phi) = \phi(s) - \int K(s, t)\phi(t) dt.$$

Then, for any complete system of functions $\phi_1, \phi_2, \cdots$, construct, by a process similar to orthogonalization, a system of functions $v_1(s), v_2(s), \cdots$, with the property that $\int v_i J(v_k) ds = \delta_{ik}$, where δ_{ik} is the Kronecker delta. Such a system is called a *complete system, polar with respect to the kernel* $K(s, t)$. Setting $a_\gamma = \int \phi J(v_\gamma) ds =$

$\int v_\gamma f ds$ immediately yields the solution $\phi(s) = \sum_{\gamma=1}^{\infty} a_\gamma v_\gamma(s)$, provided this series converges uniformly.

entire algebroidal function An algebroidal function that has no pole in $|z| < \infty$ is called an *entire algebroidal function. See also* algebroidal function.

envelope A curve which at each of its points is tangent to an element of a one-parameter family of curves is called an *envelope* of that family. The notion can be extended to a one-parameter family of surfaces, as well as to two (or more) parameters.

envelope of holomorphy Given a domain G in complex n-space, let $H(G)$ be the holomorphic functions on G. For $f \in H(G)$, the domain $\tilde{G}_f$ of holomorphy for f is defined to be the maximal domain to which f may be continued analytically. The common existence domain for all $f \in H(G)$ is called the envelope of holomorphy.

enveloping surface Consider a 1-parameter family of surfaces in 3-space given by the equation $F(x_1, x_2, x_3, t) = 0$. A surface E, not belonging to this family, is called an *enveloping surface* of the family $\{S_t\}$ if E is tangent to some S_t at each point of E. In other words, E and S_t have the same tangent plane.

epicycloid The curve in $\mathbf{R}^2$ traced out by a point P on the circumference of a circle of radius b which is rolling externally on a fixed circle of radius a. If the fixed circle is centered at the origin, $A(a, 0)$ is one of the points at which the given point P comes in contact with the fixed circle, B is the moving point of tangency of the two circles, and the parameter t is the radian measure of the angle AOB, then parametric equations for the epicycloid are given by

$$x = (a+b)\cos t - b\cos\frac{a+b}{b}t,$$

and

$$y = (a+b)\sin t - b\sin\frac{a+b}{b}t.$$

See roulette.

epitrochoid One of a class of plane curves, called *roulettes*, in which one curve C' rolls on a fixed curve C without slipping and always tangent to C. The locus Γ of a point X keeping a fixed position with the curve C' is called a roulette. When the base curve C and the rolling curve C' are both circles which are externally tangent, the resulting locus Γ is called an epitrochoid. The *epicycloid* is the special case where X lies on C'.

equation of curve An equation in two variables, x and y, such that all the points on the curve and only those points, satisfy the equation.

equation of locus An equation in n variables, $x_1, \ldots, x_n$, such that all the points on a given locus in $\mathbf{R}^n$, and only those points, satisfy the equation.

equation of surface An equation in n variables, $x_1, \ldots, x_n$, such that all the points on a given surface in $\mathbf{R}^n$, and only those points, satisfy the equation.

equivalence relation A binary relation $\sim$ defined on a set, which satisfies the three properties listed below.

(1) It is reflexive: $a \sim a$;

(2) It is symmetric: $a \sim b \Rightarrow b \sim a$;

(3) It is transitive: $a \sim b, b \sim c \Rightarrow a \sim c$.

For example, ordinary $=$ is an equivalence relation on the integers.

error function The function $y = \text{erf}(x)$ defined by the equation

$$y = \text{erf}(x) = \int_0^x e^{-t^2} dt.$$

It is of great importance in probability and statistics and is an odd function which is continuous for all x and satisfies the inequality $-1 < \text{erf}(x) < 1$ for all x.

essential singularity If a complex function $f(z)$ is analytic in a region $0 < |z-a| < \delta$, then $z = a$ is called an *isolated* singularity of $f(z)$. An isolated singularity $z = a$ such that $\lim_{z\to a} |z-a|^{\alpha}|f(z)|$ is neither 0 nor ∞ for any α is called an *essential* singularity. Thus, in the neighborhood of an essential singularity $f(z)$ is at the same time unbounded and comes arbitrarily close to 0.

essential spectrum For a linear operator T on a complex, normed vector space X, the set of complex numbers λ such that one of the following fails to hold.
(1.) the operator $\lambda I - T$ has a finite dimensional null space (kernel);
(2.) the range (image) of the operator $\lambda I - T$ is a closed subspace of finite codimension in X. *See* Fredholm operator.

For a self-adjoint operator on Hilbert space, the essential spectrum is the set of limit points of the spectrum, together with the eigenvalues of infinite multiplicity.

essential supremum Let f be a μ-essentially bounded function with domain S, i.e., there is a μ-null set N such that the restriction of f to $S\backslash N$ is bounded. The quantity

$$\inf_N \sup_{s\in S} |f(s)|,$$

where N ranges over the μ-null subsets of S is called the μ-essential supremum of $|f(\cdot)|$ or the μ-essential least upper bound of $|f(\cdot)|$.

essentially bounded function Let μ be an additive set function defined on a field of subsets of a set S. A subset N of S is said to be a μ-null set if $v^*(\mu, N) = 0$, where v^* is the extension of the total variation v of μ. A function f defined on S is said to be μ-essentially bounded, or simply essentially bounded, if there is a μ-null set N such that the restriction of f to $S\backslash N$ is bounded.

essentially self-adjoint operator A linear operator whose least closed linear extension is self-adjoint. *See* closed operator. *See also* self-adjoint operator.

Euler formula There are several formulas attributed to Leonhard Euler. Euler's formula for complex numbers is

$$e^{i\theta} = \cos\theta + i\sin\theta.$$

Euler's formula for polyhedra is

$$V - E + F = 2,$$

where V = the number of vertices, E = the number of edges, and F = the number of faces in the polyhedron. The formulas for the Fourier coefficients of a periodic function of period 2π (*see* Fourier series) are also sometimes referred to as Euler's formulas.

Euler polynomial The polynomial of degree n defined by

$$E_n(x) = a_0x^n + \binom{n}{1}a_1x^{n-1} + \binom{n}{2}a_2x^{n-2} + \cdots + a_n.$$

Euler transform To obtain solutions of a linear differential equation

$$L[u] + \mu u = 0$$

in the form of an integral representation, the method of integral transforms is often useful. In place of the unknown function $u(z)$ we introduce a new unknown function $v(\zeta)$ of the complex variable $\zeta = \xi + i\nu$ by means of an equation

$$u(z) = \int_C K(z, \zeta)v(\zeta)d\zeta,$$

where the transformation kernel $K(z, \zeta)$ (assumed analytic in each complex variable) and the path of integration C are to be suitably determined. The differential equation then takes the form

$$\int_C (L[K] + \mu K)v(\zeta)d\zeta = 0,$$

where here the differentiation process L refers to the variable z and it is assumed that the process L is interchangeable with integration. K may be specified by subjecting it to a suitable partial differential equation, and various kernels are used in practice. The kernel

$$K(z, \zeta) = (z - \zeta)^{\alpha}$$

produces the Euler transform if the path of integration is suitably chosen. *See also* Laplace transform.

Euler's angles Suppose we have two rectangular coordinate systems, (x', y', z') and (x, y, z), sharing the same origin. The correlation of the two is given by *Euler's angles* (θ, ϕ, ψ), where θ is the angle between the z axis and the z' axis, ϕ is the angle between the zx-plane and the zz'-plane, and ψ is the angle between the $z'x'$-plane and the $z'z$-plane. These angles are subject to the inequalities $0 < \theta < \pi$, and $0 < \phi, \psi < 2\pi$ and are used in the dynamics of rigid bodies.

Euler's constant The constant γ defined as

$$\gamma = \lim_{n\to\infty}(1 + \frac{1}{2} + \frac{1}{3} + \ldots + \frac{1}{n} - \ln n).$$

The value of Euler's constant is approximately .577215 and has been computed to over 10,000 decimal places. The constant is named for the Swiss mathematician Leonhard Euler. It is not known whether γ is rational or irrational.

Euler-Lagrange differential equation If $u = u(x)$ and $J[u] = \int f(u, u', x)ds$, then the condition for the vanishing of the variational derivative of J with respect to u, $\frac{\delta J}{\delta u} = 0$, is given by the Euler-Lagrange equation:

$$\left[\frac{\partial}{\partial u} - \frac{d}{dx}\frac{\partial}{\partial u'}\right]f = 0.$$

If $w = w(x)$ and $J = \int g(w'', w', w', x)ds$, then the Euler-Lagrange equation is

$$\left[\frac{\partial}{\partial w} - \frac{d}{dx}\frac{\partial}{\partial w'} - \frac{d^2}{dx^2}\frac{\partial}{\partial w''}\right]g = 0.$$

Finally, if $v = v(x, y)$ and $J = \int\int h(v_x, v_y, v, x, y)dxdy$, then the Euler-Lagrange equation is

$$\left[\frac{\partial}{\partial v} - \frac{d}{dx}\frac{\partial}{\partial v_x} - \frac{d}{dy}\frac{\partial}{\partial v_y}\right]h = 0.$$

even function A function $f : \mathbf{R} \to \mathbf{R}$ with a domain D such that if $x \in D$ then $-x \in D$ and $f(-x) = f(x)$.

evolute *See* involute evolute.

evolution equation A general term for an equation of the form

$$\frac{\partial u}{\partial t} = Au,$$

where A is a (in general unbounded) linear operator on a function space.

An example is the Schroedinger equation.

exact differential form A differential form ν which can be expressed in the form $\nu = d\omega$ where $d\omega$ is the exterior derivative of a differential form ω. (*See also* exterior derivative.)

expansion formula The infinite series appearing in an expansion in series. *See* expansion in series.

expansion in series An infinite series of functions, associated with a given function, whose values may reasonably be expected to be equal to the function at certain points. *See* Fourier series, Taylor expansion. Note, for example, that the series

$$\sum_{n=0}^{\infty}\frac{f^{(n)}(a)}{n!}(x - a)^n$$

is called the Taylor expansion of $f(x)$ at $x = a$, as long as the terms $f^{(n)}(a)$ are taken to be the derivatives of f at $x = a$, even if the series does not converge to f at any points except $x = a$.

explicit function In general, an equation such as $F(x, y) = 0$ or $F(x, y, z) = 0$ defines one variable, say y (or z) in terms of the other variable(s). Then y (or z) is sometimes called an *implicit* function of x (or of x and y) to distinguish it from a so-called *explicit* function f, where $y = f(x)$ (or $z = f(x, y)$).

exponential curve The graph of the equation $y = e^x$ is called the exponential curve.

exponential function For a, a positive number, the (real) exponential function is defined by the formula $f(x) = a^x$. For $x = \frac{m}{n}$, a positive rational number, a^x is defined as $(a^{\frac{1}{n}})^m$. In addition, $a^{-\frac{m}{n}}$ means the reciprocal of $a^{\frac{m}{n}}$. For irrational values of x, a^x is defined as the limit of a^{s_n}, where $\{s_n\}$ is a sequence of rational numbers whose limit is x. In most cases, a is taken to be the constant e, defined as

$$\lim_{n \to 0}(1+n)^{\frac{1}{n}}.$$

This irrational constant has the approximate value 2.718281829. (*See also* exponential series.)

For $z = x + iy$ a complex number, the exponential function e^z is defined by

$$e^z = e^{x+iy} = e^x(\cos y + i \sin y).$$

exponential function with base a *See* exponential function.

exponential generating function The *exponential generating function* is defined by

$$g(z) = \sum_{n=0}^{\infty} B(n)(n!)^{-1} z^n,$$

where $B(n)$ is the number of ways to divide n completely dissimilar objects into non-empty classes. The exponential generating function is convergent for all complex numbers z and equals $e^{e^z - 1}$. The Bell numbers, $B(n)$, can be defined recursively by $B(0) = 1$, and

$$B(n+1) = \sum_{k=0}^{n} \binom{n}{k} B(k).$$

exponential integral The *exponential integral* $Ei(x)$ is defined

$$Ei(x) = \int_{-\infty}^{x} \frac{e^t}{t} dt.$$

Note that a principal value must be taken at $t = 0$ if $x > 0$. $Ei(x)$ has important applications in quantum mechanics.

exponential series The expansion of the exponential function e^x into the Taylor series

$$e^x = \sum_{n=0}^{\infty} \frac{x^n}{n!},$$

which is convergent for all real x.

extended real numbers The *extended real number system* consists of the real number system to which two symbols, $+\infty$ and $-\infty$, have been adjoined with the following properties:

(1) If x is real, then $-\infty < x < +\infty$, and $x + \infty = +\infty$, $x - \infty = -\infty$, $\frac{x}{+\infty} = \frac{x}{-\infty} = 0$.

(2) If $x > 0$, then $x \cdot (+\infty) = +\infty$, and $x \cdot (-\infty) = -\infty$.

(3) If $x < 0$, then $x \cdot (+\infty) = -\infty$, and $x \cdot (-\infty) = +\infty$.

extension A function g is an extension of a function f if V, the domain of f, is a subset of the domain of g, and $g(v) = f(v)$ for all $v \in V$.

exterior algebra The algebra of the exterior product, also called an alternating algebra or Grassmann algebra. The study of exterior algebra is also called ausdehnungslehre or extensions calculus. Exterior algebras are graded algebras.

In particular, the exterior algebra of a vector space is the direct sum over $k = 1, 2, \ldots$, of the vector spaces of alternating k-forms on that vector space. The product on this algebra is then the wedge product of forms. The exterior algebra for a vector space V is constructed by forming monomials u, $v \wedge w$, $x \wedge y \wedge z$, etc., where u, v, w, x, y, and z are vectors in V and $\wedge$ is asymmetric multiplication.

The sums formed from linear combinations of the monomials are the elements of an exterior algebra.

exterior derivative If ω is a differential form of degree r written

$$\omega = (1\backslash r!)\sum a_{i_1 \ldots i_r} dx^{i_1} \wedge \cdots \wedge dx^{i_r}$$

in a coordinate neighborhood U, then the exterior derivative is the differential form $d\omega$ of degree $r+1$ defined in U by

$$d\omega = (1\backslash r!)\sum da_{i_1 \ldots i_r} \wedge dx^{i_1} \wedge \cdots \wedge dx^{i_r}.$$

exterior problem (1.) A geometric condition related to the famous Cauchy problem. It is considered to be the converse of the interior problem. *See* interior problem.
(2.) Let D be a domain in Euclidean n-space for $n \geq 2$ with compact boundary S. The classical Dirichlet problem is the problem of finding a harmonic function in D that assumes the values of a prescribed continuous function on S. If D is unbounded, this is called an exterior problem. (*See also* interior problem, Dirichlet problem.)

exterior product For ω and θ differential forms of degree r and s respectively, the exterior product $\omega \wedge \theta$ is the differential form of degree $r+s$ defined by

$$(\omega \wedge \theta)_p = \omega_p \wedge \theta_p,$$

$p \in M$, where M is an n-dimensional topological manifold. Let $X_1, X_2, \ldots, X_{r+s}$ be $r+s$ vector fields in M. Then,

$$(\omega \wedge \theta)(X_1, X_2, \ldots, X_{r+s}) = \sum sgn(i;j)\omega(X_{i_1}, \ldots, X_{i_r})\,\theta(X_{j_1}, \ldots, X_{j_s}),$$

where the summation runs over all possible partitions of $(1, 2, \ldots, r+s)$ such that $i_1 < i_2 < \ldots < i_r, j_1 < j_2 <, \ldots, < j_s$ and $sgn(i;j)$ means the sign of the permutation $(1, 2, \ldots, r+s) \to (i_1, \ldots, i_r, j_1, \ldots, j_s)$. Note that if $\omega_1, \omega_2, \ldots \omega_r$ are differential forms of degree 1, then

$$(\omega_1 \wedge \ldots \wedge \omega_r)(X_1, \ldots, X_r) = \det(\omega_i(X_j)).$$

extremal distance Let Ω be a domain in a plane, $\partial\Omega$ its boundary, and X_1, X_2 sets on the union of Ω and $\partial\Omega$. The extremal length of the family of curves in Ω connecting points of X_1 and points of X_2 is called the *extremal distance* between X_1 and X_2 (relative to Ω) and is denoted by $\lambda_\Omega(X_1, X_2)$.

extremal length Let C be a countable collection of locally rectifiable curves in a plane domain and let Γ be a family of such collections C. Define $M(\Gamma) = \inf \int \rho^2 dxdy$, where the inf runs over all admissible Baire functions in the plane. Then the extremal length of Γ is the reciprocal $\frac{1}{M(\Gamma)}$.

extremal length defined by Hersch and Pfluger Let C be a countable collection of locally rectifiable curves in a plane domain and let Γ be a family of such collections C. Define $M(\Gamma) = \inf \int \rho^2 dxdy$, where the inf runs over all continuous admissible Baire functions in the plane. Then the extremal length of Γ, as defined by Hersch and Pfluger, is the reciprocal $\frac{1}{M(\Gamma)}$. *See also* extremal length.

extreme value of a function A maximum or minimum value of a real-valued function. *See also* maximum value of a function, minimum value of a function.

extremum A maximum or a minimum value of a real-valued function. *See also* maximum value of a function, minimum value of a function.

F

F_σ-set A set which is the union of a countable collection of closed sets, in a topological space.

F. and M. Riesz Theorem Let μ be a Borel measure on the unit circle T. If $\int_T e^{in\theta} d\mu = 0$ for all positive integers n, then μ is absolutely continuous with respect to Lebesgue measure. In other words, if μ is analytic, then it is absolutely continuous with respect to Lebesgue measure.

factor analysis A data reduction technique which builds a model from data. The technique finds underlying factors, also called *latent variables* and provides models for these factors based on variables in the data. For example, suppose a market research survey is given, which asks the importance of 9 product attributes. Suppose also that three underlying factors are found. The variables which *load* highly on these factors give some information about what these factors might be. For example, if three attributes such as technical support, customer service, and availability of training courses all load highly on one factor, one might call this factor *service*.

This technique can be very helpful in finding important underlying characteristics, which might not themselves be observed, but which might be found as manifestations of variables which are observed. Another good application of factor analysis is for grouping together products based on a similarity of buying patterns. By using factor analysis one might locate opportunities for cross-selling and bundling.

The factor analysis model is similar to a multivariate model, but several of the assumptions are changed. It is assumed that each observation vector y_i has $E(y_i) = \mu$ and $\text{Cov}(y_i) = \epsilon$. For n observations, the factor analysis model is

$$Y = \mu' + XB + e,$$

where Y is $n \times q$ and X is $n \times r$. The assumptions about the rows of e are: $E(\epsilon_i) = 0$, $\text{Cov}(\epsilon_i, \epsilon_j) = 0$, $i \neq j$ and $\text{Cov}(\epsilon_i) = A$, with A diagonal. Each row of X is assumed to be an unobservable random vector with $E(x_i) = 0$, $\text{Cov}(x_i, x_j) = 0$, $i \neq j$, $\text{Cov}(x_i) = I_r$, and $\text{Cov}(x_i, \epsilon_j) = 0$ and i, j. The matrix B remains a fixed but unknown matrix of parameters. The coefficients of matrix B are called the factor loadings.

See also factor score, factor loading.

factor loading The classical factor analysis model can be written for the value of variable j for individual i as

$$z_{ji} = \sum_{p=1}^{m} a_{jp} F_{pi} + d_j U_{ji},$$

$i = 1, 2, \ldots, N$, $j = 1, 2, \ldots, n$. In this expression, F_{pi} is the value of a common factor p for an individual i, and each of the m terms $a_{jp}F_{pi}$ represents the contribution of the corresponding factor to the linear composite, while $d_j U_{ji}$ is the *residual error* in the theoretical representation of the observed measurement z_{ji}. The common factors account for the correlations among the variables, while each unique factor accounts for the remaining variance (including error) of that variable. The coefficients of the factors are called *factor loadings*.

factor of automorphy Let X be a complex analytic manifold and Γ a discontinuous group of (complex) analytic automorphisms of X. A set of holomorphic functions (without zeros) on X, $\{j_\gamma(z), \gamma \in \Gamma\}$, is called a factor of automorphy if it satisfies the condition $j_{\gamma\gamma'}(z) = j_\gamma(\gamma'(z))j_\gamma(z)$ for all $\gamma, \gamma' \in \Gamma$, and $z \in X$.

factor score Estimated best linear predictors used to check the assumption of multivariate normality of a factor analysis model. *See also* factor analysis.

0-8493-0320-6/00/$0.00+$.50

factorial series A series of the form

$$\sum_{n=1}^{\infty} \frac{n!a_n}{z(z+1)(z+2)...(z+n)},$$

$z \neq 0, -1, -2, \ldots$. It converges or diverges together with the ordinary Dirichlet series $\sum a_n/n^z$ except at $z = 0, -1, -2, \ldots$.

Factorization Theorem (1.) (Bing) Let $C = \bigcap_i T_i$ where each T_i is a compact neighborhood of C in E^n (Euclidean space) and $T_{i+1} \subset T_i$ for each i. Suppose that for each i and $\epsilon > 0$ there is an integer N and an isotropy μ_t of E^{n+1} onto E^{n+1} such that μ_0 = identity, μ_1 is uniformly continuous and
(i.) $\mu_t \mid (E^{n+1} - (T_i \times E^1)) = 1$,
(ii.) μ_t changes $(n+1)$st coordinates less than ϵ, and
(iii.) for each $w \in E^1$, diameter $\mu_1(T_N \times w) < \epsilon$.
Then,

$$(E^n/C) \times E^1 \approx E^{n+1}.$$

(2.) (Andrews and Curtis) Let α be an arc in E^n (Euclidean space). Then,

$$(E^n \mid \alpha) \times E^1 \approx E^{n+1}.$$

(3.) Any natural number can be expressed as the product of a unique set of prime natural numbers. For example $264 = 2^3 \cdot 3 \cdot 11$. This result is also called the Fundamental Theorem of Arithmetic or the Unique Factorization Theorem.

family of confocal central conics The family of curves given by

$$\frac{x^2}{a^2+\lambda} + \frac{y^2}{b^2+\lambda} = 1,$$

with a, b held fixed and λ the parameter. The differential equation of this family of plane curves is

$$(xy' - y)(yy' + x) = (a^2 - b^2)y'.$$

family of confocal parabolas A family of parabolas (*see also* family of curves) with a common focus.

family of curves A relation involving a parameter, as well as one or both of the coordinates of a point in the plane, represents a family of curves, one curve corresponding to each value of the parameter. For example, the equation

$$(x-c)^2 + (y-c)^2 = 2c^2$$

may be interpreted as the equation of a family of circles of differing radii, each having its center on the line $y = x$ and each passing through the origin.

family of frames A set whose members are frames. *See* frame.

family of functions A set whose members are functions. *See* function.

family of mappings A set whose members are mappings. *See* mapping.

family of points A set whose members are points.

family of quasi-analytic functions A set whose members are quasi-analytic functions. *See* quasi-analytic function.

family of sets A set whose members are sets.

fast Fourier transform *See* discrete Fourier transform.

Fatou's Lemma If f_n are real-valued, nonnegative, measurable functions on a measure space (X, μ), for $n = 1, 2, \ldots$, then

$$\int_X (\liminf_{n\to\infty} f_n) d\mu \leq \liminf_{n\to\infty} \int_X f_n d\mu.$$

Fatou's Theorem Let f be a bounded analytic function on the open unit disk D; that is, $f \in H^\infty(D)$. For most all t, the radial limit $\lim_{r\to 1^-} f(re^{it})$ exists.

Fefferman-Stein decomposition If $f \in BMO(\mathbf{R}^n)$, then there are $n+1$ functions

$f_0, f_1, \ldots, f_n \in L^\infty(\mathbf{R}^n)$ such that $f = f_0 + \sum_{k=1}^{n} R_k(f_k)$, where R_k are the Riesz transforms.

Fejer kernel For a nonnegative integer n, the nth Fejer's kernel is the function

$$K_n(t) = \sum_{k=-n}^{n} (1 - \frac{|k|}{n+1})e^{ikt} = \frac{1}{n+1}\left[\frac{\sin\frac{n+1}{2}t}{\sin\frac{1}{2}t}\right]^2.$$

Fejer means Also known as Cesàro means. *See* Cesàro means.

Fejer's Theorem Let $f(t)$ be a continuous, periodic function of period 2π. Then the arithmetic means $\sigma_k(t)$ of the partial sums of the Fourier series of $f(t)$ converge uniformly to $f(t)$.That is, if $\sum_{n=-\infty}^{\infty} a_n e^{int}$ is the Fourier series of $f(t)$, $s_k(t) = \sum_{n=-k}^{k} a_n e^{int}$ is the kth partial sum of the Fourier series, and $\sigma_k(t) = \frac{1}{k+1}\sum_{n=0}^{k} s_n(t)$ is the kth Cesàro mean, then the sequence $\{\sigma_k(t)\}$ converges uniformly to $f(t)$.

Feynman amplitude A concept appearing in quantum field theory, in connection with the Feynman rules for the computation of a definite Green's function or scattering amplitude.

Feynman amplitude is given by

$$i\mathcal{T}(2\pi)^4\delta^4(p_i - p_f),$$

where $(2\pi)^4\delta^4(\Sigma p)$ is a momentum conservation delta function. This expression is obtained up to renormalization, from the Feynman rules by putting the external lines on their mass shell, and providing external fermion lines with spinors according to whether the line enters or leaves the diagram and whether it belongs to the initial or final state.

Feynman diagram A diagram which depicts certain physical processes in quantum physics.

Feynman graph Also known as Feynman diagram. *See* Feynman diagram.

Feynman integral Consider the integral

$$I = \int_{-\infty}^{\infty} e^{i(2\pi/h)F(x)}\, dx,$$

where h is a small number, and $F(x)$ is a real function of the real variable x. $F(x)$ is supposed to have exactly one stationary point, which can be either a maximum or minimum, and which lies at $x = a$. Choose F to be the action integral derived from the Lagrangian L:

$$F = \int L\, dt.$$

The integral I now becomes the summation over all possible trajectories between fixed end points, and using

$$F(a) = \lim_{h\to 0} \frac{h}{2\pi i} \log_e\left\{\int_{-\infty}^{\infty} dx \exp[i(2\pi/h)F(x)]\right\}$$

(found by a Taylor-type expansion of $F(x)$ about $x = a$) yields an F value.

The Feynman integral is given by

$$I = \sum_{\text{all trajectories}} \exp\left(-(i/h_0)\int_{t_1}^{t_2} L(x)\, dt\right). \quad (1)$$

$(h_0 = \frac{h}{2\pi})$

We note that I given by (1) is used in quantum field theory in connection with the Feynman rules for the computation of a definite Green function or scattering amplitude.

Feynman rules Rules for the computation of a definite Green's function or scattering amplitude:

(i.) Draw all possible topologically distinct diagrams, connected or disconnected

but without vacuum-vacuum subdiagrams, contributing to the process under study, at the desired order.
(ii.) For each diagram, and to each internal line, attach a propagator:

$$\bullet\overset{k}{\text{—}}\bullet \qquad \frac{i}{k^2 - m^2 + i\varepsilon}$$

for a spin 0 boson,

$$\alpha \bullet\overset{p}{\rightarrow}\bullet \beta \quad \left(\frac{i}{\not{p} - m + i\varepsilon}\right)_{\beta\alpha}$$

for a spin 1/2 fermion,

$$\rho \bullet \overset{k}{} \bullet \sigma$$

$$-i\left(\frac{g_{\rho\sigma} - k_\rho k_\sigma/\mu^2}{k^2 - \mu^2 + i\varepsilon} + \frac{k_\rho k_\sigma/\mu^2}{k^2 - \mu^2/\lambda + i\varepsilon}\right)$$

$$= -i\left[\frac{g_{\rho\sigma}}{k^2 - \mu^2 + i\varepsilon} - \frac{(1-\lambda^{-1})k_\rho k_\sigma}{(k^2 - \mu^2 + i\varepsilon)(k^2 - \mu^2/\lambda + i\varepsilon)}\right]$$

for a spin 1 boson of mass μ in the Stueckelberg gauge, i.e., endowed with a kinetic Lagrangian

$$\mathcal{L} = -\tfrac{1}{4}(\partial_\mu A_\nu - \partial_\nu A_\mu)^2 - \frac{\lambda}{2}(\partial \cdot A)^2 + \frac{\mu^2}{2}A^2.$$

(iii.) To each vertex, assign a weight derived from the relevant monomial of the interaction Lagrangian. It is composed of a factor coming from the degeneracy of identical particles in the vertex, of the coupling constant appearing in $i\mathcal{L}_{\text{int}}$, of possible tensors in internal indices, and of a momentum conservation delta function $(2\pi)^4\delta^4(\Sigma p)$. To each field derivative $\partial_\mu\phi$ is associated $-ip_\mu$ where p is the corresponding incoming momentum. Vertices for the most common theories are listed below.
(iv.) Carry out the integration over all internal momenta with the measure $d^4k/(2\pi)^4$, possibly after a regularization.
(v.) Multiply the contribution of each diagram by
(a.) a symmetry factor $1/S$ where S is the order of the permutation group of the internal lines and vertices leaving the diagram unchanged when the external lines are fixed;
(b.) a factor minus one for each fermion loop; and
(c.) a global sign for the external fermion lines, coming from their permutation as compared to the arguments of the Green function at hand.

These rules appear in quantum field theory.

field A nonempty set $\mathbf{F}$ with binary operations $+$ and $\cdot$ defined on it, is said to be a field if the following axioms are satisfied. (1.) The Associative Laws: $a+(b+c) = (a+b)+c$ and $a \cdot (b \cdot c) = (a \cdot b) \cdot c$ for all $a, b, c \in \mathbf{F}$. (2.) The Commutative Laws: $a+b = b+a$ and $a \cdot b = b \cdot a$ for all $a, b \in \mathbf{F}$. (3.) The Distributive Law: $a \cdot (b+c) = a \cdot b + a \cdot c$ for all $a, b, c \in \mathbf{F}$. (4.) Existence of an Additive Identity: there is an element $0 \in \mathbf{F}$ satisfying $a + 0 = a$ for all $a \in \mathbf{F}$. (5.) Existence of a Multiplicative Identity: there is an element $1 \in \mathbf{F}$, $1 \neq 0$, satisfying $a \cdot 1 = a$ for all $a \in \mathbf{F}$. (6.) Existence of Additive Inverses: for each element $a \in \mathbf{F}$, there is an element $-a \in \mathbf{F}$ satisfying $a + (-a) = 0$. (7.) Existence of Multiplicative Inverses: for each element $a \in \mathbf{F}$, $a \neq 0$, there is an element $a^{-1} \in \mathbf{F}$ satisfying $a \cdot a^{-1} = 1$. For example, the set of all real numbers $\mathbf{R}$ with usual addition $+$ and multiplication $\cdot$ is a field.

field of scalars Let V be a vector space over the field $\mathbf{F}$. Elements of $\mathbf{F}$ are called scalars, and $\mathbf{F}$ is called the *field of scalars* of V. *See* field.

field of sets A collection $\mathcal{F}$ of subsets of a set S such that if $E, F \in \mathcal{F}$, then (i.) $E \cup F \in \mathcal{F}$, (ii.) $E \backslash F \in \mathcal{F}$, and (iii.) $S \backslash F \in \mathcal{F}$. If $\mathcal{F}$ is also closed under the taking of countable unions, then $\mathcal{F}$ is called a σ-field. Fields and σ-fields of sets are sometimes called algebras and σ-algebras of sets, respectively.

filter Let X be a nonempty set. A *filter* F in X is a family of subsets of X satisfying (1.) the empty set is not a member of F, (2.) the intersection of any two members of F is a member of F, and (3.) any subset of X which contains a member of F is a member of F.

filter base Let X be a nonempty set and F, a filter in X. A family B of subsets of X is a base for the filter F if $F = \{f : f \subset X$ and $b \subset f$ for some $b \in B\}$.

final set The range of a partial isometric operator. The *final set* of a partial isometric operator V is also called the final space of V. *See* partially isometric operator.

finite sequence A sequence which has a finite number of terms.

finite type power series space Replace infinity by a fixed (finite) positive integer in the description of *infinite type power series space*. *See* infinite type power series space.

finite-dimensional linear space A vector or linear space which has a basis consisting of a finite number of vectors. One says the space has dimension n if there are n vectors in the basis.

finiteness The property of *finiteness* (of generation) asserts that a point x is in the convex hull of a set S if and only if x is in the convex hull of a finite subset of S.

Finiteness Theorem (1.) A polynomial vector field on the real plane has a finite number of limit cycles. This is the Finiteness Theorem for limit cycles.
(2.) Let R be a compact Riemann surface and Θ be the sheaf of holomorphic functions on R. The first cohomology group $H^1(R, \Theta)$ is a finite dimensional complex vector space. This is the Finiteness Theorem for compact Riemann surface. The dimension of $H^1(R, \Theta)$ is called the genus of R.

Finsler manifold A $\mathbf{C}^r$ Banach manifold M is a *Finsler manifold* if for every $p \in M$, there is a norm $\|\cdot\|_p$ defined on the tangent space $T_p(M)$ such that $\|\cdot\|_p$ is continuous in p.

Finsler metric Let M be an n-dimensional manifold, $T(M)$ be the tangent bundle of M, $T_p(M)$ be the tangent space at $p \in M$, and $X_p \in T_p(M)$. A *Finsler metric* on M is a function $F : T(M) \to [0, \infty)$ satisfying (1.) F is continuous; (2.) for every $p \in M$, $F(X_p) = 0$ if and only if $X_p = O_p$, where O_p is the zero vector of $T_p(M)$; (3.) for every $p \in M$, every vector $X_p \in T_p(M)$ and every complex number α, $F(\alpha X_p) = |\alpha| F(X_p)$.

Finsler space A differentiable manifold M with a Finsler metric F defined on it is called a *Finsler space.*

first category A subset E of a topological space is *of the first category* if E is the countable union of nowhere dense sets.

first Cousin problem Let Ω be a domain in $\mathbf{C}^n$ and $\{G_\alpha : \alpha \in \Lambda\}$ be an open cover of Ω. For each $\alpha \in \Lambda$, let there be given a meromorphic function f_α defined on G_α such that for any $\alpha, \beta \in \Lambda$, the difference $f_\alpha - f_\beta$ is analytic in $G_\alpha \cap G_\beta$. Find a meromorphic function f defined on Ω such that $f - f_\alpha$ is analytic in G_α for all $\alpha \in \Lambda$. This is called the *first Cousin* (Pierre Cousin, 1867–1933) *problem.* A domain Ω for which the problem can be solved is called a Cousin domain of the first kind. For example, any domain Ω in $\mathbf{C}^n$ for which the Cauchy-Riemann equations on Ω can be solved is a Cousin domain of the first kind. In particular, any polydisk in $\mathbf{C}^n$ is a Cousin domain of the first kind.

first fundamental form Let F be an arbitrary smooth function of the three Cartesian coordinates. Define a surface S to be the set of all solutions of an equation

$$F(x, y, z) = 0. \qquad (1)$$

Restricting the Euclidean metric of the embedding space $\mathbf{R}^3$ to S, we obtain an expression for the square of the arc element, namely

$ds^2 = dx^2 + dy^2 + dz^2$. If we substitute the defining equations $x = x(u, v)$; $y = y(u, v)$; $z = z(u, v)$ (x, y, z smooth), equation (1) reduces to a homogeneous second degree polynomial P in the coordinate differentials du and dv. P so constructed is called the *first fundamental form*. *See* fundamental form.

First Mean Value Theorem *See* Mean Value Theorem of differential calculus, Mean Value Theorem of integral calculus.

first quadrant of Cartesian plane The portion of the Cartesian plane $\mathbf{R}^2$ consisting of points (x, y) with $x > 0$ and $y > 0$.

first-order derivative Let $f : D \to \mathbf{R}$ be a function with domain $D \subset \mathbf{R}$. We say $f(t)$ is differentiable at t if $f(t)$ is defined in a neighborhood of t and the limit $f'(t) = \lim_{h \to 0} \frac{f(t+h)-f(t)}{h}$ exists. The function $f'(t)$ defined for those t for which the limit exists is called the *first-order derivative* of $f(t)$. The first-order derivative of $f(t)$ is usually called the derivative or the first derivative of $f(t)$.

flabby resolution Let S be a sheaf (of Abelian groups) over a topological space. A resolution of the sheaf S is an exact sequence $0 \to S \to S_1 \to S_2 \to \cdots$, where each $S_i, i = 1, 2, 3, \ldots$ is a sheaf. The resolution is a flabby resolution if each sheaf S_i in the sequence is flabby.

flat function A function $f(t)$ is said to be flat at t_0 if it is infinitely differentiable at t_0 and its derivatives of all orders at t_0 are 0. That is, $f^{(n)}(t_0) = 0$ for $n = 0, 1, 2 \ldots$. In other words, if $f(t)$ is flat at t_0, then its formal Taylor expansion at t_0 vanishes. A function is a flat function if it is flat at a point of its domain. For example, the function $f(t) = e^{-1/t}$ for $t > 0$ and $f(t) = 0$ for $t \leq 0$ is a flat function for it is flat at 0.

flat point Let M be a surface and $p \in M$ a point on M. The point p is a flat (or planar) point if both principal curvatures at p are zero.

Floquet's Theorem Let $\mathbf{y}' = \mathbf{A}(t)\mathbf{y}$ be a homogeneous, first-order system of n linear differential equations in which the coefficient matrix $\mathbf{A}(t)$ is periodic of period p ($\mathbf{A}(t + p) = \mathbf{A}(t)$). Any fundamental matrix $\mathbf{Y}(t)$ of the system can be expressed as $\mathbf{Y}(t) = \mathbf{B}(t)e^{\mathbf{C}t}$, where $\mathbf{C}$ is an $n \times n$ constant matrix, and $\mathbf{B}(t)$ is periodic of period p.

flux Let $\mathbf{F}$ be a continuous vector field defined on an oriented surface S with a continuous unit normal vector $\mathbf{n}$ in $\mathbf{R}^3$. The *flux* ϕ of $\mathbf{F}$ across S in the direction of $\mathbf{n}$ is defined by the surface integral $\phi = \int\int_S \mathbf{F} \cdot \mathbf{n}\, dS$.

focal chord A chord of a conic which passes through a focus of the conic.

focal radius A line segment joining a focus and a point of a conic.

focus Let $e > 0$ be a fixed positive constant and let L and F, respectively, be a fixed line and a fixed point in a plane such that the point F is not on the line L. For any point P in the plane, let $d(P, L)$ be the (perpendicular) distance between P and L, and let $d(P, F)$ be the distance between P and F. A conic is the set of all points P in the plane satisfying $\frac{d(P,L)}{d(P,F)} = e$. The conic is called a parabola, if $e = 1$; an ellipse, if $0 < e < 1$; and a hyperbola, if $e > 1$. The fixed point F is called a focus of the conic.

foliated structure *See* foliation.

foliation Let $M \subseteq \mathbf{C}^n$ be a real $\mathbf{C}^k$ manifold. A foliation Φ of M is a set of complex submanifolds of M and a $\mathbf{C}^k$ map $\rho : M \to \mathbf{R}^q$ such that

(1.) the Frechet derivative ρ' of ρ has rank q, everywhere on M,

(2.) $\Phi = \{\rho^{-1}(S) : S \subset \mathbf{R}^q\}$, and

(3.) each member of Φ has the same dimension, as a submanifold of $\mathbf{C}^n$.

folium of Descartes The graph of the equation $x^3 + y^3 = 3cxy$, where $c > 0$. The folium has the line $x + y + c = 0$ as its asymptote.

formal Taylor expansion Let $f(x)$ be a function which is infinitely differentiable at $x = x_0$. The formal power series $\sum_{n=0}^{\infty} \frac{f^{(n)}(x_0)}{n!}(x - x_0)^n = f(x_0) + f'(x_0)(x - x_0) + \frac{f''(x_0)}{2!}(x - x_0)^2 + \ldots$, is called the formal Taylor expansion of $f(x)$ at $x = x_0$.

formal vector field on a manifold A tensor field of rank $(1, 0)$. *See also* tensor field.

four-space The set of all 4-tuples (x_1, x_2, x_3, x_4) of real numbers, usually denoted $\mathbf{R}^4$ or $\mathbf{E}^4$.

Four-vertex Theorem A simple closed smooth plane curve with positive curvature has at least four vertices, where a vertex of a plane curve is a point of the curve at which the curvature has a relative maximum or minimum.

Fourier coefficients (1.) The coefficients c_n $(n = 0, \pm 1, \pm 2 \ldots)$, or the coefficients a_0, a_n, and b_n $(n = 1, 2, 3, \ldots)$, in the Fourier series of a function $f(t)$. *See* Fourier series. (2.) The coefficients in the expansion of an element of a Hilbert space, with respect to an orthonormal basis. *See* orthonormal basis.

Fourier cosine transform Let $f(t)$ be an absolutely integrable function on the entire real line. The function $C(\lambda) = \sqrt{\frac{2}{\pi}} \int_0^{\infty} f(t) \cos \lambda t dt$ is called the Fourier cosine transform of $f(t)$.

Fourier integral Let $f(t)$ be an absolutely integrable function on the entire real line. The integral $F(t) = \int_0^{\infty} (a(\lambda) \cos \lambda t + b(\lambda) \sin \lambda t) d\lambda$, where $a(\lambda) = \frac{1}{\pi} \int_{-\infty}^{\infty} f(t) \cos \lambda t dt$ and $b(\lambda) = \frac{1}{\pi} \int_{-\infty}^{\infty} f(t) \sin \lambda t dt$, is called the *Fourier integral* of $f(t)$. The Fourier integral of $f(t)$ can also be expressed as $F(t) = \frac{1}{\pi} \int_0^{\infty} \int_{-\infty}^{\infty} f(s) \cos \lambda(s - t) ds d\lambda$, or in complex form as $F(t) = \frac{1}{2\pi} \int_{-\infty}^{\infty} \int_{-\infty}^{\infty} f(s) e^{i\lambda(s-t)} ds d\lambda$.

Fourier integral operator Let Ψ be the Schwartz class of test functions on $\mathbf{R}^n$ and S^m be the class of functions $a(x, y)$ which are $\mathbf{C}^{\infty}$ functions of $(x, y) \in \mathbf{R}^n \times \mathbf{R}^n$ and satisfy the differential inequalities $|\partial_x^{\beta} \partial_y^{\alpha} a(x, y)| \leq A_{\alpha,\beta}(1 + |y|)^{m-|\alpha|}$ for all multi-indices α and β. Let $\phi(x, y)$ be a real-valued function on $\mathbf{R}^n \times \mathbf{R}^n$ which is homogeneous of degree 1 in y. Further, on the support of $a(x, y)$, the function $\phi(x, y)$ is assumed to be smooth in (x, y), for $y \neq 0$, and satisfy $\det\left(\frac{\partial^2 \phi}{\partial x_i \partial y_j}\right) \neq 0$, for $y \neq 0$. The Fourier integral operator T on Ψ is defined for $f \in \Psi$ by the formula $(Tf)(x) = \int_{\mathbf{R}^n} e^{2\pi i \phi(x,y)} a(x, y) \widehat{f}(y) dy$, where $\widehat{f}$ is the Fourier transform of f. The function $a(x, y)$ is called the symbol of the operator T. One important question concerning the operator T is under what conditions on the symbol $a(x, y)$ will T admit extension to a bounded operator on a broader class of functions, such as the class $L^p(\mathbf{R}^n)$, than Ψ.

Fourier Integral Theorem Let $f(t)$ be an absolutely integrable function on the entire real line. If $f(t)$ is piecewise smooth on every bounded interval of the real line, then, for each real number t, the Fourier integral of $f(t)$ converges to the value

$$\frac{1}{2}[f(t - 0) + f(t + 0)]$$

$$= \frac{1}{2}\left[\lim_{x \to t^-} f(x) + \lim_{x \to t^+} f(x)\right].$$

In particular, the Fourier integral of $f(t)$ converges to $f(t)$ if $f(t)$ is continuous at t.

Fourier Inversion Formula Let f be an absolutely integrable function on the real line $\mathbf{R}$; that is $f \in L^1(\mathbf{R})$, and $\widehat{f}$ be the Fourier transform of f. If $\widehat{f} \in L^1(\mathbf{R})$, then, for almost all $t \in \mathbf{R}$, $f(t) = \frac{1}{\sqrt{2\pi}} \int_{\mathbf{R}} \widehat{f}(x) e^{ixt} dx$. The equality which holds for almost all t is called the Fourier inversion formula. (In case the Fourier transform is defined by $\widehat{f}(x) = \int_{\mathbf{R}} f(t)e^{-ixt}dt$, then the inversion formula takes the form $f(t) = \frac{1}{2\pi}\int_{\mathbf{R}} \widehat{f}(x)e^{ixt}dx$.) *See* Fourier transform.

Fourier kernel The function e^{-ixt} in the Fourier transform. *See* Fourier transform.

Fourier series Let $f(t)$ be a complex-valued function of a real variable t. If $f(t)$ is periodic of period $2p$, then the formal series $\sum_{n=-\infty}^{\infty} c_n e^{i\pi nt/p}$, where $c_n = \frac{1}{2p}\int_{-p}^{p} f(t)e^{-i\pi nt/p}dt$, $n = 0, 1, 2, \ldots$, is called the *Fourier series* of $f(t)$. Alternatively, the Fourier series of $f(t)$ may be defined as the formal series $\frac{a_0}{2} + \sum_{n=1}^{\infty}\left(a_n \cos\frac{\pi nt}{p} + b_n \sin\frac{\pi nt}{p}\right)$, where $a_n = \frac{1}{p}\int_{-p}^{p} f(t)\cos\frac{\pi nt}{p} dt$, and $b_n = \frac{1}{p}\int_{-p}^{p} f(t)\sin\frac{\pi nt}{p}dt$.

Fourier sine transform Let $f(t)$ be an absolutely integrable function on the entire real line. The function $S(\lambda) = \sqrt{\frac{2}{\pi}}\int_0^{\infty} f(t)\sin\lambda t dt$ is called the *Fourier sine transform* of $f(t)$.

Fourier transform Let f be an absolutely integrable function on the real line $\mathbf{R}$; that is $f \in L^1(\mathbf{R})$. The *Fourier transform* $\widehat{f}$ of f is defined by $\widehat{f}(x) = \frac{1}{\sqrt{2\pi}}\int_{\mathbf{R}} f(t)e^{-ixt}dt$, $x \in \mathbf{R}$. (Some authors define $\widehat{f}$ by $\widehat{f}(x) = \int_{\mathbf{R}} f(t)e^{-ixt}dt$.)

Fourier–Laplace transform Consider the Fourier transform pair

$$F(w) = \frac{1}{2\pi}\int_{-\infty}^{+\infty} f(x)e^{-iwx}\,dx,$$

$$f(t) = \int_{-\infty}^{+\infty} F(w)e^{iwt}\,dw.$$

One has, under appropriate hypotheses,

$$f(t) = \int_{-\infty}^{\infty}\frac{1}{2\pi}\left(\int_{-\infty}^{+\infty} f(x)e^{-iwx}\,dx\right)e^{iwt}\,dw.$$

Setting $f(t) = g(t)e^{-at}$, where $g(t) = 0$ for $t < 0$ and making the change of variables $s = a + iw$, the third equation can be written as

$$g(t) = \frac{1}{2\pi i}\int_{a-i\infty}^{a+i\infty}\left[\int_0^{\infty} g(x)e^{-sx}\,dx\right]e^{st}\,dt$$

or

$$g(t) = \int^{-1}(G(s)) = \frac{1}{2\pi i}\int_{a-i\infty}^{a+i\infty} G(s)e^{st}\,ds,$$

where $G(s)$ is the Laplace transform of $g(t)$. The above is called the Fourier–Laplace transform or the Bromwich integral for the inversion of Laplace transforms.

Fourier-Stieltjes transform Let μ be a finite Borel measure on the real line $\mathbf{R}$. The *Fourier-Stieltjes transform* $\widehat{\mu}$ of μ is the function $\widehat{\mu}(x) = \int_{\mathbf{R}} e^{-ixt}d\mu(t)$, $x \in \mathbf{R}$.

fourth quadrant of Cartesian plane The portion of Cartesian plane $\mathbf{R}^2$ consisting of points (x, y) with $x > 0$ and $y < 0$.

fractal A set F in $\mathbf{R}^n$ whose topological dimension is strictly less than its Hausdorff dimension. For example, the Cantor ternary set is a fractal for it has topological dimension 0, and Hausdorff dimension log 2/log 3. It can be shown that if the Hausdorff dimension of F is a non-integer, then F is a fractal.

frame Let M be an n-dimensional manifold and $p \in M$. A *frame* of M at p is a set $\{V_1, V_2, \ldots, V_n\}$ of n linearly independent tangent vectors at p.

Frechet derivative Let X and Y be Banach spaces, G be an open set in X, and p be a point in G. A function $F : G \to Y$ is said to be Frechet differentiable at p if there is a bounded linear operator $L : X \to Y$ such that $\lim_{h\to 0}\frac{\|F(p+h)-F(p)-L(h)\|}{\|h\|} = 0$. The operator L is usually denoted $dF|_p$ and is called the *Frechet derivative* of F at p. If F is Frechet differentiable at p, then its Frechet derivative is unique. For example, let $X = \mathbf{R}^n$, $Y = \mathbf{R}^m$ and $F(\mathbf{x}) = \mathbf{y}$, where $\mathbf{x} = (x_1, x_2, \ldots, x_n) \in \mathbf{R}^n$and $\mathbf{y} = (y_1, y_2, \ldots, y_m) \in \mathbf{R}^m$. If F is Frechet differentiable at $\mathbf{p}$, then all first-order partial derivatives $\partial y_i/\partial x_j$ ($i = 1, 2, \ldots, m$ and $j = 1, 2, \ldots n$) exist at $\mathbf{p}$, and $dF|_\mathbf{p}$ has the $m \times n$ matrix whose ij-entry is the value of $\partial y_i/\partial x_j$ at $\mathbf{p}$.

Frechet distance Let C be a curve and z be a point in the complex plane. The distance $d(z, C)$ between z and C is defined by $d(z, C) = \inf\{|z - w| : w \in C\}$. Now, let C_1 and C_2 be two curves in the complex plane. The *Frechet distance* $D_F(C_1, C_2)$ between the curves C_1 and C_2 is defined by $D_F(C_1, C_2) = \sup\{d(z, C_2) : z \in C_1\}$. Frechet distance D_F possesses the following properties: (1.) $D_F(C_1, C_2) \geq 0$; (2.) $D_F(C_1, C_2) = 0$ if and only if $C_1 = C_2$; (3.) $D_F(C_1, C_2) = D_F(C_2, C_1)$; and (4.) $D_F(C_1, C_2) \leq D_F(C_1, C_3) + D_F(C_3, C_2)$.

Frechet space A linear (vector) space that is complete under a quasi-norm. That is, such that every Cauchy sequence has a limit under the quasi-norm.

Fredholm integral equation of the first kind Let $g(x)$ and $k(x, y)$ be given functions and μ, a and b be constants. An equation of the form $g(x)+\mu \int_a^b k(x, y)f(y)dy = 0$ with unknown $f(x)$ is called a *Fredholm integral equation of the first kind.* The function $k(x, y)$ is called the kernel of the equation. If $\mu \neq 0$ and one writes $(Tf)(x) = \int_a^b k(x, y)f(y)dy$, then the equation may be written as $Tf = h$, where $h(x) = -\mu^{-1}g(x)$.

Fredholm integral equation of the second kind Let $g(x)$ and $k(x, y)$ be given functions and μ, a and b be constants. An equation of the form $g(x) + \mu \int_a^b k(x, y)f(y)dy = f(x)$ with unknown $f(x)$ is called a *Fredholm integral equation of the second kind.* The function $k(x, y)$ is called the kernel of the equation. If $\mu \neq 0$ and one writes $(Tf)(x) = \int_a^b k(x, y)f(y)dy$ and $\lambda = \mu^{-1}$, then the equation may be written as $(T - \lambda)f = h$, where $h(x) = -\mu^{-1}g(x)$.

Fredholm integral equation of the third kind Let $g(x), h(x)$ and $k(x, y)$ be given functions and μ, a and b be constants. An equation of the form $g(x) + \mu \int_a^b k(x, y)f(y)dy = h(x)f(x)$ with unknown $f(x)$ is called a *Fredholm integral equation of the third kind.* The function $k(x, y)$ is called the kernel of the equation.

Fredholm mapping *See* Fredholm operator.

Fredholm operator A bounded linear operator $T : X \to Y$ from a Banach space X to a Banach space Y such that: (1) the range ran$T$ of T is a closed subspace of Y, (2) the dimension of the kernel ker$T$ of T is finite, and (3) the dimension of the kernel kerT' of T', the Banach space adjoint of T, is finite. If X and Y are Hilbert spaces, then the operator T' in (3) is to be replaced by the operator T^*, the Hilbert space adjoint of T.

Fredholm's Alternative Theorem Let $T : X \to X$ be a compact operator on a Banach space X, $\lambda \neq 0$, a nonzero complex number, and $A = T - \lambda I$. Then, the operator A is a Fredholm operator, and the dimension of the kernel ker$A$ of A and the dimension of the kernel ker$A'$of A' are equal. In other words, the operator A is a Fredholm operator and either ran$A = X$ and ker$A = \{0\}$, or ran$A \neq X$ and ker$A \neq \{0\}$. If X is a Hilbert space, then the operator A' is to be replaced by the operator A^*. *See* Fredholm operator.

Fredholm's determinant Let $k(x, y)$ be a continuous function for $(x, y) \in [a, b] \times$

$[a, b]$, and let $d_0 = 1$. For $n = 1, 2, \ldots$, let

$$\mathbf{K}\begin{pmatrix} x_1, x_2, \ldots, x_n \\ x_1, x_2, \ldots, x_n \end{pmatrix} = \begin{vmatrix} k(x_1,x_1) & k(x_1,x_2) & \ldots & k(x_1,x_n) \\ k(x_2,x_1) & k(x_2,x_2) & \ldots & k(x_2,x_n) \\ \ldots & \ldots & \ldots & \ldots \\ k(x_n,x_1) & k(x_n,x_2) & \ldots & k(x_n,x_n) \end{vmatrix},$$

and

$$d_n = \frac{(-1)^n}{n!}\int_a^b\int_a^b \cdots \int_a^b \mathbf{K}\begin{pmatrix} x_1, x_2, \ldots, x_n \\ x_1, x_2, \ldots, x_n \end{pmatrix} dx_1 dx_2 \ldots dx_n.$$

The power series $\sum_{n=0}^{\infty} d_n\mu^n$ is convergent for all $\mu \in \mathbf{C}$, and its sum $D(\mu) = \sum_{n=0}^{\infty} d_n\mu^n$ is called the *Fredholm's determinant* of $k(x, y)$.

Fredholm's first minor Let $k(x, y)$ be a continuous function for $(x, y) \in [a, b] \times [a, b]$, and let $m_0(x, y) = k(x, y)$. For $n = 1, 2, \ldots$, let

$$\mathbf{K}\begin{pmatrix} x, x_1, \ldots, x_n \\ y, x_1, \ldots, x_n \end{pmatrix} = \begin{vmatrix} k(x,y) & k(x,x_1) & \ldots & k(x,x_n) \\ k(x_1,y) & k(x_1,x_1) & \ldots & k(x_1,x_n) \\ \ldots & \ldots & \ldots & \ldots \\ k(x_n,y) & k(x_n,x_1) & \ldots & k(x_n,x_n) \end{vmatrix},$$

and

$$m_n(x, y) = \frac{(-1)^n}{n!}\int_a^b\int_a^b \cdots \int_a^b \mathbf{K}\begin{pmatrix} x, x_1, \ldots, x_n \\ y, x_1, \ldots, x_n \end{pmatrix} dx_1 dx_2 \ldots dx_n.$$

For each compact set K in $\mathbf{C}$, the series $\sum_{n=0}^{\infty} m_n(x, y)\mu^n$ converges absolutely and uniformly for $(x, y, \mu) \in [a, b] \times [a, b] \times K$, and for each $\mu \in \mathbf{C}$, the sum $M_\mu(x, y) = \sum_{n=0}^{\infty} m_n(x, y)\mu^n$, which is a continuous function of $(x, y) \in [a, b] \times [a, b]$, is called *Fredholm's first minor* of $k(x, y)$.

Fredholm's method Consider the Fredholm integral equation of the second kind $g(x) + \mu\int_a^b k(x, y)f(y)dy = f(x)$, where the function $g(x)$ is continuous on $[a, b]$, and the kernel $k(x, y)$ is continuous on $[a, b] \times [a, b]$. For each $\mu \in \mathbf{C}$ for which Fredholm's determinant $D(\mu) \neq 0$, the integral equation has a unique solution $f(x)$ continuous on $[a, b]$. The solution is given by $f(x) = g(x) + \frac{\mu}{D(\mu)}\int_a^b M_\mu(x, y)g(y)dy$, where $M_\mu(x, y)$ is the Fredholm's first minor. *See* Fredholm's determinant, Fredholm's first minor.

Frenet frame Let $\mathbf{r}(t) = f(t)\mathbf{i} + g(t)\mathbf{j} + h(t)\mathbf{k}, a \le t \le b$, be a smooth curve in $\mathbf{R}^3$, $s(t) = \int_a^t |\mathbf{r}'(t)|dt$ be the arc length function, $\mathbf{T}(s)$ be the unit tangent vector, $\mathbf{N}(s)$ be the unit normal vector, and $\mathbf{B}(s)$ be the binormal unit vector. The triple $\{\mathbf{T}(s), \mathbf{N}(s), \mathbf{B}(s)\}$ is called the *Frenet frame* of the curve $\mathbf{r}(t)$.

Frenet's formulas Let $\mathbf{r}(t) = f(t)\mathbf{i} + g(t)\mathbf{j} + h(t)\mathbf{k}, a \le t \le b$, be a smooth curve in $\mathbf{R}^3$, $s(t) = \int_a^t |\mathbf{r}'(t)|dt$ be the arc length function, $\mathbf{T}(s)$ be the unit tangent vector, $\mathbf{N}(s)$ be the unit normal vector, $\mathbf{B}(s)$ be the binormal unit vector, $\kappa(s)$ be the curvature, and $\tau(s)$ be the torsion. The following formulas are called *Frenet's formulas*. (1.) $\frac{d\mathbf{T}}{ds} = \kappa\mathbf{N}$, (2.) $\frac{d\mathbf{N}}{ds} = -\kappa\mathbf{T} + \tau\mathbf{B}$, (3.) $\frac{d\mathbf{B}}{ds} = -\kappa\mathbf{N}$.

frequency of periodic function Let $f(t)$ be a periodic function of fundamental period p. The number $1/p$ is called the frequency of $f(t)$. *See* fundamental period.

Fresnel integral Either of the improper integrals $\int_0^\infty \sin x^2 dx$ and $\int_0^\infty \cos x^2 dx$. Both integrals are convergent and $\int_0^\infty \sin x^2 dx = \int_0^\infty \cos x^2 dx = \frac{\sqrt{\pi}}{2\sqrt{2}}$.

Friedrich's extension An (unbounded) semi-bounded symmetric operator T defined on a subspace $D(T)$ of a Hilbert space $\mathbf{H}$ admits a semi-bounded self-adjoint extension to the entire space $\mathbf{H}$.

Frobenius Theorem Let $\mathcal{V}$ be a subbundle of the tangent bundle TM of a manifold M. In order for a unique integral manifold of $\mathcal{V}$ to pass through every point of M, it is necessary and sufficient that $\mathcal{V}$ be involutive. *See* integral manifold.

Fubini's Theorem Let (X, A, μ) and (Y, B, ν) be σ-finite measure spaces, and $f(x, y)$ be a function defined on $X \times Y$. For each $x \in X$, let $f_x(y) = f(x, y)$, and for each $y \in Y$, let $f_y(x) = f(x, y)$. If $f \in L^1(\mu \times \nu)$, then (1) $f_x \in L^1(\nu)$ for almost all x, and $f_y \in L^1(\mu)$ for almost all y, (2) the function $g(x)$ defined for almost all x by $g(x) = \int_Y f_x(y)d\nu$, and the function $h(y)$ defined for almost all y by $h(y) = \int_X f_y(x)d\mu$ are such that $g \in L^1(\mu)$ and $h \in L^1(\nu)$, and (3) $\int_X \int_Y f(x, y)d\nu d\mu = \int_{X\times Y} f(x, y)d(\mu \times \nu) = \int_Y \int_X f(x, y) d\mu d\nu$.

Fuchs relation Consider the second-order linear homogeneous differential equation $y'' + p(z)y' + q(z)y = 0$. Assume that the equation has only isolated regular singular points at $z_1, z_2, \ldots, z_n$ and ∞. The Fuchs relation states that $\sum_{i=0}^{n} (\alpha_i + \beta_i) = n - 1$, where α_0 and β_0 are the roots of the indicial equation at ∞, and α_i and β_i are the roots of the indicial equation at $z_i, i = 1, 2, \ldots, n$.

function For two sets X and Y, a subset G of $X \times Y$, with the properties (1) $(x, y_1) \in G$ and $(x, y_2) \in G$ imply $y_1 = y_2$ and (2) for every $x \in X$, there is $y \in Y$ such that $(x, y) \in G$. Instead of the set G, a function $f : X \to Y$ is usually thought of as the rule that assigns x to y, where $(x, y) \in G$, and one writes $f(x) = y$.

function element A pair (f, D) where D is an open disk (or region) in the complex plane and f is an analytic function on D.

function of a complex variable A function $f(z)$ whose argument z is a complex number.

function of a real variable A function $f(x)$ whose argument x is a real number.

function of bounded variation A complex-valued function $f(x)$ defined on the real interval $[a, b]$, for which there is a positive constant C such that $\sum_{i=1}^{n} |f(x_i) - f(x_{i-1})| \le C$ for all partitions $P = \{a = x_0 < x_1 < \cdots < x_n = b\}$ of the interval $[a, b]$.

function of class $\mathbf{C}^r$ Let r be a positive integer. A function f of a single variable is of class $\mathbf{C}^r$ if it is r times continuously differentiable. If f is a function of several variables, then it is of class $\mathbf{C}^r$ if all of its partial derivatives up to and including order r exist and are continuous. This class of functions is also denoted by the symbol $\mathbf{C}_r$.

function of confluent type A function which is a solution of a confluent differential equation. *See* confluent differential equation.

function of Gevrey class *See* Gevrey class.

function space A set of functions endowed with a topology. For example, the space $H(G)$ of analytic functions on an open set G with the compact-open topology is a function space.

function-theoretic null set **(1.)** A set of measure zero.
(2.) The empty set.

functional Let V be a vector space over the field $\mathbf{F}$. A functional f on V is a function (usually linear) from V to $\mathbf{F}$.

functional analysis A branch of mathematical analysis which studies the analytical structure of a vector space endowed with a topology (such as topological vector space and normed linear space) and the functionals and operators defined on such spaces.

functional calculus A formula or definition for taking functions of an operator or class of operators on Hilbert space. Strictly speaking, a multiplicative homomorphism of an algebra of functions into the operators from a Hilbert space to itself, sending the function $f(z) = z$ to a given operator.

For a bounded operator T, with spectrum K and the class of functions analytic in a neighborhood of K, one may define a functional calculus by defining

$$f(T) = \frac{1}{2\pi i}\int_C (\lambda I - T)^{-1} d\lambda,$$

where C is a simple closed curve with K in its interior (Dunford calculus).

For a self-adjoint operator T, with spectral resolution $E(\lambda)$, and the bounded measurable functions on $\mathbf{R}$, one may define

$$f(T) = \int_{-\infty}^{\infty} f(\lambda) dE(\lambda).$$

A similar construction works for any normal operator (Von Neumann calculus).

For a completely nonunitary contraction operator T, one may use the minimal unitary dilation U to define $f(T)$, for $f(z)$ analytic and bounded in $|z| < 1$, by

$$f(T) = Pf(U).$$

(Sz.-Nagy-Foias calculus).

functional differential equation A differential equation which depends on the past state of its variable. For example, let $t_0 > 0$ be a fixed positive constant. The differential equation $\frac{dy(t)}{dt} = F(t, y(t), y(t - t_0))$ is a *functional differential equation* for it depends on the past state $y(t - t_0)$ of the variable y.

functional equation A relation satisfied by a function. For example, the functional equation $f(x + \lambda) = f(x)$ expressing periodicity of $f(x)$.

functional model A structure, including a function space, with which an abstract mathematical object is realized. For example, if N is a normal operator on a separable Hilbert space $\mathbf{H}$, then there is a finite measure space (X, μ) and a function $\varphi \in L^\infty(X, \mu)$ such that N is unitarily equivalent to the multiplication operator M_φ on $L^2(X, \mu)$ defined by $M_\varphi f = \varphi f$. Thus, the *functional model* $\{M_\varphi, L^2(X, \mu)\}$ is a realization (through unitary equivalence) of the pair $\{N, \mathbf{H}\}$.

functional value A value of a function f; thus, each element of the range of the function f.

fundamental form Associated with a Finsler metric F, with Hessian G_{ab}: the form

$$\Phi = iG_{\alpha\bar{\beta}} dz^\alpha \wedge \overline{dz^\beta}$$

on $\tilde{M}$, the tangent bundle minus the 0-section of a real finite-dimensional manifold M.

fundamental invariant (for an equivalence relation) Let E be an equivalence relation on a set X. If T is another set, a function $f: X \to T$ is called a *fundamental* (or complete) *invariant* for E if $xEy \Leftrightarrow f(x) = f(y)$.

Fundamental Lemma in the Calculus of Variations Let $\varphi(t)$ and $\psi(t)$ be two fixed piecewise continuous functions defined on the interval $[a, b]$. Let $\wp$ be the set of all continuous functions $f(t)$ on $[a, b]$ having $f(a) = f(b) = 0$ and having piecewise continuous derivatives. Then, $\int_a^b (\varphi(t)f(t) + \psi(t)f'(t))\, dt = 0$ for all $f \in \wp$ if and only if there is a constant c such that $\psi(t) = \int_a^t \varphi(x)dx + c$ $(t \in [a, b])$.

fundamental period Let $f(x)$ be a periodic function. The smallest $p > 0$ for which $f(x + p) = f(x)$ for all x in the domain of f is called the *fundamental period* of $f(x)$.

fundamental period parallelogram Let $f(z)$ be a doubly periodic complex-valued function defined on the entire complex plane $\mathbf{C}$. Thus, there is a nonzero $\alpha \in \mathbf{C}$ of least modulus satisfying $f(z + \alpha) = f(z)$ for all

$z \in \mathbf{C}$. Moreover, there is another nonzero $\beta \in \mathbf{C}$ of least modulus such that β/α is not a real number and $f(z+\alpha) = f(z)$ for all $z \in \mathbf{C}$. The pair $\{\alpha, \beta\}$ is called a fundamental period pair of the function $f(z)$. For each complex number $a \in \mathbf{C}$, the parallelogram with vertices $a, a+\alpha, a+\beta, a+\alpha+\beta$ is called a *fundamental period parallelogram.*

fundamental solution Let P be a complex polynomial of n variables and $P(D)$ be the corresponding linear partial differential operator. A distribution φ in $\mathbf{R}^n$ is called a *fundamental solution* of $P(D)$ if φ satisfies the partial differential equation $P(D)\varphi = \delta$, where δ is the Dirac delta "function."

fundamental space A complete metric space. That is, a metric space X where each Cauchy sequence in X converges to a point in X.

fundamental system A set of solutions which forms a basis of the solution space of a linear homogeneous differential equation or of a first order linear homogeneous system. Thus, the set $\{y_1, y_2, \ldots, y_n\}$ of solutions of such an equation or system is a *fundamental system* if the solutions $y_1, y_2, \ldots, y_n$ are linearly independent and every solution of the equation or system can be expressed as a linear combination of these solutions.

fundamental system of solutions A homogeneous linear differential equation of order n has at most n linearly independent solutions. A set of n linearly independent solutions constitute a *fundamental system of solutions* of such an equation.

fundamental system of solutions A set of solutions (over a common interval) of a homogeneous linear differential equation such that every solution of the equation can be represented as a linear combination of solutions from this set. Furthermore, no proper subset of this set will allow the representation of every solution.

One can show that an equation of order n has exactly n linearly independent functions in each such fundamental system. For first order systems over $\mathbf{R}^n$, the fundamental system consists of n linearly independent n-vector solutions.

A condition that ensures that a set of n solutions forms a fundamental system is the non-vanishing of the *Wronskian Determinant* of this set of functions.

fundamental tensor of Finsler space The notion of Finsler space arises in an attempt to provide a notion of *length of arc* in a differentiable manifold which is independent of the parametrization and of the sense of traversal.

A differentiable manifold is called a *metric space* if there is a continuous function $F(x, \xi)$ defined for all points x and all tangent vectors ξ and having the following properties: a) $F(x, \xi)$ is positive definite in ξ; b) $F(x, \lambda\xi) = |\lambda| F(x, \xi)$, i.e., $F(x, \lambda\xi)$ is positive homogeneous of degree 1 in ξ. The length of the arc $x(t)$, $t \in [a, b]$ is defined to be $| \int_a^b F(x, \dot{x})dt |$.

A metric space is called a *Finsler Space* if

$$\sum_{i,k} g_{ik}(x, \xi)\eta_i\eta_k > 0, \quad for \ \ \xi \neq 0, \eta \neq 0,$$

where

$$g_{ik}(x, \xi) \equiv \frac{1}{2}\frac{\partial^2 F^2}{\partial \xi_i \, \partial \xi_k}(x, \xi),$$

i.e., if the metric tensor is positive definite and is obtained from the *metric* function. Every Riemannian space is a Finsler Space.

Fundamental Theorem of Calculus Assume that f is a Riemann integrable function on the closed real interval $[a, b]$. If there is a real-valued differentiable function F on $[a, b]$ such that $F' = f$, then

$$\int_a^b f(x)dx = F(b) - F(a).$$

More general versions of this theorem, e.g., for Lebesgue theory, can be formulated and proved. The Lebesgue version requires that F be differentiable at every point of $[a, b]$, and not just *almost everywhere*.

Extensions of this theorem to higher dimension are known as *Green's Theorem* (in dimension 2), the *Divergence Theorem* (in dimension 3) and *Stokes' Theorem* in general.

Fundamental Theorem of the Theory of Curves Let (a, b) be an open interval containing 0.

Let $\kappa(s)$ and $\tau(s)$ be real-valued functions defined on (a, b), with $\kappa(s)$ positive and with continuous first derivative there, while $\tau(s)$ is continuous. Then there exists one and (up to *rigid motions*) only one curve in $\mathbf{R}^3$ for which $\kappa(s)$, $\tau(s)$, and s are *curvature, torsion* and *arc-length*, respectively.

Fundamental Theorem of the Theory of Surfaces Also known as *Bonnet's Fundamental Theorem*. Let Ω be an open, simply connected subset of $\mathbf{R}^2$ and suppose that $\alpha(u)$ and $\beta(u)$ are quadratic forms on the tangent space to $\mathbf{R}^2$ at u, $u \in \Omega$. Assume further that their coefficients $\alpha_{ij}(u)$ and $\beta_{ij}(u)$ are sufficiently differentiable functions of u. If $\alpha(u)$ is positive definite and the *Gauss curvature equation* and the *Codazzi-Mainardi equations* are satisfied, then:

a) There exists a surface $f : \Omega \to \mathbf{R}^3$ whose *first* and *second fundamental forms* are α and β, respectively;

b) given any two surfaces f and g defined on Ω which have the same first and second fundamental forms, there exists an *isometry* $\Phi : \mathbf{R}^3 \to \mathbf{R}^3$ such that $f = \Phi \circ g$.

Fundamental Theorems of Stein Manifolds Let X be a Stein manifold (a noncompact complex manifold which admits an embedding into $\mathbf{C}^n$) of dimension n; let Ω be the sheaf of germs of holomorphic functions on X; and let F be an analytic coherent sheaf of Ω-modules on X.

Theorem A: $H^0(X, F)$ generates the stalk F_x as an Ω_x-module at every point x of X. This is known as *Theorem A of Cartan*.

Theorem B: $H^q(X, F) = 0$ for all $q > 0$. This is known as *Theorem B of Cartan*.

G

G_δ set A set which is the intersection of a countable collection of open sets, in a topological space.

G-space Let G be a topological group, and X be a topological space. We say that X is a *G-space* if G acts on X and the map $G \times X \to X$ is continuous.

G-surface A G-space whose underlying topological space is two-dimensional.

gamma function Let z be a complex variable. We define

$$\Gamma(z) = \int_0^\infty t^{z-1} e^{-t} dt,$$

for all z with positive real part. The definition can be extended via the use of some standard theorems from complex variable theory, so that Γ is *meromorphic* on the whole complex plane, with an *entire* reciprocal.

The gamma function is familiar in the context of the *factorial function,* since

$$\Gamma(n+1) = n! \;\; for \;\; n = 0, 1, 2, \ldots,$$

which is a special case of the more general relation $\Gamma(z+1) = z\Gamma(z)$. Several other formulae are derivable. One explicit representation is given by

$$\Gamma(z) = \frac{e^{-\gamma z}}{z} \prod_{n=1}^{\infty} (1 + \frac{z}{n})^{-1} e^{z/n}$$

where

$$\gamma = \lim_{n \to \infty} (\sum_{i=1}^{n} \frac{1}{i} - \log(n)),$$

is the *Euler-Mascheroni constant.* Another well-known formula satisfied by the Gamma function is *Gauss's formula.*

gamma structure (1.) The study of *schlicht* functions leads to the differential equation $\frac{P(w)dw^2}{w^2} = \frac{Q(z)dz^2}{z^2}$ which is solved by $w = f(z)$, the schlict function under study. Any zero of $Q(z)$ inside the unit circle is mapped to a zero of $P(w)$.

Any arc on which $\frac{Q(z)dz^2}{z^2}$ has a constant argument is mapped on an arc on which $\frac{P(w)dw^2}{w^2}$ has a constant argument. These arcs are called *trajectories,* vertical if $\frac{Q(z)dz^2}{z^2} \leq 0$, horizontal if $\frac{Q(z)dz^2}{z^2} \geq 0$.

Let Γ_w be the set of *maximal vertical trajectories* of $\frac{P(w)dw^2}{w^2}$ which do not begin and end at the origin. This set is finite and each trajectory that is part of this set connects a zero to a zero, a zero to a pole, the pole at ∞ to the pole at 0, and only a finite number of trajectories issue from the zeros and from the pole at ∞.

In the z-plane, one carries out the same construction, but limited to $|z| \leq 1$, and denotes by Γ_z the set of trajectories contained in the circle. Γ_z and Γ_w make up the Γ-structure.

(2.) Let E and M be topological spaces, $\phi : E \to M$, be a homeomorphism of an open set $U \subset E$ to an open set $V \subset M$. ϕ is called a *local coordinate system of M with respect to E.* A set Σ of local coordinate systems of M with respect to E is said to define a Γ-structure on M if Σ satisfies the conditions: a) the totality of the images of local coordinate systems belonging to Σ covers M; b) if two local coordinate systems ϕ_1 and ϕ_2 of Σ have a transformation of local coordinates (i.e., the homeomorphism $\phi_2^{-1} \circ \phi_1 : \phi_1^{-1}(V_1 \cap V_2) \to \phi_2^{-1}(V_1 \cap V_2)$), this transformation belongs to Σ.

gap value A complex number a is said to be a *gap* or *lacunary value* of a complex-valued function $f(z)$ if $f(z) \neq a$ in the region where f is defined. For example, 0 is a gap value of e^z in the complex plane. Gap

0-8493-0320-6/00/$0.00+$.50

values were investigated by E. Picard, Hallstrom, and Kametani.

Gårding's inequality Let Ω be any open set in $\mathbf{R}^n$; let $\mathbf{C}_0^m(\Omega)$ be the $\mathbf{C}^m$ functions with compact support in Ω, and let α and β denote multi-indices. Assume that

$$B(\phi, \psi) = \sum_{|\alpha|\le m, |\beta|\le m} \int_\Omega a_{\alpha\beta}(x)\, D^\alpha \phi \overline{D^\beta \psi}\, dx$$

is a uniformly strongly elliptic quadratic form on Ω with ellipticity constant E_0, that is

$$\Re\{ \sum_{|\alpha|=m, |\beta|=m} a_{\alpha\beta}(x)\xi^{\alpha+\beta}\} \ge E_0 \mid \xi \mid^{2m}$$

for all real ξ and all $x \in \Omega$, and E_0 is the largest such constant.

Assume further that

(1.) $a_{\alpha\beta}(x)$ is *uniformly continuous* on Ω, for $| \alpha | = | \beta | = m$;

(2.) $a_{\alpha\beta}(x)$ is *bounded* and *measurable* for $|\alpha| + |\beta| \le 2m$.

Then there are constants $\gamma > 0$ and $\lambda \ge 0$ such that

$$\Re(B(\phi)) \ge \gamma E_0 \|\phi\|^2_{m,\Omega} - \lambda\|\phi\|^2_{0,\Omega}$$

for all $\phi \in C_0^m(\Omega)$, where γ depends only on n and m; λ depends only on n, m, E_0, the *modulus of continuity* of $a_{\alpha\beta}(x)$ for $| \alpha | | \beta | = m$, and $\sup_{x\in\Omega} \sum_{|\alpha|+|\beta|\le 2m} | a_{\alpha\beta}(x) |$. Furthermore

$$\| \phi \|_{m,\Omega} \equiv \{ \sum_{|\alpha|\le m} \int_\Omega | D^\alpha \phi(x) |^2\, dx\}^{1/2}.$$

Gateaux derivative Let $x, \eta \in X$, where X is a vector space, and let Y be a normed linear space. We define the *Gateaux derivative* of $u : X \to Y$ at x in the direction η as the limit

$$Du(x)\eta = \lim_{\alpha\to 0} \frac{u(x + \alpha\eta) - u(x)}{\alpha}$$

if such limit exists. We say that u is Gateaux differentiable at x if it is Gateaux differentiable in every direction. In this case, the operator $Du(x) : X \to Y$ is called the *Gateaux derivative* at x.

gauge function Let A and B be subsets of a vector space X over $\mathbf{R}$. We say that A *absorbs* B if there exists $\lambda > 0$ such that $B \subseteq \mu A$ for all μ with $|\mu| \ge \lambda$. If A absorbs every finite subset of X, then A is said to be *absorbing*. For an absorbing set A, the function

$$p_A(x) = \inf \{\lambda > 0 : x \in \lambda A\}$$

for any $x \in X$, is called the *gauge* (or *Minkowski*) function of A.

gauge transformation In electromagnetic theory, the *electric field vector* $\vec{E} = (E_1, E_2, E_3)$ and the *magnetic field vector* $\vec{H} = (H_1, H_2, H_3)$ are related, in the absence of electric charges, by Maxwell's equations. It is convenient to express $\vec{E}$ and $\vec{H}$ in terms of a four-dimensional *electromagnetic potential* (A_0, A_1, A_2, A_3), by setting

$$\vec{E} = \text{grad} A_0 - \frac{\partial}{\partial x_0}\vec{A},$$

where grad $A_0 = (\frac{\partial A_0}{\partial x_1}, \frac{\partial A_0}{\partial x_2}, \frac{\partial A_0}{\partial x_3})$, $\vec{A} = (A_1, A_2, A_3)$, and $\vec{H} = \text{curl}\vec{A}$.

The potential $(A_0, \ldots, A_3)$ is not uniquely determined by the vectors $\vec{E}$ and $\vec{H}$, since they do not change under any transformation of the type

$$A'_j(x) = A_j(x) + \frac{\partial f}{\partial x_j}(x),$$

for $j = 0, 1, 2, 3$, where f is an arbitrary function. This transformation is called a *gauge transformation*.

Gauss equations Several equations fall under this general name.

(1.)

$$D_X Y = \bar{D}_X Y - (L(X), Y)N;$$

where M is a hypersurface in $\mathbf{R}^n$; Y is a $\mathbf{C}^\infty$ field on M; $\bar{D}$ is the natural connection on

$\mathbf{R}^n$; N is a unit normal field that is $\mathbf{C}^\infty$ on M; $L(X) \equiv \bar{D}_X N$ for X tangent to M and $(\cdot,\cdot)$ is the Riemannian metric tensor. L is also known as the *Weingarten map*. The importance of this equation resides in the fact that D can be shown to be the Riemannian connection on M, and that $\bar{D}_X Y$ can be decomposed into its tangent and normal components relative to M.

(2.) *Gauss curvature equation.* Let the previous notation hold; let X, Y and Z be $\mathbf{C}^\infty$ fields on some open set Ω in M. Then

$$\begin{aligned} &D_X D_Y Z - D_Y D_X Z - D_{[X,Y]} Z \\ &\quad = (L(Y), Z)L(X) - (L(X), Z)L(Y). \end{aligned}$$

(3.) *The Hypergeometric Equation.*

$$\begin{aligned} w'' &+ \frac{(-\gamma + (1+\alpha+\beta)z)}{z(z-1)} w' \\ &+ \frac{\alpha\beta}{z(z-1)} w = 0, \end{aligned}$$

where α, β and γ are (generally) real numbers. The solutions of this equation can be expressed in terms of *Riemann's P-function.*

Gauss map *See* Gaussian frame.

Gauss's circle problem (1.) In his study of the cyclotomic equation $x^p - 1 = 0$, which is important both in the study of the solvability of arbitrary degree equations by radicals and also in the (related) construction of regular polygons with p sides, Gauss showed that the roots of the equation can be expressed as rational functions of the roots of a *sequence* of equations, $f_i(x) = 0$, $i = 1, 2, \ldots$, whose coefficients are rational functions of the roots of the preceding equations of the sequence. The degrees of the equations $f_i = 0$ are the prime factors of $p - 1$; there is an f_i for each factor, even if repeated; each of the $f_i = 0$ can be solved by radicals so that the original cyclotomic equation can be also so solved. This leads to the conclusion that a regular polygon of p sides can be constructed by ruler and compass if $p = 2^{2^n} + 1$ is a prime, extending the number of polygons known to be constructible from classical times.

(2.) Let $r(n)$ denote the number of integer solutions of the equation $x^2 + y^2 = n$, where n is an integer. The problem of estimating the sum $\sum_{n \le x} r(n)$ is the problem of counting the number of integer lattice points inside the circle of radius $\sqrt{x}$, and is called *Gauss's circle problem.*

Gauss's formula (1.) Formulae relating the sides and angles of a spherical triangle. Let A, B and Γ denote the length of the three sides of a spherical triangle, with α, β and γ denoting the opposite angles. We have:

$$\begin{aligned} \sin\tfrac{\Gamma}{2} \cdot \sin\tfrac{\alpha-\beta}{2} &= \cos\tfrac{\gamma}{2} \cdot \sin\tfrac{A-B}{2}, \\ \sin\tfrac{\Gamma}{2} \cdot \cos\tfrac{\alpha-\beta}{2} &= \sin\tfrac{\gamma}{2} \cdot \sin\tfrac{A+B}{2} \\ \cos\tfrac{\Gamma}{2} \cdot \sin\tfrac{\alpha+\beta}{2} &= \cos\tfrac{\gamma}{2} \cdot \cos\tfrac{A-B}{2}, \\ \cos\tfrac{\Gamma}{2} \cdot \cos\tfrac{\alpha+\beta}{2} &= \sin\tfrac{\gamma}{2} \cdot \cos\tfrac{A+B}{2}. \end{aligned}$$

(2.) The multiplication formula for the Gamma function:

$$\begin{aligned} \Gamma(nx) &= (2\pi)^{(1-n)/2} n^{nx-1/2} \Gamma(x) \\ &\quad \Gamma(x + 1/n) \ldots \Gamma(x + (n-1)/n). \end{aligned}$$

(3.) Gauss's Theorem, also known as the *Divergence Theorem* or *Ostrogradsky's Theorem.*

Gauss's Theorema Egregium This theorem states that two isometric surfaces in $\mathbf{R}^3$ must have, in some sense, the *same geometry:* in particular, the curvature $K(p)$ can be computed from the first fundamental form and its first and second partial derivatives. In more formal terminology, we have:

Let M be a surface in $\mathbf{R}^3$, with M_p the tangent space at the point p, and let X and Y be an orthonormal basis for M_p. Then the total curvature $K(p) = \det(L_p) = (L(Y), Y)(L(X), X) - (L(X), Y)(L(Y), X) = (R(X, Y)Y, X)$, where L is the *Weingarten map,* $(\cdot,\cdot)$ denotes the Riemannian metric on $\mathbf{R}^3$ and $R(X, Y)Z \equiv D_X D_Y Z - D_Y D_X Z - D_{[X,Y]} Z$, where $[\cdot,\cdot]$ is the *bracket operator,* and $D_X Y$ is the induced *covariant derivative* on the surface.

The theorem can be generalized to the case where M is a k-dimensional submanifold of a Riemannian manifold.

Gauss's Theorema Elegantissimum This is a corollary to the Gauss-Bonnet Theorem that assumes the boundary arcs are geodesic segments and thus with curvature 0. The result follows from the Gauss-Bonnet formula, which loses those integral components.

Gauss-Bonnet formula We start with a simple form of the result, according to Bonnet (1848). Gauss had earlier proved a similar result for geodesic triangles.

Let M denote a connected, oriented Riemannian 2-manifold. Let $A \subseteq U \subset M$, where U is a coordinate patch and ∂A is a simple closed curve with exterior corner angles $\alpha_1, \ldots, \alpha_r$. Then

$$\int_{\partial A} k = 2\pi - \sum_{j=1}^{r} \alpha_j - \int_A K$$

where k is the geodesic curvature function on ∂A and K is the Riemannian (Gaussian) curvature function on A.

A more general version frees the result from the requirement that $A \subseteq U$ (a coordinate domain), by requiring that A be a union of images of squares (with disjoint interiors) under orientation-preserving diffeomorphisms.

Let $A \subset M$ be as described; let ∂A be a finite disjoint union of simple closed curves; let $\alpha_1, \ldots, \alpha_r$ be the exterior angles of ∂A. Then

$$\int_{\partial A} k = 2\pi\chi(A) - \sum_{j=1}^{r} \alpha_j - \int_A K,$$

where $\chi(A)$ is the Euler characteristic of A.

If M is a compact connected oriented Riemannian 2-manifold with Riemannian curvature function K, $\int_M K = 2\pi\chi(M)$. Furthermore, if M is a compact even-dimensional hypersurface in $\mathbf{R}^{k+1}$, then $\int_M K = \frac{1}{2}\pi_k\chi(M)$, where π_k is the volume of the unit k-sphere S^k.

Gaussian frame Let the surface $f : \Omega \to \mathbf{R}^3$, where Ω is a domain in $\mathbf{R}^2$ be given; let $(x, y) \in \Omega$; let f_x and f_y denote the tangent vectors to the surface at (x, y); let $n = (f_x \times f_y)/|f_x \times f_y|$ denote the *unit normal field along f*. The 3-frame (f_x, f_y, n) is called the *Gaussian frame* of the surface $f : \Omega \to \mathbf{R}^3$. The mapping $n : \Omega \to \mathbf{S}^2 \subset \mathbf{R}^3$ is called the *Gauss Map*.

Gel'fand-Fuks cohomology Let $\mathcal{X}(M)$ denote the space of all smooth vector fields on a smooth manifold M. Under the bracket operation $[X, Y] = XY - YX$, where the vector fields are regarded as endomorphisms of the space $\mathbf{C}^\infty(M)$ of smooth functions on M, acting as first order differential operators, $\mathcal{X}(M)$ has the structure of a Lie algebra. We introduce on $\mathcal{X}(M)$ the topology of uniform convergence of the components of vector fields and all their partial derivatives on compact sets of M. We define a cochain complex : $C^0 = \mathbf{R}$, $C^p = \{\phi : \phi : \mathcal{X}(M) \times \ldots \times \mathcal{X}(M) \to \mathbf{R}, p \geq 1\}$, where the product is taken p times. We define a coboundary operator d on C^p as

(1) $d\phi = 0$, for $\phi \in C^0$;

(2) $d\phi(X_1, \ldots, X_{p+1}) = \sum_{i<j}(-1)^{i+j} \phi([X_i, X_j], X_1, \ldots, \check{X}_i, \ldots \check{X}_j, \ldots, X_{p+1})$ for $\phi \in C^p$, $p \geq 1$.

The cochain complex is $\oplus\{C^p, d\}$ and the cohomology group $H^*(\mathcal{X}(M))$ of this complex is called the *Gel'fand-Fuks cohomology* group of M. The result according to Gel'fand and Fuks is that, if M is a compact oriented manifold of dimension n, the dimension of $H^p(\mathcal{X}(M))$ is finite for all $p \geq 0$, and $H^p(\mathcal{X}(M)) = 0$ for $0 \leq p \leq n$.

Gel'fand-Shilov generalized function These are the continuous linear functionals on function spaces on $\mathbf{R}^n$ or $\mathbf{C}^n$ which have (at least) the properties (1) they are countably normed spaces or countable unions of such spaces; (2) their topology is stronger than that of pointwise convergence.

By *countably normed* we mean that the space has a topology induced by a countable set of compatible norms. *See* Generalized function for an example of such a countable set of norms.

Gelfand transform Let B be a commutative Banach algebra and let Δ be the set of all continuous, multiplicative homomorphisms from B to the complex numbers. For $x \in B$ and $h \in \Delta$, denote by $\hat{x}(h)$ the homomorphism h evaluated at x:

$$\hat{x}(h) = h(x).$$

The map $x \to \hat{x}$ from B to the algebra of complex functions on Δ is called the *Gelfand transform. See* Banach algebra.

general analytic space A ringed space $(X, \mathcal{O}_X)$ is a *general analytic space* if
(1.) it is locally isomorphic to a ringed space $(A, \mathcal{H}_A)$, where A is an analytic set in a domain $G \subset \mathbf{C}^n$;
(2.) $\mathcal{H}_A = (\mathcal{O}_G/\mathcal{J}) \mid_A$ for some coherent analytic subsheaf $\mathcal{J}$ of $\mathcal{J}(A)$ such that $\text{supp}(\mathcal{O}_G/\mathcal{J}) = A$.
The notion of *analytic space* can also be generalized by going to infinite dimensions (analytic sets in complex Banach spaces).

general curve A one-dimensional continuum. In $\mathbf{R}^2$ this coincides with the Cantor definition: a curve in $\mathbf{R}^2$ (two-dimensional Euclidean plane) is a continuum that is nowhere dense in $\mathbf{R}^2$.

A continuum is a connected compact metric space consisting of more than one point. A topological space is said to be connected if there are no proper closed subsets A and B of X such that $A \cap B = \emptyset$ and $A \cup B = X$.

general geometry of paths This refers to an alternative definition of connection on a Finsler space, based on curves satisfying appropriate second order differential equations and called *paths*.

general rational function A quotient of polynomials in the form:

$$R(z) = A\frac{(z-a_1)^{\lambda_1}(z-a_2)^{\lambda_2}\dots(z-a_n)^{\lambda_n}}{(z-b_1)^{\mu_1}(z-b_2)^{\mu_2}\dots(z-b_m)^{\mu_m}}$$

where λ_j and μ_j are positive integers, A, a_j and b_j are complex numbers, and z is a complex variable.

General Tauberian Theorem *See* Tauberian Theorems. The version given here is *Wiener's General Tauberian Theorem:*

Let $K \in L^1(\mathbf{R})$ and $f \in L^\infty(\mathbf{R})$. Assume $\hat{K}(\xi)$, the Fourier transform of $K(x)$, does not vanish for any ξ, and

$$\lim_{x\to\infty} \int K(x-y)f(y)dy = H \int K(t)dt,$$

for some constant H. Then

$$\lim_{x\to\infty} \int K'(x-y)f(y)dy = H \int K'(t)dt,$$

for all $K' \in L^1(\mathbf{R})$.

general term This refers to sequences and series. The *general term* of a sequence is sometimes known as the *nth term* and refers to the pair (n, S_n) in the graph of the sequence as a function from the (non-negative) integers to its range space. Since a series is the sum of a sequence, the general term of the series refers to the *nth* summand.

generalized absolute continuity Let $E \subset \mathbf{R}$, let $F(x)$ be a real-valued function whose domain contains E. $F(x)$ is a *generalized absolutely continuous* function on E if: a) F is continuous on E; b) E is the sum of a countable sequence of sets E_n such that F is absolutely continuous on E_n for all n.

generalized absolute continuity in the restricted sense This refers to the same generalization of the notion of absolute continuity in the restricted sense that was introduced for *generalized absolute continuity.*

generalized conformal mapping Let Ω be a domain in $\mathbf{R}^2$ and $f : \Omega \to \mathbf{R}^3$ a continuous surface. The mapping $f(u, v) = (x(u, v), y(u, v), z(u, v))$ is a *generalized conformal mapping* if the partial derivatives of x, y and z with respect to u and v exist almost everywhere in Ω, are square integrable, and angles are preserved (almost everywhere) by the mapping.

generalized convolution (1.) A generalization of the operation of convolution. Let $(X, \mathcal{X}, m_X)$, $(Y, \mathcal{Y}, m_Y)$ and $(Z, \mathcal{Z}, m_Z)$ be σ-finite measure spaces. Assume further that $l(\cdot, \cdot)$ maps pairs of measurable functions f and g, defined on X and Y respectively, to measurable functions defined on Z. We say that l is a *convolution structure* if

(i.) for $f \in L^1(X)$, $g \in L^1(Y)$,

$$\|l(f, g)\|_1 \leq \|f\|_1 \|g\|_1;$$

(ii.) for $f \in L^1(X)$, $g \in L^\infty(Y)$,

$$\|l(f, g)\|_\infty \leq \|f\|_1 \|g\|_\infty;$$

(iii.) for $f \in L^\infty(X)$, $g \in L^1(Y)$,

$$\|l(f, g)\|_\infty \leq \|f\|_\infty \|g\|_1.$$

(2.) An extension of the usual operation, meaningful on spaces of generalized functions. If u is a distribution and ϕ is a $\mathbf{C}^\infty$ function of compact support, the convolution $u * \phi$ is a function f of class $\mathbf{C}^\infty$, $f(x) \equiv u_{(y)}(\phi(x - y))$. For distributions u and v, at least one of which has compact support, there exists a unique distribution w such that $w * \phi = u * (v * \phi)$ for all $\phi \in \mathbf{C}_0^\infty$. This distribution is called the *generalized convolution* of u and v and is denoted by $u * v$.

generalized Fourier integral *See* generalized Fourier transform. (The Fourier integral of a function and the Fourier transform are either the same or differ by a normalization constant.)

generalized Fourier transform (1.) The extension of the notion of *Fourier transform* from Euclidean (or locally Euclidean), spaces to more general spaces, e.g., *topological groups*. In this context, a *generalized Fourier transform* is an invertible integral transform whose kernel, and the kernel of its inverse, are of the form $K(xy)$, i.e., a function of a product.

(2.) The Fourier transform of a generalized function. One needs to find conditions on the *generalized functions* that suffice for the *product theorem:* the Fourier transform of a convolution is the product of the individual Fourier transforms. For *distributions of compact support*, the function $\hat{u}(\xi) = u_x(e^{-i<x,\xi>})$ is defined for every complex vector $\xi \in \mathbf{C}^n$, and is an entire analytic function of ξ, called the *Fourier-Laplace transform* of u. For *temperate distributions*, we can define the Fourier transform as $\hat{v}(\phi) \equiv v(\hat{\phi})$ for all functions ϕ which, along with their derivatives, vanish at infinity faster than any (reciprocal) polynomial. One can show that, for such u and v, the Fourier transform $u*v$ is a temperate distribution, and its Fourier transform is precisely the temperate distribution $\hat{u}\,\hat{v}$.

generalized function The *linear functionals* on some function space. They were introduced in order to understand and formulate the behavior of solutions of partial differential equations, and thus the functions have multiple derivatives and convergence usually requires convergence of the derivatives. We give one example, observing that the imposition of different conditions on the underlying function space gives rise to different spaces of generalized functions. *See also* Gel'fand-Shilov generalized function.

Let Ω denote an open set in $\mathbf{R}^n$; let $\mathbf{C}_0^k(\Omega)$ denote the set of all functions u defined and with compact support in Ω, whose partial derivatives of order $\leq k$ all exist and are continuous. Let α denote a multi-index, $\alpha = (\alpha_1, \ldots, \alpha_n)$.

A *generalized function* (or *distribution*) u in Ω is a linear form on $\mathbf{C}_0^\infty(\Omega)$ such that for every compact set $\mathcal{K} \subset \Omega$ there exist constants C and k such that

$$|\, u(\phi) \,| \leq C \sum_{|\alpha| \leq k} \sup |\, D^\alpha \phi \,|,$$

for all $\phi \in \mathbf{C}_0^\infty(\mathcal{K})$.

generalized helix A curve in $\mathbf{R}^3$ for which the ratio torsion/curvature is a constant. It can be easily shown that general (or cylindrical) helices are precisely those curves whose tangents make a fixed angle with a given direction.

generalized isoperimetric problem A natural generalization of the *isoperimetric problem:* among all closed curves of a given length l, find the curve enclosing the greatest area. The length of the enclosing curves is the fixed, equal (= *iso*) perimeter. The problem is sometimes known as *Dido's problem,* since the mythical founder of Carthage was offered *as much land as she could encompass in a cow's skin.*

We can state the generalization as follows: find the curve $\mathbf{y}(x) = (y_1(x), \ldots, y_n(x))$ in $\mathbf{R}^n$ for which the functional

$$\mathbf{F}(\mathbf{y}) = \int_a^b F(x, y_1(x), \ldots, y_n(x), y_1'(x), \ldots, y_n'(x))dx$$

has an extremum, and where the admissible curves satisfy the boundary conditions $y_i(a) = y_{ia}$, $y_i(b) = y_{ib}$, for $i = 1, \ldots, n$, and $a < b$, and are such that another functional

$$\mathbf{G}(\mathbf{y}) = \int_a^b G(x, y_1(x), \ldots, y_n(x), y_1'(x), \ldots, y_n'(x))dx$$

takes a fixed value, say l.

generalized Jacobian Let (T, A) be a continuous plane mapping, where A is a domain (e.g., a simply or multiply connected region bounded by one or more simple Jordan curves) in $\mathbf{R}^2$ and T is a continuous map from A to $\mathbf{R}^2$. Using notation from *Geocze area,* we have a function $v(T, \pi)$ (there is only one projection) defined over the class of all simple polygonal regions $\pi \subset A$.

Denote by $v^+(T, \pi)$ and $v^-(T, \pi)$ the functions

$$\int_A (1/2)(|\gamma(z; C)| \pm \gamma(z; C))dxdy,$$

respectively. From these, we have functions $V(T, A)$, $V^+(T, A)$ and $V^-(T, A)$, which can be identified to the total variations of v, v^+ and v^-.

If the mapping (T, A) is of *bounded variation,* the derivatives of V, V^+ and V^- at a point can be defined (by using squares with sides parallel to the coordinate axes, containing the point and of diameter tending to 0), and it can be shown that such derivatives exist and are finite almost everywhere. The generalized Jacobian at $x \in A$ is the function $\mathcal{J}(x) \equiv (V^+)'(x) - (V^-)'(x)$.

generalized Lamé's differential equation A linear second order complex differential equation with rational function coefficients. More precisely, let $a_1, a_2, \ldots, a_{n-1}, \infty$ be the n *elementary singularities,* i.e., *regular singular points* with exponent difference $\frac{1}{2}$. The equation is:

$$\frac{d^2 w}{d z^2} + \{\sum_{j=1}^{n-1} \frac{\frac{1}{2}}{z - a_j}\}\frac{d w}{d z} + \frac{A_0 + A_1 z + \ldots + A_{n-3} z^{n-3}}{\prod_{j=1}^{n-1}(z - a_j)} w = 0,$$

where $A_{n-3} = \dfrac{(n-2)(n-4)}{16}$.

generalized trigonometric polynomial Functions of the form $\sum_{n=1}^m \alpha_n e^{i\lambda_n x}$, where $\{\lambda_n : n = 1, 2, \ldots\}$ are real numbers, and $\{\alpha_n : n = 1, 2, \ldots\}$ are complex numbers. Recall that $e^{i\lambda_n x} = \cos(\lambda_n x) + i \sin(\lambda_n x)$. They are almost periodic functions in the sense of Bohr.

generalized trigonometric series Let $S_m(x) = \sum_{n=1}^m \alpha_n e^{i\lambda_n x}$, $n = 1, 2, \ldots$ be a sequence of *generalized trigonometric polynomials,* and assume that $\lim_{m\to\infty} S_m(x)$ exists uniformly in x. Then $\sum_{n=0}^\infty \alpha_n e^{i\lambda_n x}$ is a *generalized trigonometric series,* and is also almost-periodic in the sense of Bohr.

generating curve A moving curve which traces out a surface. Usually, the curve is rotated around an axis in $\mathbf{R}^3$, and the surface generated is a *surface of revolution.*

generating function (1.) Given a sequence of numbers or functions $P_n(x)$, where x ranges over some common domain, and $n = 0, 1, \ldots$, the formal power series

$$P(x,t) = \sum_{n=0}^{\infty} P_n(x)t^n$$

is called a *generating function* for the $P_n(x)$.

For example, the Legendre polynomials have generating function

$$P(x,t) = (1 - 2xt + t^2)^{-1/2}.$$

Generating functions arise naturally in the theory of eigenfunction expansions, in number theory and also in probability theory, where, for example, P_n is the probability that the random variable x will take the value n.
(2.) A similar idea arises with regard to *canonical transformations* and *contact transformations:* the *generating function* is one whose (partial) derivatives provide the components of the desired transformation.

generating line A moving line which traces out a surface. A special case of *generating curve.* Surfaces that possess a generating line are known as *ruled surfaces.*

generic property Let X be a topological space, and let $\mathcal{P}$ denote a property which is either true or false of each element of X. We say that property $\mathcal{P}$ is *generic* in X if the set $\{x \in X : \mathcal{P} \text{ is true for } x\}$ contains a set which is open and dense in X.

Geocze area Let (T, A) be a surface, where A is a domain (e.g., a simply or multiply connected region bounded by one or more simple Jordan curves) in $\mathbf{R}^2$ and T is a continuous map from A to $\mathbf{R}^3$; let E_1, E_2 and E_3 denote the coordinate planes in $\mathbf{R}^3$, and let $T_i, i = 1, 2, 3$, denote the composition of T with the projection on E_i. Let $\partial\pi$ be the positively oriented boundary of a polygonal domain π in A and C_i be the oriented image of $\partial\pi$ by T_i. The rotation number $\gamma(z; C_i)$ of z with respect to the closed plane curve C_i, is a measurable function of $z (= x + iy)$. We can define $v_i = v(T_i; \pi) = \int_{E_i} |\gamma(z; C_i)| dxdy$, and $v(T; \pi) = (v_1^2 + v_2^2 + v_3^2)^{1/2}$. Let S denote any finite collection of non-overlapping polygonal domains contained in A. The Geocze area V of (T, A) is defined as $V(T, A) = \sup_S \sum_{\pi \in S} v(T; \pi)$.

There is a variant, the Geocze area U of (T, A), where $u_i = u(T_i; \pi) = \int_{E_i} \gamma(z; C_i) dxdy$, and $u(T; \pi) = (u_1^2 + u_2^2 + u_3^2)^{1/2}$, with $U(T, A) = \sup_S \sum_{\pi \in S} u(T; \pi)$.

Geocze problem This is the problem of determining whether the Lebesgue area of a surface (T, A), where A is a domain in $\mathbf{R}^2$ and T is a continuous map from A to $\mathbf{R}^3$, is the same as that obtained by using the set of all sequences $\{(F_n, O_n) : n = 1, 2, \ldots\}$ such that (F_n, O_n) converges to (T, A) and each (F_n, O_n) is inscribed in (T, A). The answer is positive if A is a Jordan domain and the Lebesgue area is finite.

geodesic (1.) The shortest curve connecting two given points on a surface and lying on the surface. On the surface of a sphere, a *geodesic* is an arc of a great circle; in a Euclidean space it is a line segment.
(2.) More generally, let M denote a smooth n-dimensional manifold, let D denote a connection (or covariant differentiation) on M, and let σ be a smooth curve on M with tangent field Σ. We say that σ is a *geodesic* if $D_\Sigma \Sigma = 0$, (i.e., the tangent field of σ is *parallel* along σ).

geodesic arc An arc of curve on a smooth manifold which satisfies the condition of being a geodesic.

geodesic coordinates Let M be an n-dimensional smooth manifold, let N_x be an open neighborhood of $x \in M$. Let (U, ϕ) be a chart containing N_x, $\phi : U \to \mathbf{R}^n$. We call ϕ a *geodesic* or *normal chart* if it maps geodesics through x to line segments through $0 \in \mathbf{R}^n$.

geodesic correspondence Two surfaces, with a diffeomorphism between them, which maps geodesics to geodesics. This map is also called *geodesic preserving* or *connection preserving.* A fairly immediate conclusion is that a surface is locally geodesically correspondent with a plane if it is a surface of constant curvature.

geodesic curvature Let M be a Riemannian manifold with Riemannian connection D. Let σ be a $\mathbf{C}^\infty$ curve in M with tangent vector field $\Sigma = \sigma_*(d/dt)$. Assuming that Σ does not vanish on the domain of σ, we define the unit tangent vector $T(t) = \Sigma(t)/|\Sigma(t)|$, where the Riemannian requirement allows for a notion of length of tangent vectors. The *geodesic curvature vector field* of σ is defined to be $D_T T$, and its length is the *geodesic curvature* of σ.

geodesic line Given two points on a Riemannian manifold, a curve of least length joining the two points. For points on an embedded hypersurface (dim = 2), the principal normal of this curve coincides with the normal to the surface at each point on the curve.

geodesic polar coordinates Let M be a two-dimensional smooth manifold, let D_ϵ be a disk of radius ϵ centered at $0 \in \mathbf{R}^2$. A map $\phi : D_\epsilon \to M$ is called a *geodesic polar coordinate system* if it maps line segments in D_ϵ passing through the origin into geodesics on M, and it maps the orthogonal curves $r = constant$ and $\theta = constant$ on D_ϵ to orthogonal curves on M.

geodesics Curves that satisfy the condition that their tangent vector be parallel along them. *See* geodesic.

geometric mean For n real numbers $x_1, \ldots, x_n$, the number $(x_1 x_2 \cdots x_n)^{1/n}$.

More generally, for G a locally compact, Abelian group and w a nonnegative function in $L^1(G)$, the geometric mean of w is

$$\Delta(w) = \exp \int_G \log w(x)dx.$$

geometric progression A sequence a_0, $a_1, \ldots$, such that $a_{n+1}/a_n = d$, for some number d and all $n \geq 0$.

geometric series A series such that the successive terms form a geometric progression.

geometry on a surface Beside Euclidean geometry, which is based on properties of the Euclidean plane, and, in particular, the parallel postulate, one can introduce geometries on other two-dimensional surfaces. Such surfaces can be spheres, leading to the well-known spherical geometry, hyperboloids, leading to the geometry of N. I. Lobatchevski, or any other surface one desires. The properties of the geometric objects will depend on the geometry of the surface on which the objects are defined. *See* non-Euclidean geometry.

germ of a set Let $\mathcal{T}$ be a topological space, A a subset of $\mathcal{T}$, a a point of $\mathcal{T}$. The *germ* of A at a, denoted by A_a, is the set of all subsets Ω of $\mathcal{T}$ which coincide with A on some $\mathcal{T}$-neighborhood of a.

germ of analytic set A set A is an analytic subset of a complex analytic manifold M if for every $a \in A$ there exists a neighborhood Ω of a and a finite number of analytic functions $\phi_1, \ldots, \phi_n$ on Ω such that

$$A \cap \Omega = \{x \in \Omega : \phi_1(x) = \ldots = \phi_n(x) = 0\}.$$

We define A_a, the germ of A at a, as a germ of a set. *See* germ of a set. Each germ A_a is associated with an ideal in the ring of germs of holomorphic functions at a.

germ of holomorphic function Let z_0 be a point in the complex plane. We define an equivalence class of holomorphic functions at z_0: two functions belong to the class if there exists an open neighborhood of z_0 on which the two functions coincide. This equivalence class is the *germ of the holomorphic function* at z_0. The notion can be extended in the obvious way to higher dimensional manifolds, and holomorphic functions of several complex variables. *See* sheaf of germs of holomorphic functions.

Gevrey class The notion of *Gevrey class* provides a description of the growth of the derivatives of a function. Let β be a nonnegative number. A function $f(x)$ defined on the real interval $[a, b]$ is said to belong to the *Gevrey class* $G_\beta[a, b]$ if it has derivatives of all orders on $[a, b]$ and satisfies the inequalities $|f^{(q)}(x)| \leq CB^q q^{q\beta}, q = 0, 1, 2, \ldots$ for some constants B and C, depending, possibly, on f.

Some consequences are that, for $\beta \leq 1$, an $f(x)$ in $G_\beta[a, b]$ is an *analytic function.* Furthermore, an analytic function in $[a, b]$ is in $G_1[a, b]$. When $\beta > 1$, the class G_β contains functions which are not analytic.

In the case of n variables, let Ω be a *closed* region, and let $G_\beta(\Omega)$ be the collection of all functions defined, continuous and infinitely differentiable on Ω, which satisfy the inequality:

$$\left| \frac{\partial^{q_1+\cdots+q_n} u(x)}{\partial x_1^{q_1} \ldots \partial x_n^{q_n}} \right| \leq C B_1^{q_1} \ldots B_n^{q_n} q_1^{q_1\beta} \ldots q_n^{q_n\beta},$$

for $q_1, \ldots, q_n = 0, 1, 2, \ldots$ The class G_1 consists of functions analytic in each of their variables. The class G_β is closed under addition, multiplication by scalars and differentiation.

Gibbs' phenomenon A phenomenon that arises in the Fourier (or any continuous eigenfunction family) approximation of discontinuous functions. As the number of terms in the Fourier representation increases, the approximation to the discontinuous function near the discontinuity both overshoots and undershoots the actual functional values, and continues to do so, with a percentage of error bounded away from 0, regardless of the number of terms in the representation. More specifically, the function

$$f(x) = \begin{cases} +1, \text{ for } 0 < x < \pi, \\ -1, \text{ for } \pi < x < 2\pi \end{cases}$$

can be represented by a series periodic in x with period 2π. One can show that, if $\sigma_n(x)$ represents the partial sums of the series containing all terms up to $\sin(nx)$, then $\lim_{n\to\infty} \sigma_n(\pi/(2n + 1)) \approx 1.179$, thus indicating an overshoot of about 18%.

global analysis The determination of maxima and minima of functions over Euclidean spaces is part of what is known as *local analysis,* i.e., the study of *local* properties of geometric or functional objects. When we examine properties of objects defined on *manifolds,* which are *locally* Euclidean spaces, we move into the domain of *global analysis.* It includes the study of differential operators on manifolds, and their extensions to Banach manifolds, Morse theory and the study of critical points, Lie groups, dynamical systems and structural stability, singularities, and others.

global concept A mathematical concept that is meaningful on a manifold. A certain property or statement is assumed dependent on the whole space, considered as a single object.

global dimension Let A be an analytic subset of a complex analytic manifold G. The global dimension of A is $\sup_{z\in A} \dim_z A$, where $\dim_z A$ is the *local dimension* of A. The local dimension of A is given by the dimension of the range of the covering mapping that exists due to the nature of A as a k-sheeted ramified covering space.

Global Implicit Function Theorems A group of theorems which provide conditions for the existence of *global* homeomorphisms between topological spaces. Recall that the standard Implicit Function Theorem provides a *local* homeomorphism only.

(i.) *Global Inverse Function Theorem (with differentiation):* let $\phi : \mathbf{R}^n \to \mathbf{R}^n$ be $\mathbf{C}^1$ and proper (inverse images of compact sets are compact). Assume further that $\det(D\phi(x)) \neq 0$ for each $x \in \mathbf{R}^n$. Then ϕ is a global diffeomorphism.

(ii.) *Global Inverse Function Theorem (with differentiation):* let X and Y be Banach spaces, $\phi \in C^1(X, Y)$, $[D\phi(x)]^{-1}$ exists and there is a constant K such that $|[D\phi(x)]^{-1}| \leq K$ for all $x \in X$. Then ϕ is a homeomorphism of X onto Y.

(iii.) *Global Implicit Function Theorem (without differentiation):* let X, Y be metric spaces and assume that $\phi : X \to Y$ is continuous and proper. If $X \setminus \phi^{-1}(\phi(W))$ (W is the set of critical points of ϕ, i.e., those points at which ϕ is not invertible) is non-empty, arcwise connected and $Y \setminus \phi(W)$ is simply connected, then ϕ is a homeomorphism from $X \setminus \phi^{-1}(\phi(W))$ onto $Y \setminus \phi(W)$.

(iv.) *Global Inverse Function Theorem (without differentiation):* assume the conditions of (iii.) hold. If $W = \emptyset$ and Y is simply connected, then ϕ is a homeomorphism from X onto Y.

Goursat's Theorem Let D be any finite, simply connected domain in the complex plane $\mathbf{C}$. Let γ be any closed, rectifiable curve contained in D. If the function $f(z)$ is differentiable in D, then $\int_\gamma f(z)dz = 0$.

Notice that this theorem requires only that the function f be differentiable, rather than holomorphic.

gradient In classical vector analysis, the gradient of a function f on $\mathbf{R}^n$ is defined to be the vector field

$$\sum_{i=1}^{n} \frac{\partial f}{\partial x_i}(x) \frac{\partial}{\partial x_i}.$$

The notion can be extended to Riemannian manifolds.

Gramian determinant Let $v_1, \ldots, v_n$ be vectors in a unitary or Euclidean space; let $x_1, \ldots, x_n$ be indeterminates, and consider the vector equation $v_1x_1 + \ldots + v_nx_n = 0$. Performing *scalar* multiplications by $v_1, \ldots, v_n$ in succession, leads to the scalar system

$$\begin{aligned}(v_1 \cdot v_2)x_1 + \ldots + (v_1 \cdot v_n)x_n &= 0,\\ \vdots \qquad\qquad & \vdots \;\; \vdots\\ (v_n \cdot v_1)x_1 + \ldots + (v_n \cdot v_n)x_n &= 0.\end{aligned}$$

The determinant

$$G(v_1, \ldots, v_n) = \det \begin{bmatrix} (v_1 \cdot v_1), \ldots, (v_1 \cdot v_n) \\ \vdots \\ (v_n \cdot v_1), \ldots, (v_n \cdot v_n) \end{bmatrix}$$

is called the *Gramian* of the vectors $v_1, \ldots, v_n$. It can be shown that the vectors $v_1, \ldots, v_n$ are linearly independent if and only if their Gramian determinant does not vanish.

graph Several slightly different notions pervade the mathematical literature. We shall indicate the most common.

(1.) A *graph* G is a pair (V, E), where V (the set of vertices) is a finite set and E (the set of edges) is a subset of the collection of subsets of V of cardinality two. If E is a set of *ordered* pairs of elements of V, then G is called a *directed* graph (or *digraph*). This definition is used in combinatorics, in particular in the analysis of the time and space complexity of algorithms.

(2.) A graph is a one-dimensional simplicial complex. This terminology is used in topology, and coincides with the definition in **(1.)**.

(3.) In mathematical analysis, we define the notion of *graph* in terms of sets and functions. Let f be a function from a set X to a set Y (the two sets could coincide). The collection of all pairs $(x, f(x))$ in $X \times Y$ is called the *graph* of the function f. Conversely, we can define a function in terms of a graph, i.e., a subset of $X \times Y$ such that no two elements of this set have the same first entry with different second entries.

(4.) More colloquially, a graph is a *pictorial representation* of a relation among variables.

In particular, a function can be given a visual geometric meaning by plotting the points of its graph on an appropriate coordinate system.

graph norm Let $f : X \to Y$, where X and Y are normed spaces. The *graph norm* on X with respect to f is given by:

$$\|x\|_f = \|x\|_X + \|f(x)\|_Y.$$

(Thus it is defined only for elements of X in the domain of f.)

graph of equation *See* graph.

graph of function *See* graph.

graph of inequality The set of points satisfying the inequality. For example, the inequality $x > a$, for $a \in \mathbf{R}$, is the set of points $\{x : x \in \mathbf{R}, x > a\}$. For the inequality $f(x_1, \ldots, x_n) \geq a$, $x_i \in X_i$ for $i = 1, 2, \ldots, n$, the graph consists of the subset of the product $X_1 \times \ldots \times X_n$ consisting of all n-tuples $(x_1, \ldots, x_n)$ satisfying the inequality.

graph of relation Given sets X and Y (where Y and Y could be the same set), *a relation from X to Y* (or *on* X, if $Y = X$) is a subset of the product set $X \times Y$. This subset, a collection of pairs, is also known as the graph of the relation.

graphing The process of obtaining a geometric pictorial representation of a relation among two or more variables by using its graph to determine points in a coordinate system.

Grauert and Remmert's Theorem Let M and N be real analytic manifolds. Then the set of real analytic maps from M to N is dense in $\mathbf{C}^r_S(M, N)$, for $0 \leq r \leq \infty$, the space of functions with continuous derivatives up to order r, where S denotes the strong (graph) topology over the underlying set.

Grauert's Theorem Let X be a complex manifold of dimension n and $\mathcal{F}$ a coherent analytic sheaf over X. Let $f : X \to Y$ be a holomorphic map of complex manifolds. If f is a proper map, then $f^q_* \mathcal{F}$, the qth direct image of $\mathcal{F}$, is coherent for all $q \geq 0$.

greatest lower bound Given a set of real numbers Σ, the greatest lower bound of Σ is a number b with the properties:
a) $x \geq b$ for every $x \in \Sigma$;
b) if $x \geq c$ for every $x \in \Sigma$, then $b \geq c$.

Green line Assume u is a harmonic function over some domain D. The set $\{w : u(w) = constant\}$ is called a *level* or *equipotential surface* for u. A point where grad(u) vanishes is called *critical.* For each noncritical point there exists an analytic curve through it such that grad(u) is tangent to it at every point of the curve. A maximal curve with this property is called an *orthogonal trajectory.* When u is a Green's function, this orthogonal trajectory is called a *Green line.*

Green measure Let $u(w)$ be a Green's function with pole w_0. Let E be the family of orthogonal trajectories issuing from the pole. Let Σ_a be a closed level surface for u, which, for sufficiently large a, is an analytic closed surface homeomorphic to a spherical surface. If $A = E \cap \Sigma_a$ is a measurable set, then the harmonic measure of A at w_0 with respect to the interior of Σ_a is called the *Green measure* of E. M. Brelot and G. Choquet proved that all orthogonal trajectories issuing from the pole, except for a set of Green measure 0, are Green lines issuing from the pole, and u goes to 0 along them.

Green space A generalization of a Riemann surface on which a Green's function exists.

Green's formula (1.) In the context of ordinary differential equations. Let $A(t)$ be a continuous $n \times n$ matrix-valued function on a t-interval $[a, b]$. *Let* $A^*(t)$ denote the complex conjugate transpose of $A(t)$; let $f(t)$

and $g(t)$ be continuous n-vector-valued functions on $[a, b]$. Let $y(t)$ denote a solution of $y' = A(t)y + f(t)$, while $z(t)$ denotes a solution of $z' = -A^*(t)z - g(t)$ on $[a, b]$. Then for all $t \in [a, b]$,

$$\int_a^t [f(x) \cdot z(x) - y(x) \cdot g(x)]dx$$
$$= y(t) \cdot z(t) - y(a) \cdot z(a).$$

(2.) In the context of complex variable theory, partial differential equations or differential geometry. Let D be a bounded domain whose boundary ∂D consists of a finite number of closed surfaces (or arcs) that are piecewise of class $\mathbf{C}^1$. Let u and v have continuous first and second order partial derivatives and suppose the partial derivatives u_x, u_y, v_x, v_y have finite limits at every boundary point. Let $\mathbf{n}$ be the unit outer normal on ∂D, let $d\mu$ denote the infinitesimal volume element on D and let $d\sigma$ denote the infinitesimal area element on ∂D. Then we have, as a special case of Stokes' Theorem:

$$\int_D (u\Delta v - v\Delta u)d\mu$$
$$= \int_{\partial D} (u\frac{\partial v}{\partial \mathbf{n}} - v\frac{\partial u}{\partial \mathbf{n}})d\sigma,$$

where Δ is Laplace's operator.

If u and v are also harmonic in D, we have

$$\int_{\partial D} (u\frac{\partial v}{\partial \mathbf{n}} - v\frac{\partial u}{\partial \mathbf{n}})d\sigma = 0.$$

Green's function Green's functions arise in the solution of boundary value problems.
(1.) A typical such function for ordinary differential equations: let $A(t)$ be a continuous $n \times n$ matrix-valued function on a t-interval $[a, b]$. Let $f(t)$ be a continuous n-vector-valued function on $[a, b]$.

Consider the system $x' = A(t)x + f(t)$, on which we impose appropriate boundary conditions.

The Green's function for the boundary value problem is a function $G(t, \tau)$, $t, \tau \in [a, b]$, such that the function

$$u(t) = \int_a^b G(t, \tau)f(\tau)d\tau$$

exists on $[a, b]$ and solves the boundary value problem.
(2.) From complex variable theory, let D be a region of finite connectivity in the complex plane whose boundary is a finite sum of analytic contours. Let $z_0 \in D$, and let $G(z, z_0)$ be the solution of the Dirichlet problem in D with boundary values $\log|\zeta - z_0|$. The Green's function of D with pole at z_0 is the function $g(z, z_0) = G(z, z_0) - \log|z - z_0|$. This function is harmonic in D, except at z_0 and vanishes on the boundary.
(3.) In general, if the solution of a boundary value problem has an integral representation, the Green's function for the boundary value problem is defined to be the kernel function of the integral representation. *See* Green's operator.

Green's operator Consider a (partial) differential equation (usually elliptic) as an operator between pairs of function spaces, say $A : X \to Y$, with appropriate boundary conditions. A *Green's operator* (relative to the boundary conditions), if one exists, is the mapping $A^{-1} : Y \to X$. Note that the domain of A in X will depend on the boundary conditions.

If a Green's operator admits an integral representation, the kernel function associated with the representation is called the Green's function for the boundary value problem.

Green's tensors Green's functions that arise in the solution of Stokes's differential equation in $\mathbf{R}^3$, or the equation $\Delta\ u + a$ grad div $u = 0$ in a bounded domain D with smooth boundary ∂D. They satisfy the condition $\mathbf{G}(P, Q) = (g_{ik}(P, Q)) = 0$ for $Q \in D$ and $P \in \partial D$.

Green's Theorem Let E be an open set in $\mathbf{R}^2$, with α and β two continuously differentiable functions on E. Let Ω be a closed subset of E, with positively oriented boundary, $\partial\Omega$. Let dA denote the infinitesimal area element in $\mathbf{R}^2$. Then

$$\int_{\partial\Omega}(\alpha dx + \beta dy) = \int_{\Omega}(\frac{\partial\beta}{\partial x} - \frac{\partial\alpha}{\partial y})dA.$$

The theorem is often stated in terms of differential forms and chains, and is one of the many generalizations of the fundamental theorem of calculus.

Gronwall's inequality Let $u(t)$ and $v(t)$ be non-negative continuous functions on $[a, b]$; let $C \geq 0$ be a constant; assume that

$$v(t) \leq C + \int_a^t v(\tau)u(\tau)d\tau$$

for $a \leq t \leq b$. Then $v(t) \leq C \exp\int_a^t u(\tau)d\tau$ for $a \leq t \leq b$. In particular, $C = 0$ implies that $v(t) \equiv 0$.

Gross area Let X be a Borel set in $\mathbf{R}^3$, let E be any plane in $\mathbf{R}^3$; divide $\mathbf{R}^3$ into a mesh of half-open cubes $M_1, M_2, \ldots$ with uniform diameter $d > 0$; let m_j denote the supremum over all planes E of the Lebesgue measure of the orthogonal projection of $M_j \cap X$ to E. $\lim_{d\to 0}\sum_j m_j$ is called the *Gross area* (the two-dimensional *Gross measure*) of X.

Gross measure Starting from the family F of all Borel subsets of $\mathbf{R}^n$, define a function $\zeta(S)$ as the least upper bound of the m-dimensional Lebesgue measures of the orthogonal projections of the Borel set S over all m-dimensional hyperplanes of $\mathbf{R}^n$.

The *Caratheodory construction* applied to this $\zeta(S)$ gives rise to $\mathcal{G}^m$ the m-dimensional *Gross measure* over $\mathbf{R}^n$, where $m \leq n$.

Gross's Theorem Let $w = f(z)$ be a meromorphic function, and let $P(w, w_0)$ be a function element of the inverse function centered at w_0. Consider the set of half-lines centered at w_0 and parametrized by their argument. Then the set of arguments of half-lines along which the analytic continuation meets a singularity at a finite point is of measure 0.

Grothendieck group Let A be an operator algebra (usually assumed to be an AF-algebra) and $D(A)$ the projections in A modulo the equivalence relation: $e \sim f \iff e = v^*v, f = vv^*$, for some $v \in A$. $D_\infty(A)$ denotes the direct limit of $D_n(A) = D(M_n(A))$, where $M_n(A)$ denotes the $n \times n$ matrices, with entries in A. The *Grothendieck group*, $K_0(A)$, is the quotient group of A. That is, $K_0(A)$ is the set of all pairs of elements $(\alpha, \beta) \in K_0(A) \times K_0(A)$, modulo the equivalence relation $(\alpha, \beta) \sim (\alpha', \beta')$ if $\alpha + \beta' = \alpha' + \beta$.

Gudermann function Setting $x = \log(\sec\theta + \tan\theta)$, we define θ to be the *Gudermannian* of x, $gd(x)$. With it we can express many relationships between trigonometric and hyperbolic functions; for example: $\sinh(x) = \tan(gd(x))$.

H

Haar measure A translation-invariant Borel measure on a topological group.

Haar space A locally compact topological group X with a Borel measure μ, such that $\mu(U) > 0$ for every non-empty Borel open set U, and $\mu(xE) = \mu(E)$ for every Borel set E and $x \in X$. Such a measure is called a Haar measure.

Hadamard manifold A complete, simply connected, Riemannian manifold of nonpositive sectional curvature.

Hadamard's Composition Theorem Let $f(z) \equiv \sum_{n=0}^{\infty} a_n z^n$, $g(z) \equiv \sum_{n=0}^{\infty} b_n z^n$ and $h(z) \equiv \sum_{n=0}^{\infty} a_n b_n z^n$, where $a_n, b_n, z \in \mathbf{C}$. For each of the functions one can define the *principal* or *Mittag-Leffler star*, $\mathbf{A}[f]$, as the set of all those points $a \in \mathbf{C}$ such that f can be continued analytically along the line segment $[0, a]$. We define $\mathbf{A}[f] * \mathbf{A}[g] \equiv \{\mathbf{A}[f]^c \cdot \mathbf{A}[g]^c\}^c$, where $\mathbf{A}[f]^c$ is the complement of $\mathbf{A}[f]$ and the product set is $\{z_1 z_2 : z_1 \in Z_1 \text{ and } z_2 \in Z_2\}$. *Hadamard's Composition Theorem* says that $\mathbf{A}[h] \subset \mathbf{A}[f] * \mathbf{A}[g]$.

Hadamard's Factorization Theorem If $f(z)$ is an entire function of finite order ρ, then

$$f(z) = e^{g(z)} z^k P(z),$$

where $g(z)$ is a polynomial of degree $q \leq \rho$, k is a non-negative integer, and $P(z)$ is a canonical product of order $\sigma \leq \rho$. If ρ is not an integer, then $\sigma = \rho$ and $q \leq \lfloor \rho \rfloor$. If ρ is an integer, then at least one of q and σ equals ρ.

Hadamard's Gap Theorem One of many results about the noncontinuability of a power series beyond its circle of convergence. Such series are called *gap* or *lacunary series*.

Given the series $\sum_{n=0}^{\infty} a_n z^{j_n}$, $a_n \neq 0$, where $\lim_{n\to\infty} n/j_n = 0$, if there exists a fixed $\lambda > 1$ such that for all n, $j_{n+1}/j_n \geq \lambda$, then the series has its circle of convergence as its natural boundary.

Hadamard's Theorem (1.) This is also known as the *Prime Number Theorem*, originally conjectured by L. Euler: let x be a nonnegative real number; let $\pi(x)$ be the number of prime numbers $\leq x$. Then $\lim_{x\to\infty} \pi(x)/(x/log(x)) = 1$. The theorem was proven by Jacques Hadamard in 1896.

(2.) If a particle is free to move on a surface which is everywhere regular and has no infinite sheets, the potential energy function being regular at all points of the surface and having only a finite number of maxima and minima on it, either the part of the orbit described in the attractive region is of length greater than any assignable quantity, or the orbit tends asymptotically to one of the positions of unstable equilibrium.

Hadamard's Three Circles Theorem Let $f(z)$ be holomorphic in $|z| < R$, and set $M(r) = M(r; f) = \max_\theta |f(re^{i\theta})|$. If $0 < r_1 \leq r \leq r_2 < R$, then

$$\log M(r) \leq \frac{\log r_2 - \log r}{\log r_2 - \log r_1} \log M(r_1) + \frac{\log r - \log r_1}{\log r_2 - \log r_1} \log M(r_2).$$

Hahn-Banach Theorem If E is a subspace of the linear space X, if f is a subadditive functional on X such that $f(ax) = af(x)$ for $a \geq 0$, and if g is an additive, homogeneous functional on E such that $g(x) \leq f(x)$ for every $x \in E$, then there exists an additive homogeneous extension h of g to all of X such that $h(x) \leq f(x)$ for every $x \in X$.

half-open interval An interval in $\mathbf{R}$, closed (i.e., containing the endpoint) at one end and open at the other; hence, $[a, b) =$

0-8493-0320-6/00/$0.00+$.50

$\{x : a \leq x < b\}$, where $a > -\infty$ and $b \leq \infty$ or $(a, b] = \{x : a < x \leq b\}$, where $-\infty \leq a$ and $b < \infty$.

half-plane In $\mathbf{R}^2$, the collection of all those points that lie on one side of an infinite straight line. The half-plane can be closed, thus containing all the points of the defining line, or open, thus containing no points of the defining line. Other conditions, where part of the line belongs to the half-plane, are also possible.

half-space The portion of a linear space bounded by a hyperplane. A number of generalizations of this notion are possible, for example: let Y be the domain of positivity of a formal real Jordan algebra A and let H be the subset $A \oplus \sqrt{-1}\, Y = \{x + \sqrt{-1}\, y : x \in A, y \in Y\}$ of the complexification $A_{\mathbf{C}}$ of A. H is called the *half-space* belonging to A. If $A = \mathbf{R}$ and Y is the set of positive real numbers, then H is the upper half-plane in $\mathbf{C}$, so it is a *half space* in more common terms.

Hallstrom-Kametani Theorem A gap-value theorem, complementing, for meromorphic functions, those that E. Picard proved for entire functions.

A single-valued meromorphic function with a set of singularities of logarithmic capacity zero takes on every value infinitely often in any neighborhood of each singularity except for at most an F_σ-set of values of logarithmic capacity zero.

Hamburger's moment problem This problem can be stated as the theorem: A sequence $\{s_n : n = 0, 1, \ldots, s_n \in \mathbf{R}\}$ is the sequence of moments of a measure $\mu \in E_+(\mathbf{R}) \equiv \{\mu \in M_+(\mathbf{R})$: the Radon measures on $\mathbf{R}$ with $\int |x^n| d\mu(x) < \infty$ for all $n \geq 0\}$, i.e., $\{s_n\}$ is of the form $s_n = \int x^n d\mu(x)$, if and only if $\sum_{j,k=0}^{n} s_{j+k}\, \alpha_j \bar{\alpha}_k \geq 0$ for all $n \geq 0$ and for all sets $\{\alpha_0, \ldots, \alpha_n\} \subset \mathbf{C}$. The set of measures $\mu \in E_+(\mathbf{R})$ for which this holds is a nonempty compact set in $M_+(\mathbf{R})$ in the coarsest topology in which the mappings $\mu \to (\mu, f)$ are continuous when f ranges over continuous functions with compact support on $\mathbf{R}$.

Hamilton's equations *See* Hamiltonian function.

Hamiltonian function In mechanics, the behavior of a system with n degrees of freedom can be expressed in terms of a function of both position (q_i) and momentum (or velocity; p_i). Such a function, usually denoted by $H(q_1, \ldots, q_n, p_1, \ldots, p_n, t)$, is called *the Hamiltonian of the system* and satisfies the system of Hamiltonian differential equations:

$$\frac{d\,p_i}{d\,t} = -\frac{\partial H}{\partial q_i}, \quad \frac{d\,q_i}{d\,t} = \frac{\partial H}{\partial p_i},$$

for $i = 1, \ldots, n$.

In the Lagrangian formulation, with Lagrangian function $L(q_1, \ldots, q_n, \dot{q}_1, \ldots, \dot{q}_n, t)$, we let $p_i = \partial L/\partial \dot{q}_i$, for $i = 1, 2, \ldots, n$. The function $\sum_{i=1}^{n} p_i \dot{q}_i - L$ is also called the Hamiltonian of the system.

Hamiltonian vector field Let (M, ω) be a symplectic manifold and $H : M \to \mathbf{R}$ a $\mathbf{C}^r$ function, $r \geq 1$. The vector field X_H given by $\omega(X_H, Y) = dH \cdot Y$, is called the Hamiltonian vector field with energy function H.

We restate this in more classical terminology: Let $(q_1, \ldots, q_n, p_1, \ldots, p_n)$ be canonical coordinates for the symplectic 2-form ω, so that $\omega = \sum dq_i \bigwedge dp_i$. Then, in these coordinates, the Hamiltonian vector field is given by

$$X_H = (\frac{\partial H}{\partial p_i}, -\frac{\partial H}{\partial q_i}) = J \cdot dH,$$

where $J = \begin{bmatrix} 0 & I \\ -I & 0 \end{bmatrix}$. Thus $(q(t), p(t))$ is an integral curve of the vector field X_H if and only if Hamilton's equations hold.

Hammerstein's integral equation A nonlinear integral equation of the form

$$\psi(x) + \int_0^1 K(x,y)f[y,\psi(y)]dy = 0.$$

The following conditions are sufficient for its solvability by successive approximations: assume $f(y,0)$ is square integrable, and $f(y,u)$ is uniformly Lipschitz continuous in u with Lipschitz coefficient $C(y)$. Assume further that (at least) $A^2(x) = \int_0^1 K^2(x,y)dy$ exists almost everywhere in (0, 1) and is integrable there, and that $\int_0^1 A^2(x)C^2(x)\,dx < 1$.

A number of other results on existence and uniqueness are known.

Hankel determinant Starting from the power series $a_0+a_1z+a_2z^2+\ldots a_nz^n+\ldots$, the Hankel determinants are

$$A_n^r = \det\begin{bmatrix} a_n & \cdots & a_{n+r-1} \\ \vdots & \vdots & \vdots \\ a_{n+r-1} & \cdots & a_{n+2r-2} \end{bmatrix}.$$

They arise in the power series expansion of rational functions. Observe further that all backward diagonals of the matrix consist of constant entries.

Hankel form Let $s_0, s_1, \ldots, s_{2n-2}$, be a sequence of numbers. The *Hankel form* is the quadratic form $S(x,y) = \sum_{i,k=0}^{n-1} s_{i+k}x_iy_k$.

Hankel matrix (1.) An $n \times n$ matrix corresponding to a *Hankel form,* or to a *Hankel determinant,* hence having the form

$$\begin{bmatrix} a_n & \cdots & a_{n+r-1} \\ \vdots & \vdots & \vdots \\ a_{n+r-1} & \cdots & a_{n+2r-2} \end{bmatrix}.$$

(2.) An infinite matrix of the form $(a_{i+j}), i, j = 0, 1, \ldots$; thus, whose entries are constant along the backward diagonals $(a_{i,j}, i = j + c)$.

See also Toeplitz matrix.

Hankel operator An operator $H_\phi : H^2 \to H^{2\perp}$ defined by

$$H_\phi f = (I - P)\phi f,$$

where P is the orthogonal projection from L^2 to H^2, ϕ is bounded and $f = f(e^{it}) \in H^2$. For more general ϕ, H_ϕ can be defined, as above, for f a trigonometric polynomial, and extended, when possible, to a bounded operator defined on all of H^2. The matrix of H_ϕ, with respect to the orthonormal basis $\{e^{ijt}\}$ of H^2, is the Hankel matrix (a_{i+j}), where $\{a_j\}, j = 0, 1, \ldots$, are the (forward) Fourier coefficients of ϕ. Often the change of variables $e^{it} \to e^{-it}$ is applied to the right side of the definition of H_ϕ, when an operator from H^2 to itself is desired.

Hankel operators can also be defined on the Bergman space B^2 and on H^2 spaces in several variables. In these cases, there are two possibilities for the projection $I - P$ above. When $I - P$, for P the projection from L^2 to B^2 [resp. H^2] is used, the resulting H_ϕ is called a *large* Hankel operator. When, instead, $I - P$ is replaced by the projection of L^2 on the space of *conjugates* of B^2 [resp. H^2] functions, then H_ϕ is called a *small* Hankel operator.

Hankel transform For a function f, piecewise continuous on $[0, \infty)$ and absolutely integrable there,

$$H[f] = \int_0^\infty x\, f(x)\, J_\nu(tx)\, dx,$$

for $\nu \geq -1/2$, where $J_\nu(z)$ is the Bessel function of the first kind of order ν.

Hardy class *See* Hardy space.

Hardy space The class H^p $(0 < p < \infty)$ of functions $f(z)$, analytic for $|z| < 1$, and satisfying

$$\sup_{0 \leq r < 1} \int_0^{2\pi} |f(re^{it})|^p dt < \infty.$$

If $p = \infty$ the integral is replaced by the essential sup. If $1 \leq p < \infty$, H^p is a Banach space with norm equal to the pth root of the above integral (and if $p = \infty$, with the sup-norm).

One may define H^p spaces on other open subsets of the plane. For example, for the half plane $\Re(z) > 0$, the integrals in the above definition are replaced by

$$\sup_{x>0} \int_{-\infty}^{\infty} |f(x+iy)|^p dy.$$

Hardy's Theorem Let $f(z) = \sum_0^\infty a_n z^n$ be an element of the Hardy space H^1. Then $\sum_1^\infty | a_n | n^{-1} < \infty$.

Hardy-Lebesgue class *See* Hardy space.

Hardy-Littlewood Theorem If $\lim_{x\to 1} \sum_{n=0}^\infty a_n x^n = A$, and $a_n = O(1/n)$, then $\sum_{n=0}^\infty a_n = A$.

This is a generalization of the original theorem of Tauber's (and the first *officially Tauberian* theorem; it was in this context that Hardy and Littlewood introduced the term), which assumed that $a_n = o(1/n)$.

harmonic (1.) Any function which satisfies Laplace's equation.
(2.) In the representation of functions in terms of sums of periodic functions, any function whose frequency is an integral multiple of a given one is called a *harmonic* of the given one. Terms associated with this interpretation are *first harmonic* or *fundamental frequency*, where the *nth harmonic* has n times the frequency of the first.

harmonic analysis The study of mathematical objects like functions, measures, etc., defined on topological groups. The group structure is used to place the objects in a translation-invariant context. The simplest such context is provided by the circle group and Fourier analysis.

harmonic boundary Let R be a Riemann surface, let R^* be a compactification of R, $\Delta = R^* \setminus R$, an *ideal boundary* of R. We define a set Γ to be the *harmonic boundary* of R with respect to R^* in the following manner: Γ consists of those points $p^* \in \Delta$ such that $\liminf_{p\to p^*, p\in R} P(p) = 0$, for every positive superharmonic function P for which the class of non-negative harmonic functions smaller than P consists only of the constant function 0. Γ is a compact subset of R^* and is the desired harmonic boundary.

If R is not compact, the non-uniqueness of the compactification of R implies that the harmonic boundary is dependent on the compactification chosen.

harmonic continuation Let H be a *schlicht* domain in $\mathbf{R}^2$ containing the domain G. Let u be a harmonic function in G. If there exists a harmonic function in H which coincides with u on G, this function is called the *harmonic continuation* of u. A fundamental result is that, if such a continuation exists, it is unique.

harmonic differential form Let M be a Riemannian manifold of dimension n. A k-form α on M satisfying $\Delta\alpha = 0$ is called *harmonic*. Δ is the Laplace-deRahm operator, $\Delta = d\delta + \delta d$.

harmonic division A polynomial function in the variable x can be uniquely divided (decomposed) into a sum of harmonic polynomials times powers of $|x|$.

harmonic flow Let u be a harmonic function on a domain D; let Σ_a be an equipotential surface $u(w) = a$, and let Σ_a^0 be the complement in Σ_a of the critical points of u. The family of orthogonal trajectories passing through a subdomain of Σ_a^0 is called a *harmonic flow*.

harmonic function A complex-valued function $f(x_1, \ldots, x_n)$, defined on a domain D in $\mathbf{R}^n$ and satisfying Laplace's differential equation there. That is,

$$\Delta f = \sum_{j=1}^{n} \frac{\partial^2}{\partial x_j^2} f = 0$$

in D.

harmonic integral *See* harmonic differential form. Both refer to the same theory.

harmonic kernel function These are kernel functions (or reproducing kernels) relative to Hilbert spaces of harmonic functions.

harmonic majorant Given two functions $f(z)$ and $g(z)$ defined on a domain G (and on its boundary), $g(z)$ is a *harmonic majorant* of $f(z)$ if g is harmonic on G and $g(z) \geq f(z)$ at all points of G. It can be shown that, if f is *subharmonic* and $g(z)$ majorizes $f(z)$ at all but a finite number of points on the boundary of G, then $g(z)$ is a majorant of $f(z)$ at all points in G, thus extending the majorization to the interior.

harmonic mapping A map $f : M \to N$ between Riemannian manifolds M and N such that $\partial f_* = 0$, or, equivalently, such that the trace of its second fundamental form $\beta(f)$ vanishes. If f is a solution of Laplace's equation, then it is harmonic.

harmonic mean Let $\{p_1, \ldots, p_n\}$ be a set of positive numbers. The *harmonic mean* of these numbers is given by

$$\left(\frac{\sum_{i=1}^{n} \frac{1}{p_i}}{n}\right)^{-1}.$$

harmonic measure Let D be a bounded domain whose boundary consists of a finite set Γ of rectifiable Jordan curves. Let $\Gamma = \alpha \cup \beta$, where α and β are finite sets of Jordan arcs with $Int(\alpha) \cap Int(\beta) = \emptyset$. The function $\omega(z, \alpha; D)$, which is harmonic in D and assumes the value 1 on α and the value 0 on β, is called the *harmonic measure* of α with respect to D, evaluated at the point z.

harmonic progression A sequence of the form $p_n = 1/(a + nd)$, $n = 0, 1, \ldots$, where $a \neq -n\,d$ for all n. In particular, the *reciprocals* of the p_n form an *arithmetic progression*.

harmonic series The series

$$\sum_{j=1}^{\infty} \frac{1}{j},$$

which diverges and is often given as an example of a divergent series with nth term tending to 0. Its partial sums are known as the *harmonic numbers.*

More generally, any series whose terms are in *harmonic progression*.

Harnack's condition A condition arising in the constructive definition of integrals, in particular Denjoy integrals.

Let S be an integral operator, $K(S, I)$ a set of real valued functions on a closed interval $I = [a, b]$ which is in the domain of S. Let E be a closed subset of $I_0 \subset \mathbf{R}$, $\{I_k\}$ be a sequence of intervals contiguous to the set consisting of the points of E and the endpoints of I_0, and f be a function on I_0 satisfying the conditions:

(1.) let $f_E(x) = f(x)\chi(E)$, where $\chi(E)$ is the characteristic function of E. Then $f_E \in K(S, I_0)$;

(2.) $f_k = f_{I_k} \in K(S, I_k)$ for each k;

(3.) $\sum_k |S(f_k; I_k)| < +\infty$ and $\lim_{k\to\infty} O(S; f_k; I_k) = 0$ when the sequence $\{I_k\}$ is infinite.

$O(S; f_k; I_k)$ is the variation of $S(f_k)$ on I_k, i.e., the least upper bound of the numbers $|S(f_J; J)|$, where J denotes any subinterval of I_k.

Harnack's First Theorem Let $\{w_n : n = 1, 2, \ldots\}$ be a sequence of real-valued functions, harmonic in a bounded domain $\Omega \subset \mathbf{R}^2$, continuous in $\bar{\Omega}$, and uniformly convergent on the boundary of Ω. Then the sequence $\{w_n\}$ is uniformly convergent throughout $\bar{\Omega}$ and the limit function w is harmonic in Ω. Furthermore, if we let δ denote any differentiation (in $\mathbf{R}^2$), the sequence

$\{\delta w_n\}$ converges uniformly to δw on every compact subset of Ω.

Harnack's inequality *See* Harnack's Lemma.

Harnack's Lemma This is also known as *Harnack's inequality*.

Let u be harmonic and non-negative in a disk D of radius R with center at z_0. For any point z of D, and $r = |z - z_0|$, the following inequalities hold:

$$\frac{R-r}{R+r}u(z_0) \leq u(z) \leq \frac{R+r}{R-r}u(z_0).$$

Harnack's Second Theorem Let $\{u_j\}$ be a sequence of harmonic functions on a bounded domain D in the plane such that $\{u_n(p)\}$ is increasing for each point $p \in D$ and that $\{u_n(p_0)\}$ is bounded for some point $p_0 \in D$, then $\{u_n\}$ converges uniformly on compact subsets of D, to some harmonic function u.

Hartogs' Continuation Theorem Let D be a domain in $\mathbf{C}^n$ of the form $D = D_1 \times D_2$ with $D_1 \subset \mathbf{C}^k$ and D_2 a bounded domain in $\mathbf{C}^{n-k}$ having boundary S. Then any function holomorphic in a neighborhood of $[(cl(D_1) \times S) \cup (\{a\} \times cl(D_2)]$, for some $a \in D_1$, can be holomorphically extended to D. Here $cl(X)$ denotes the closure of a set X.

Hartogs' Theorem of Continuity Let G be a domain of holomorphy in $\mathbf{C}^n$ containing domains $S_j, T_j, j = 1, 2, \ldots$. Suppose that
(i.) $cl[S_j \cup T_j]$ is contained in G,
(ii.) $\sup\{|f(z)| : z \in T_j\} = \sup\{|f(z)| : z \in S_j \cup T_j$ for all f holomorphic in $G\}$
(iii.) $S_0 = \lim S_j$ is bounded.
Then for $T_0 = \lim T_j$, $cl(T_0) \subset G$ implies $cl(S_0) \subset G$. Here $cl(S)$ denotes the closure of a set S.

Hartogs' Theorem of Holomorphy Let f be a complex-valued function defined in a neighborhood of $z_0 \in \mathbf{C}^n$. Then f is holomorphic in the n-dimensional variable $z = (z_1, \ldots, z_n)$ if it is holomorphic with respect to each coordinate variable z_j separately. A major part of the theorem is that the continuity of f follows from the existence of the n partial derivatives.

Hartogs-Osgood Theorem Let G be a bounded simply connected domain in $\mathbf{C}^n, n > 1$, having connected boundary. Then every function holomorphic in a neighborhood of the boundary of G extends to a function holomorphic in G. This is very different from the behavior in the case of one complex variable.

Haupt's Theorems Theorems concerning the stability of solutions of the differential equation

$$u'' + (\lambda + \gamma F(x))u = 0,$$

where λ, γ are parameters and $F(x+2\pi) = F(x)$. (Note that this is a special case of Hill's equation and a generalization of Mathieu's equation.) For fixed γ, there are two countably infinite sequences $\{\lambda_j\}, \{\lambda'_j\}$ corresponding to periodic and half-periodic solutions respectively with

$$\lambda_0 < \lambda'_1 \leq \lambda'_2 < \lambda_1 \leq \lambda_2 < \lambda'_3 \ldots.$$

For $\lambda \in [(\lambda_{2j-1}, \lambda_{2j}) \cup (\lambda'_{2j-1}, \lambda'_{2j})]$ the equation has unstable solutions and for λ in the complementary open intervals, it has stable solutions. These results are useful when applied to the Mathieu functions arising in problems in oscillation theory.

Hausdorff dimension For the Hausdorff measures m_p and A a Borel set in the metric space X, there exists a unique p_0 such that

$$m_p(A) = \infty \quad \text{if} \quad p < p_0$$

and

$$m_p(A) = 0 \quad \text{if} \quad p > p_0.$$

(*See* Hausdorff measure.) This unique p_0 is the *Hausdorff dimension* of A. This coincides with the natural concept of dimension for nice sets in Euclidean spaces.

Hausdorff measure A collective name for a class of measures on a metric space X defined in the following manner: For $l = l(A)$ a non-negative function defined on a class of open sets in X, let $\lambda(B, \varepsilon) = \inf\{\Sigma l(A_i) : B \subset \cup_i A_i, l(A_i) < \varepsilon\}$. The *Hausdorff measure* $\lambda(B)$, of B, is defined by

$$\lambda(B) = \lim_{\varepsilon \to 0} \lambda(B, \varepsilon).$$

Most frequently, l is taken to be a function of the diameter defined on the class of open balls in X. In the most important examples, we have $l_p(A) = \text{diam}(A)^p, 0 < p < \infty$, to yield the Hausdorff measures m_p. *Hausdorff measures* appear in many different parts of analysis related to capacity, minimal surfaces, fractals, etc..

Hausdorff moment problem In one dimension, to find necessary and sufficient conditions, given a sequence $\{\mu_n\}$, for there to exist a measure μ with $\mu_n = \int_0^1 t^n d\mu(t), n = 0, 1 \ldots$ This extends to k dimensions in the obvious manner, using multi-indices and multiple integrals on $[0, 1]^k$.

Hausdorff-Young Theorem Let $1 \leq p \leq 2$ and $f \in L^p(\mathbf{T})$, $\mathbf{T}$ the unit circle. Then $\sum |\hat{f}(j)|^q < \infty$, where $1/p + 1/q = 1$. In fact

$$(\sum |\hat{f}(j)|^q)^{1/q} \leq \|f\|_{L^p}.$$

Heaviside's function A seemingly innocuous function defined by

$$H(x) = \begin{cases} 1 & \text{if} \quad x \geq 0 \\ 0 & \text{if} \quad x < 0. \end{cases}$$

This corresponds to the case of a constant unit input or output, in a system beginning at time 0. Obviously $H'(x) = 0$ if $x \neq 0$. This function has played a central role in motivating the theory of distributions, in which $H'(x) = \delta_0$, the Dirac delta function.

Heine's series A generalization of the hypergeometric series defined by

$$\phi(a, b, c; q, z) = 1 + \frac{(1-q^a)(1-q^b)}{(1-q)(1-q^c)} q^z + \frac{(1-q^a)(1-q^{a+1})(1-q^b)(1-q^{b+1})}{(1-q)(1-q^2)(1-q^c)(1-q^{c+1})} q^{2z} + \ldots$$

Taking $q = 1 + \varepsilon, z = \frac{1}{\varepsilon} \log x$, and letting ε tend to 0, we obtain the hypergeometric series in the limit.

helicoidal surface A surface generated by a curve simultaneously rotated about a straight line with a constant angular velocity and also translated in the direction of the line at a constant linear velocity.

helix A curve in three-dimensional space on the surface of a cylinder that intersects all generators of the cylinder at constant angles. Often the circular case, i.e., the spiral staircase, is simply referred to as a *helix* and the cases of more general cylinders are called generalized helices. Occasionally, the analogous curve on the surface of a cone is also called a helix. All the above are special cases of curves of constant slope.

Helly's Theorem If $\sigma(x)$ and $\sigma_n(x), n = 1, 2, \ldots$, are functions of bounded variation, whose total variations are uniformly bounded and $\lim \sigma_n(x) = \sigma(x)$ at every point of continuity of σ, then, for the Riemann-Stieltjes integrals,

$$\lim_{n \to \infty} \int_a^b f(x) d\sigma_n(x) = \int_a^b f(x) d\sigma(x),$$

for all continuous functions f. There is also a theorem of Helly on intersections of convex sets and also the Helly Selection Theorem, concerning convergence of sequences of monotone functions, both of which are sometimes referred to as *Helly's Theorem.*

Helmholtz differential equation The partial differential equation $(\Delta + k^2)u(x) = 0$, where Δ is the Laplacian operator, $\Delta u = (u_{11} + u_{22} + \ldots + u_{nn})$. A generalization of Laplace's equation, this equation occurs, among other places, in the solution of higher dimensional heat and wave equations. It is usually accompanied by boundary conditions on u and can be solved explicitly via Fourier series, in the cases of simple geometries for the domain of u.

Helson set A compact subset K of a locally compact Abelian group G with a thinness condition. Specifically, K is a *Helson set* if there is a constant C such that the norm of μ is bounded by C times the sup norm of its Fourier transform, for all regular Borel measures μ on K, i.e., $\|\mu\| \leq C\|\hat{\mu}\|_{\infty}$. Here the Fourier transform is considered as a function on the dual group of G. There are several other equivalent conditions for a Helson set.

Herglotz integral representation A representation (based on the Poisson integral representation) for functions $f(z)$, holomorphic in $|z| < 1$ and having $0 \leq \Re(f)$ and $f(0) = 1$, given by

$$f(z) = \int_0^{2\pi} \frac{e^{it} + z}{e^{it} - z} dm(t), \quad |z| < 1.$$

Here m is a probability measure on $[0, 2\pi]$. The representation is often attributed to M. Riesz.

Herglotz Theorem Let $\{a_n\}$ be a positive definite sequence of complex numbers, i.e., $\Sigma a_{j-k}\zeta_j\bar{\zeta}_k \geq 0$, for any $\zeta_1, \ldots, \zeta_n \in \mathbf{C}$. Then there is a monotonically increasing bounded function $\alpha(t)$ on $(-\pi, \pi]$, such that $a_n = \int_{-\pi}^{\pi} e^{int} d\alpha(t)$, $n = 0, 1, \ldots$ The converse is clearly true (a sequence with such an integral representation is positive definite).

Hermite polynomial The polynomials, orthogonal on $(-\infty, \infty)$, with respect to the weight function $w(x) = e^{-x^2}$. The nth *Hermite polynomial* satisfies the Hermite equation $y'' + y' + 2ny = 0$ and can be given by the formula

$$H_n(x) = (-1)^n e^{x^2} \frac{d^n}{dx^n}(e^{-x^2}).$$

See orthogonal polynomials.

Hermitian form A mapping $f(x, y)$ from $L \times L$ into $\mathbf{C}$, where L is a vector space, such that, for each $x, y \in L$, $f_x(y) = f(x, y)$ is antilinear and $f^y(x) = f(x, y)$ is linear and $f(x, y) = \overline{f(y, x)}$. These are the properties necessary to define an inner product by $(x, y) = f(x, y)$.

Hermitian inner product *See* inner product.

Hermitian kernel A complex-valued function $K(x, y)$ which is square integrable on $[a, b] \times [a, b]$ and satisfies $K(x, y) = \overline{K(y, x)}$ for almost all $(x, y) \in [a, b] \times [a, b]$.

Hermitian linear space A vector space on which a Hermitian linear product is defined; generally referred to simply as an inner product space, a pre-Hilbert space or, sometimes, as a vector space if the inner product is understood.

Hermitian metric A metric on a Hermitian linear space generated by its inner product, i.e., $d(x, y) = (x - y, x - y)^{\frac{1}{2}}$. This is the standard definition of distance in a vector space.

higher plane curve A somewhat archaic term for a curve in the xy-plane that is the graph of a two-variable polynomial $p(x, y)$ of degree strictly greater than 2, as compared with a conic section which is the graph of a polynomial of degree 2 or less.

higher-order derivative A derivative of order greater than or equal to 2, where a function $f(x)$ is successively differentiated n times to yield a derivative of order n, defined inductively by $f^{(n)}(x) = \frac{d}{dx} f^{(n-1)}(x)$.

Hilbert manifold A Banach manifold, based on a Banach space which is also a Hilbert space. *See* Banach manifold.

Hilbert matrix The infinite Hankel matrix $H = (\frac{1}{1+i+j})$, $i, j = 0, 1, \ldots$. That is, the matrix

$$H = \begin{pmatrix} 1 & \frac{1}{2} & \frac{1}{3} & \cdots \\ \frac{1}{2} & \frac{1}{3} & \frac{1}{4} & \cdots \\ \frac{1}{3} & \frac{1}{4} & \frac{1}{5} & \cdots \\ \cdot & \cdot & \cdot & \end{pmatrix}.$$

See Hankel matrix.

Hilbert modular form A holomorphic function on π_+^n that is compatible with the action of the Hilbert modular group on π_+^n, analogous to the case of modular forms. Here π_+^n denotes the n-fold product of the upper half-plane. *See also* modular form.

Hilbert modular function A function meromorphic on π_+^n, invariant under the Hilbert modular group. (*See* Hilbert modular group.) Here π_+^n denotes the n-fold product of the upper half-plane. *See also* modular function.

Hilbert modular group The group SL $(2, K) = \{2 \times 2$ matrices, whose entries are algebraic integers of K, with determinant $1\}$, where K is an algebraic number field of finite degree n over the rational number field. This can be considered as a group of transformations on π_+^n, the n-fold product of the upper half-plane. *See also* modular group.

Hilbert module A Hilbert space H, together with the action of a function algebra A on H, denoted $(a, x) \to ax$ for $a \in A$ and $x \in H$, satisfying

(i.) $(a + b)x = ax + bx$, $\quad a(bx) = (ab)x$,
(ii.) $a(x + y) = ax + by$, and
(iii.) the map $(a, x) \to ax$ is continuous from $A \times H$ to H.

A consequence of (iii.) is that

$$\|ax\| \leq C\|a\|\|x\|,$$

for all $a \in A$ and $x \in H$, for some constant C. If the constant C can be taken to be 1, we call the Hilbert module *contractive* and if H is similar to a contractive Hilbert module (if the Hilbert module $K = LHL^{-1}$ is contractive, for some bounded invertible operator L, where $aLx = Lax$ for the Hilbert module K), then H is called *cramped*.

Hilbert space A complete normed vector space H, i.e., a Banach space, in which the norm is given by an inner product. That is, the norm of an element $x \in H$ is given by $\|x\| = (x, x)^{\frac{1}{2}}$. This is the most natural infinite-dimensional generalization of Euclidean space, maintaining the concept of angle, and is the setting for a very wide variety of both theoretical and applied work.

Hilbert transform The singular integral operator defined by

$$Hf = \frac{1}{\pi}\int_{-\infty}^{\infty} \frac{f(x+y)}{y} dy.$$

Because of the singular nature of the integral, a limit procedure must be used to define H. Accordingly, one defines

$$Hf(x) = \frac{1}{\pi} \lim_{\epsilon \to 0, N \to \infty} \int_{\epsilon}^{N} \frac{f(x+t) - f(x-t)}{t} dt.$$

For $f \in L^1$, $Hf(x)$ exists a.e.. The notation $Hf(x) = \frac{1}{\pi} p.v. \int_{-\infty}^{\infty} \frac{f(x+t)}{t} dt$ indicates the limit process. This transform is useful in connection with Fourier transforms and studying relationships between real and imaginary parts of analytic functions.

Hilbert's problem Any one of 23 problems posed by David Hilbert (1862–1943) at the International Congress of Mathematicians, held in Paris, in 1900. Hilbert was generally considered to be the leading mathematician of his time and these problems, ranging over a wide spectrum of mathematical endeavor, served as a stimulus and guideline for much subsequent mathematical research.

The term is also used to refer to a generalized isoperimetric problem in the calculus of variations, namely, that of finding the curve of fixed length bounding a maximum area, subject to an auxiliary condition that the variation of some functional stay fixed; the constraint also involves some differential equations.

Various other problems are associated with Hilbert's name, but they usually contain an additional noun. Hilbert's problem can refer to the problem of finding a sectionally holomorphic function $F(z)$, defined on the set Γ, the complex plane $\mathbb{C}$ cut along a set of arcs F_1, F_2. F is required to be of order z^n at ∞ and at each z in Γ, the limiting values of $F(z)$ from opposite sides of F_1 and F_2, satisfy $F_1(z) - aF_2(z) = f(z)$, for some fixed $f(z)$ and some constant a.

Hilbert-Schmidt Expansion Theorem An eigenfunction expansion theorem for a class of operators on an L^2space. Let

$$(Tf)(x) = \int_X K(x, y)f(y)d\mu(y), \quad f \in L^2(X, \mu),$$

where $K \in L^2(X \times X, \mu \times \mu)$ and $K(x, y) = \overline{K(y, x)}$, for almost all (x, y). Then

$$(Tf)(x) = \Sigma_n \lambda_n (f, \phi_n)\phi_n$$

and $K(x, y) = \Sigma \lambda_n \phi_n(x)\phi_n(y)$, where $\{\phi_n\}$ is an orthonormal system of eigenfunctions of T corresponding to eigenvalues $\lambda_n \neq 0$.

Hilbert-Schmidt kernel A measurable function $K(x, y)$, such that the integral operator

$$(Tf)(x) = \int_X K(x, y)f(y)d\mu(y)$$

is a Hilbert-Schmidt operator on $L^2(\mu)$. *See* Hilbert-Schmidt operator. An equivalent definition to this is that $K(x, y) \in L^2(X \times X, \mu \times \mu)$.

Hilbert-Schmidt operator An operator T on a Hilbert space H such that T is compact and, if $\lambda_1, \lambda_2, \ldots$ are the eigenvalues of $(TT^*)^{\frac{1}{2}}$, then $\sum \lambda_n^2 < \infty$.

Hill's determinant An infinite determinant occurring in the solution of Hill's differential equation involving the Fourier coefficients of the coefficient function. *See* Hill's differential equation. Equating this determinant to 0 yields Hill's determinantal equation. Simplifying this equation is the key step in Hill's method of solution.

Hill's differential equation An equation of the form $u'' + F(x)u = 0$, with $F(x + 2\pi) = F(x)$. Note that this includes Mathieu's equation as a special case. *See* Mathieu's differential equation.

Hill's functions Stable solutions of Hill's equation satisfying normalized initial conditions. *See* Hill's differential equation.

Hodge manifold A compact manifold, necessarily a Kahler manifold, having a Hodge metric. *See* Hodge metric. An important result is that a compact, complex manifold is projective algebraic if and only if it is a Hodge manifold. (*See also* Kahler manifold.)

Hodge metric A Kahler metric g on a compact manifold, satisfying the additional condition that the associated fundamental form for g be an integral differential form, i.e., that the integral of the fundamental form on a 2-cycle with integral coefficients be an integer. (*See also* Kahler metric.)

Hölder condition A function $f(x)$ is said to satisfy the *Hölder condition of order p* if $|f(x) - f(y)| \leq M|x - y|^p$, for some constant M. Also known as the *Lipschitz condition*, this can obviously be extended to more general metric spaces.

Hölder inequality If μ is a positive measure on X, $f \in L^p(\mu)$ and $g \in L^q(\mu)$ $\frac{1}{p} + \frac{1}{q} = 1$, $1 < p, q < \infty$, then $fg \in L^1(\mu)$ and

$$\int_X |fg| d\mu \leq \left(\int_X |f|^p d\mu\right)^{1/p} \left(\int_X |g|^q d\mu\right)^{1/q}.$$

If $p = 1$, the theorem is true with L^q replaced by L^∞ and with the final integral replaced by ess $\sup_{x \in X} |g(x)|$.

Holmgren-type theorem The *Holmgren Uniqueness Theorem* states that an initial-value problem involving a linear partial differential equation of first order with coefficients of class $\mathbf{C}^\omega$ has a unique $\mathbf{C}^1$ solution. Holmgren-type theorems deal with extensions of this result to cases not having $\mathbf{C}^\omega$ coefficients. Note that higher order equations can be reduced to first order.

holomorphic function *See* analytic function.

holomorphic hull For a compact set K contained in $\mathbf{C}^n$, the *holomorphic hull* $\hat{K}$, is the intersection over all f holomorphic on K of $\{z \in \mathbf{C}^n : |f(z)| \leq \sup |f(w)|, w \in K\}$.

holomorphic mapping A function f mapping an open set $G \subset \mathbf{C}^n$ into $\mathbf{C}^p$ such that $u \circ f$ is holomorphic for every continuous linear mapping $u : \mathbf{C}^p \to \mathbf{C}$. This can be generalized to the case of an open G contained in an analytic space or a complex manifold.

holomorphic modification An analytic space X is called a *holomorphic modification* of an analytic space Y if there exist two analytically thin sets (a technical condition for "small" sets) $M \subset X$ and $N \subset Y$ and a map ϕ taking $X \backslash M$ isomorphically onto $Y \backslash N$.

holomorphic part For the Laurent series expansion

$$f(z) = \sum_{j=1}^{\infty} c_{-j}(z-a)^{-j} + \sum_{j=0}^{\infty} c_j(z-a)^j,$$

defined in some annulus $r < |z-a| < R$, centered at the isolated singularity a, we call $\Sigma_0^\infty c_j(z-a)^j$ the holomorphic part of f at a. Note that such a Laurent series exists whenever f is holomorphic in such an annulus.

holomorphically complete domain A domain G in $\mathbf{C}^n$, which is equal to the common existence domain of the holomorphic extensions of all functions holomorphic on G. That is, there is no larger domain G_1 with the property that every holomorphic function on G extends holomorphically to G_1.

holomorphically convex domain A subset of $\mathbf{C}^n$ equal to its holomorphically convex hull. *See* polynomially convex region.

holonomic system A system of N particles in $\mathbf{R}^3$ whose $3N$ position coordinates at time t satisfy constraints of the form $f_j(x_1, \ldots, x_{3N}, t) = 0$, $j = 1, \ldots k$, with $f_j(x, t) \in \mathbf{C}^2$. These equations are the only constraints on the motion of the particles and do not explicitly contain any velocities or accelerations. However, by differentiating the constraints, we see that velocity and acceleration also influence the constraints.

homogeneous integral equation An integral equation in which the sum of all terms containing the unknown function is equal to 0. For example, an equation of the form

$$f(t) + \int_X f(x)k(x,t)d\mu(x) = 0.$$

Alternatively defined as an integral equation in which every scalar multiple of a solution is also a solution. The concepts are equivalent for linear equations.

homogeneous linear differential equation A differential equation of the form

$$y^{(n)} + p_1(x)y^{(n-1)} + \ldots + p_n(x)y = 0,$$

where the functions p_j are continuous and $y^{(j)} = d^j y/dx^j$. The term could also refer to a partial differential equation of the form $Ly = 0$, where L acts on a function y by multiplications and partial derivatives of various orders and satisfies

$$L(c_1y_1 + c_2y_2) = c_1Ly_1 + c_2Ly_2.$$

homology Homology theory is based on associating a certain sequence of groups to a topological space and using algebraic properties of these groups and their homomorphisms to study topological properties of the space and its mappings. There are several different types of homology theories. Homology is particularly relevant to questions concerning dimensions and degrees of mappings.

homothety A transformation of Euclidean space, depending upon a point O, such that each point P is mapped to a P' such that $\text{dist}(O, P') = k\text{dist}(O, P)$ for some real $k > 1$. The concept can be extended to more general contexts such as affine spaces.

horizontal space *See* vertical space.

horizontal vector field *See* vertical space.

Hörmander's Theorem Any one of several theorems in partial differential equations due to L. Hörmander. The most common reference is to an existence theorem. Let $P(D)$ be a partial differential operator with constant coefficients. The solutions of $P(D)u = 0$ are all $\mathbf{C}^\infty$ if and only if $P(D)$ is hypoelliptic, i.e., if $P(\zeta + i\eta) = 0$ and $|\zeta + i\eta| \to \infty$, then $|\eta| \to \infty$.

hyperbola A conic section equal to the intersection of a double cone and a plane parallel to the axes. Equivalently, it can be defined as the set of all points, the difference of whose distances from two fixed points (the *foci*) is constant. An example is the graph of $x^2 - y^2 = 1$.

hyperbolic differential operator A linear partial differential operator in the variables $t, x_1, \ldots, x_n$ of second order.

$$L(u) = u_{tt} - \sum a_{0i}u_{ti} - \sum a_{ij}u_{ij} - a_0u_t - \sum a_iu_i - au,$$

with coefficients $a_{ij}(t, x)$, is called *hyperbolic* if

$$H(t, x, \lambda, \zeta) = \lambda^2 - \sum a_{0i}\lambda\zeta_i - \sum a_{ij}\zeta_i\zeta_j$$

has two distinct real roots $\lambda_1(t, x, \zeta)$ and $\lambda_2(t, x, \zeta)$ for any $\zeta = (\zeta_1, \ldots, \zeta_n)$. The classic example is the wave equation, for which the above condition simplifies greatly. For higher order equations, there are several different definitions of hyperbolic; the concept is most often defined in terms of uniqueness of solutions of Cauchy problems.

hyperbolic equation A partial differential equation of the form $L(u) = f(t, x_1, \ldots, x_n)$, where L is a hyperbolic differential operator. *See* hyperbolic differential operator.

hyperbolic function Any of the six real or complex functions sinh, cosh, tanh, coth, sech, or csch defined by

$$sinh(x) = \frac{e^x - e^{-x}}{2},$$
$$cosh(x) = \frac{e^x + e^{-x}}{2},$$
$$tanh(x) = \frac{sinh(x)}{cosh(x)},$$
$$coth(x) = \frac{1}{tanh(x)},$$
$$sech(x) = \frac{1}{cosh(x)},$$
$$csch(x) = \frac{1}{sinh(x)}.$$

They are related to hyperbolas, as the trigonometric functions are related to circles and the multitudinous trigonometric identities each tend to hold in the corresponding hyperbolic case, except for some occasional minus signs. A further connection to the trigonometric functions is given in the complex case by $sinh(ix) = i\sin(x)$ and $cosh(ix) = \cos(x)$. The hyperbolic functions appear in a wide variety of analysis problems, especially involving differential equations.

hyperbolic paraboloid A quadric surface in $\mathbf{R}^3$, two of whose coordinate cross-sections are parabolas and the third of which is a hyperbola. The simplest example is the graph of $z = x^2 - y^2$, whose critical point at the origin is the classic example of a saddle point.

hyperbolic point A point on a surface where the Gaussian curvature is strictly negative, which is equivalent to the condition that the approximating quadric surface at that point be a hyperbolic paraboloid.

hyperbolic spiral A plane curve such that, in polar coordinates, each point has a radius, inversely proportional to the polar angle; that is, $r = \frac{k}{\theta}$ is the polar equation of the curve.

hyperboloid A quadric surface in $\mathbf{R}^3$ such that the cross-sections in two coordinate directions are hyperbolas and, in the third, ellipses. For example, the graph of $\frac{x^2}{a^2} + \frac{y^2}{b^2} - \frac{z^2}{c^2} = 1$. A *hyperboloid* can consist of either one or two sheets accordingly as the hyperbolas do not or do intersect the axis of the hyperboloid.

hyperelliptic integral An integral of the form $\int_{z_0}^{z_1} R(z, w)dz$, where R is a rational function of the variables z and w, which are related by $w^2 = P(z)$, where P is a polynomial of degree at least 5, having no multiple roots.

hyperelliptic Riemann surface A Riemann surface arising as a two-sheeted covering surface for the function $F(z, w) = w^2 - P(z)$. (*See* hyperelliptic integral.) An alternative definition is a Riemann surface of a function having the form $f(z) = (z - a_1)^{\frac{1}{2}} \dots (z - a_n)^{\frac{1}{2}}$ for distinct complex numbers $a_1, \dots, a_n$.

hyperfunction A member of the largest class of generalized functions that are boundary values on $\mathbf{R}$ of holomorphic functions. *Hyperfunctions* can equivalently be considered as locally finite germs of continuous linear functionals defined on real-analytic functions or formal derivatives of continuous functions of infinite order differential operators of local type.

hypergeometric differential equation An ordinary differential equation of the form

$$z(1-z)w''+(\gamma-(\alpha+\beta+1)z)w'-\alpha\beta w = 0,$$

for α, β, γ constants, $w = w(z)$ and $' = d/dz$.

hypergeometric function An analytic continuation of the hypergeometric series that is single-valued in some domain in $\mathbf{C}$. (*See* hypergeometric series.) It is a solution of the hypergeometric differential equation. (*See* hypergeometric differential equation.) These functions form a well-known class of special functions with a wide variety of applications and can be generalized in many different contexts.

hypergeometric function of confluent type A solution of the differential equation

$$z^2w'' + (\gamma - (\alpha + \beta)z)w' - \alpha\beta w = 0.$$

This corresponds to the hypergeometric differential equation with the regular singular point considered to be at ∞, i.e., a confluence of singularities. Its solutions, hypergeometric functions of confluent type, can be expressed as

$$F(\alpha, \gamma, z) = \sum \frac{\alpha}{n!\gamma} \frac{\alpha+1}{\gamma+1} \cdots \frac{\alpha+n-1}{\gamma+n-1} z^n,$$

for $|z| < \infty$.

hypergeometric integral An integral over a suitable path C in the complex plane of the form

$$\int_C t^{\alpha-\gamma}(t-1)^{\gamma-\beta-1}(t-x)^{-\alpha}dt$$

that is used for expressing hypergeometric functions.

hypergeometric series Also called the Gauss series, a solution of the form

$$w(z) = \frac{\Gamma(\gamma)\Gamma(\beta)}{\Gamma(\alpha)} \sum \frac{\Gamma(\alpha+n)\Gamma(\beta+n)}{n!\Gamma(\gamma+n)} z^n$$

for the hypergeometric differential equation. (*See* hypergeometric differential equation). It converges in $|z| < 1$, for all α, β, γ.

hyperinvariant subspace A closed linear subspace, most often in a Hilbert space or Banach space, that is invariant for (i.e., mapped into itself by) all operators commuting with a given linear operator.

hypersurface A higher dimensional analogue of a standard surface in three dimensions. That is, a subset of $n+1$-dimensional space, for $n > 2$, consisting of points $(x_1, \ldots, x_n, x_{n+1})$, such that $x_i = f_i(u_1, \ldots, u_n)$, $i = 1, \ldots, n+1$, where the f_i are continuous functions defined on a common n-dimensional domain.

hypersurface element For $x = (x_1, \ldots, x_n)$ and $y \in \mathbf{R}$, the point (y, x), together with the n-dimensional hyperplane $y^* - y = \sum c_j(x_j^* - x_j)$ in $\mathbf{R}^{n+1}$, for constants $c_1, \ldots, c_n$.

hypocycloid A curve traced in the plane by a fixed point on a circle that rolls without slipping on the inside of another circle.

hypoelliptic differential operator A differential operator $P(D)$, ordinary or partial, with constant coefficients such that, if $P(\zeta + i\eta) = 0$ and $|\zeta + i\eta| \to \infty$, then $|\eta| \to \infty$. More generally, if $P(D) = \sum a_\alpha(x)D^\alpha$, where D is either an ordinary or partial derivative and α is an integer or multi-index, respectively, of order at most m, having $\mathbf{C}^\infty$ coefficients, then $P(D)$ is called *hypoelliptic* if the solutions of $P(D)u = g(x)$ are of class $\mathbf{C}^\infty$, for all $\mathbf{C}^\infty$functions $g(x)$. (*See also* Hörmander's Theorem).

hypofunction A generalization of a function, with a range which includes infinite values, that arises in constructing solutions of the Dirichlet problem using Perron's method. *See* Dirichlet problem, Perron's method.

hyponormal operator A bounded linear operator T, defined on a Hilbert space, whose self-commutator $(T^*T - TT^*)$ is positive. This generalizes the concept of normal operator, for which the self-commutator is 0. An extensive theory for hyponormal operators has been developed, which includes spectral theory and singular integral models.

hypotenuse The side of a right triangle opposite the right angle; the *hypotenuse* is best known for its central role in the Pythagorean Theorem.

hypotrochoid A curve traced in the plane by a point rigidly attached to a circle, not at the center, where the circle rolls without slipping on the inside of a fixed circle.

I

ideal boundary For a compact Hausdorff space X, a set of the form $X_c \backslash X$, where X_c is a compactification of X, (i.e., a compact Hausdorff space containing X as a dense subset).

idempotent measure A regular, bounded measure μ on a locally compact, Abelian group G such that $\mu \star \mu = \mu$, where $\star$ denotes convolution of measures, defined by $\lambda \star \mu(E) = \int_G \lambda(E - y) d\mu(y)$. By duality, this is equivalent to the condition that the Fourier transform $\hat{\mu}$ be idempotent under multiplication, (i.e., $\hat{\mu}(\gamma) = 0$ or 1, for all $\gamma \in G$).

identity function A function mapping a set to itself assigning each element of its domain to itself as an image, (i.e., $f(x) = x$).

image (1.) The set of values assumed by a function $f : A \to B$; that is, $\{f(a) : a \in A\}$. (2.) The value assumed by a function at a point a, by a function $f : A \to B$ may be called the *image of a* (under f).

imaginary axis The y-axis in the complex plane. Namely, the vertical axis, under the standard representation of the complex numbers as points in the plane, via the Cartesian coordinate system, where (x, y) corresponds to the complex number $z = x + iy$, with the points on the vertical axis corresponding to the pure imaginary numbers.

imbedded submanifold The image, in a manifold M, of an imbedding $F : N \to M$ having a manifold N as domain. *See* imbedding.

imbedding A homeomorphism of a topological space M *into* a topological space N, so that M is homeomorphic to a subspace of N. If M and N are differentiable manifolds, an imbedding is also required to be differentiable. The term is also spelled *embedding.*

immersed submanifold A manifold M, together with an immersion of M into another manifold N. Frequently it is the case that M is a subset of N with a not necessarily compatible structure, and the immersion is the inclusion map. A famous result is that an n-dimensional manifold can be immersed in $\mathbf{R}^{2n-1}$ (and imbedded in $\mathbf{R}^{2n+1}$). *See also* Whitney's Theorem.

immersion A continuous mapping f from a topological space X to a topological space Y, for which each point $x \in X$ has a neighborhood mapped homeomorphically onto its image $f(U)$; hence, a local homeomorphism. In the case of differentiable manifolds, f is required to be differentiable and have maximal rank at each point $x \in X$.

implicit differentiation A process by which an implicitly defined function is differentiated via the chain rule without first solving for the function explicitly. For example, if y is a function of x, satisfying $y^4 + y - x - x^4 = 0$, implicit differentiation yields $dy/dx = (1 + 4x^3)/(1 + 4y^3)$, without having to solve for y in terms of x.

implicit function A function $y = f(x)$, defined by a relation of the form $F(x, y) = 0$. Often additional conditions, such as a value (x_0, y_0) through which $y = f(x)$ passes, is required to determine the function uniquely.

Implicit Function Theorem A theorem establishing the existence of a function $y = g(x_1, \ldots, x_n)$, from an implicit relation of the form $f(x_1, \ldots, x_n, y) = 0$. In its simpler case, it states that if $f(x_1, \ldots, x_n, y)$ is continuously differentiable on a domain $G \subset \mathbf{R}^{n+1}$ and if $f(P_0) = 0$ and the partial derivative $f_y(P_0) \neq 0$, then, in a neighborhood of $P_0 = (x^0, y^0)$ there exists a unique

0-8493-0320-6/00/$0.00+$.50

function $g(x_1, \ldots, x_n)$, such that

$$f(x_1, \ldots, x_n, g(x_1, \ldots, x_n)) = 0$$

and $g(x^0) = y^0$.

improper integral An expression, having the form of a definite integral, in which either the domain is unbounded, or the integrand is unbounded on the region of integration. The improper integral is defined to be a limit of appropriate definite integrals. For example,

$$\int_0^\infty \frac{1}{(1+x)^2}dx = \lim_{R\to\infty} \int_0^R \frac{1}{(1+x)^2}dx = 1.$$

inclination of a line The angle that a straight line makes with some fixed reference line. For lines in the xy-plane, the reference line is generally taken to be the x axis.

inclination of a plane The smallest of the dihedral angles that a plane makes with some fixed reference plane. Somewhat of an old-fashioned concept, the direction determined by the normal vector to the plane frequently is used to describe the inclination of a plane.

incomplete beta function The function defined by

$$B_x(a, b) = \int_0^x t^{a-1}(1-t)^{b-1}dt,$$

for $0 \le x < 1, a, b > 0$. $B_1(a, b) = B(a, b)$ is the standard beta function. B_x is frequently defined with a factor $\frac{1}{B(a,b)}$ in front of the integral. *See also* beta function.

incomplete gamma function The function defined by

$$\Gamma_x(m) = \int_0^x e^{-t}t^{m-1}dt,$$

for $0 \le x, 0 < m$. $\Gamma_\infty(m) = \Gamma(m)$ is the standard gamma function. Γ_x is sometimes defined with a factor of $\frac{1}{\Gamma(m)}$ in front of the integral. (*See also* gamma function.)

increasing function A real-valued function, defined on a set D of real numbers, such that, for any $x_1, x_2 \in D$ with $x_1 < x_2$, we have $f(x_1) < f(x_2)$. Often the condition $f(x_1) \le f(x_2)$ is used in place of the strict inequality above. The former case is called *strictly increasing* and the latter *nondecreasing*. The concept is easily extended to the case of a function mapping any directed set into a directed set.

increasing sequence A sequence $\{a_n\}$ of real numbers, such that $a_n < a_{n+1}$ for all n. As with the term *increasing function*, the inequality $a_n \le a_{n+1}$ is often used in place of strict inequality. Note that an increasing sequence can be defined as an increasing function defined on the positive, or nonnegative, integers. *See* increasing function.

increment A change in the argument or values of a function. For example, for a function $y = f(x)$, $x_2 - x_1 = \Delta x$ is called the *increment in* x and $f(x_2) - f(x_1) = y_2 - y_1 = \Delta y$ is the corresponding *increment in* y.

indefinite D-integral A generalization of the Lebesgue integral due to Denjoy. For a real-valued function f on $[a, b]$, if there exists a function F, of generalized absolute continuity in the restricted sense, whose approximate derivative $ADF(x) = f(x)$, a.e., then F is an *indefinite D-integral* for f. (*See* Denjoy integral.)

indefinite integral An integral $\int f(x)dx$, without limits of integration, defined up to an additive constant, often considered to be the whole class of functions $F(x) + C$, where $F' = f$ and C is an arbitrary constant. This can also be written as $F(x) = \int_a^x f(t)dt$, where a is an arbitrary real number. Also called *antiderivative* or *primitive function*.

index of an eigenvalue For a linear operator $T : X \to Y$, where X and Y are Banach spaces, the number

$$\nu(T, \lambda) = \dim Ker(T - \lambda I) - \dim Ker(T - \lambda I)^*.$$

The index ν is especially important when the operator $T - \lambda I$ is Fredholm. *See* Fredholm operator.

index set For a family of objects $\{a_\lambda : \lambda \in \Lambda\}$, Λ is called the *index set* for the family. Λ can be considered to be the domain of a function f, mapping Λ into $\{a_\lambda\}$ with $f(\lambda) = a_\lambda$. In the simplest case, Λ is the set of natural numbers, in which case, $\{a_\lambda\}$ is a sequence.

Index Theorem A theorem which establishes a relationship between the analytic index of a linear operator D mapping L_0 into L_1, where the analytic index

$$i_a(D) = \dim ker(D) - \dim coker(D),$$

and a topological index, namely some topological dimensions related to D, L_0 and L_1. The most famous index theorem is the Atiyah-Singer Index Theorem, involving elliptic pseudo-differential operators and sections of vector-bundles on differentiable manifolds.

indicial equation (1.) For the differential equation $x^r \frac{dy}{dx} = A(x)y$, $A(x)$ an $n \times n$ matrix function, analytic at 0, with $A(0) \neq 0$, the equation $\det(A(0) - \rho I) = 0$ is called the *indicial equation* at $x = 0$. The roots of the indicial equation determine the behavior of the solution of the above equation at the singular point $x = 0$.

(2.) If x_0 is a regular singular point for a linear nth order differential equation

$$p_n(x)\frac{d^n y}{dx^n} + p_{n-1}(x)\frac{d^{n-1} y}{dx^{n-1}} + \dots + p_0(x) = 0,$$

with polynomial coefficients, and if $y = (x - x_0)^r(a_0 + a_1(x - x_0) + \dots)$ is substituted in the equation, then the coefficient of the lowest power of $x - x_0$ is called the *indicial equation*. It is a polynomial in r of degree n. *See* regular singular point.

inferior limit *See* lower limit.

infinite determinant For an infinite matrix $A = (a_{ij})$, with $i, j = 1, 2, \dots$, a limit in the natural sense of the determinants of $A_n = (a_{ij})$, $i, j = 1, 2, \dots, n$. Another meaning involves compact linear operators of trace class, defined on an infinite-dimensional Hilbert space. For T a trace class operator, with eigenvalues $\{\lambda_j\}$, counted with multiplicities, the *infinite determinant* of $I + T$ is defined by $\det(I + T) = \prod(1 + \lambda_j)$. The product can be shown to be either finite or convergent; if T has no eigenvalues, the product is taken to be 1.

infinite sequence A function f, defined on the natural numbers, $\{1, 2, \dots\}$, into a given set, where we denote $f(n)$ by a_n. Usually, a sequence is thought of as a progression of terms $a_1, a_2, \dots$, or a countable set $\{a_n\}$, $n = 1, 2, \dots$, indexed by the natural numbers.

infinite series A formal expression of the form

$$\sum_{j=1}^{\infty} a_j = a_1 + a_2 + \dots,$$

generated by the sequence $a_1, a_2, \dots, a_n, \dots$ of real or complex numbers. The elements a_n are referred to as terms and the elements

$$s_n = (a_k + a_{k+1} + a_{k+2} + \dots + a_n)$$

are partial sums of the infinite series.

If the sequence $s_1, s_2, \dots$, is convergent, its limit is referred to as the sum of the infinite series. It is conventional to denote the sum of the series by the same expression or symbol as the series itself. For example, either

$$\sum_{k}^{\infty} a_n = \mathcal{M}$$

or $a_k + a_{k+1} + a_{k+2} + \dots + a_n + \dots = \mathcal{M}$ means that the series is convergent and its sum is $\mathcal{M}$.

The concept of series is not restricted to series of real or complex numbers. In abstract, a series can be generated by a sequence of the elements of a set X in which "addition" and "closeness" are defined.

infinite type power series space Let K be any field with a principal valuation v and uniformizer π. Taking v to be normalized, we have $v(\pi) = 1$. We denote by $\mathcal{O}_v$ the valuation ring, (π) the maximal ideal and $k = \mathcal{O}_v/(\pi)$ the residue class field. Let us take a set of representatives A for k in $\mathcal{O}_v$; thus each element of A is congruent (mod π) to exactly one element of k. For convenience we assume that $O \in A$, so that O is represented by itself.

Given $c \in K^*$, suppose that $v(c) = r$. Then $c\pi^{-r}$ is a unit, so there exists a unique $a_r \in A$ such that $c\pi^{-r} \equiv a_r \pmod{\pi}$; moreover, $a_r \neq 0$. It follows that $v(c - a_r\pi^r) \geq r + 1$, so there exists a unique $a_{r+1} \in A$ such that

$$(c - a_r\pi^r)\pi^{-r-1} \equiv a_{r+1} \pmod{\pi}.$$

Now an induction shows that for any $n \geq r$ we have a unique expression

$$c \equiv \sum_r^{\infty} a_i\pi^i$$

$$\text{where } a_i \in A,\ r = v(i),\ a_r \neq 0. \qquad (1)$$

Here r can be any integer, positive, negative or zero. From the definition of r as $v(c)$ it is clear that $r \geq 0$ precisely when $c \in \mathcal{O}_v$. The structure (field) of K is entirely determined by expression (1) for its elements, together with addition and multiplication tables for the transversal. By this we mean: given $a, a' \in A$, the elements $a + a'$ and aa' lie in $\mathcal{O}_v$ and so are congruent (mod π) to uniquely determined elements of A:

$$a + a' \equiv s, \quad aa' \equiv p \pmod{\pi},$$
$$\text{where } s, p \in A.$$

The space with the structure described above is an *infinite type power series space.*

infinite-dimensional linear space A vector space having an infinite linearly independent set of vectors.

infinite-dimensional manifold A manifold such that each point has a neighborhood that is homeomorphic to an open subset of an infinite-dimensional topological vector space. *See also* Banach manifold, Hilbert manifold.

infinitely differentiable function (1.) A function $f(x)$, of one variable, such that $\frac{d^n f}{dx^n}$ exists and is differentiable, for all n.
(2.) A real or complex-valued function $f(x_1, \ldots, x_n)$ such that the partial derivatives of f of all orders exist and are differentiable. The notation $\mathbf{C}^\infty(D)$ is used for the infinitely differentiable functions with domain D.

infinitesimal (1.) The differentials dx, dy were regarded as infinitely small, and often called *infinitesimal,* at an early stage of the development of calculus. Using this (not formally justified) idea, the tangent line to a curve was considered to be a line passing through the two points (x, y) and $(x + dx, y + dy)$ which were considered to be infinitely close to each other with the slope $\frac{dx}{dy}$. This ambiguity was resolved later, using the concept of limits, by mathematicians such as Cauchy, Riemann, and Weierstrass.
(2.) A function f together with a point a such that $\lim_{x\to a} f(x) = 0$. For example, $\Delta x = x - a$ is an infinitesimal.
(3.) Let $(x_1, \ldots, x_n)$ be the coordinates of a point in $\mathbf{R}^n$. Also, let $\mathrm{a}(t) = (a_1(t), \ldots, a_n(t))$, $\mathrm{a}(0) = \mathrm{e}$ be a differentiable curve in $\mathbf{R}^n$. The tangent vector $\alpha = \frac{d\mathrm{a}}{dt}|_{t=0}$ to $\mathrm{a}(t)$ at e is called the infinitesimal vector of $\mathrm{a}(t)$.

All vectors $\alpha \in \mathbf{R}^n$ are infinitesimal vectors of some differentiable curves, say $\mathrm{a}(t) = \alpha t$. Therefore, the set of infinitesimal vectors is an n-dimensional vector space.

infinitesimal generator For a semi-group $T(t)$ of bounded linear operators on a Banach space, satisfying $T(s+t) = T(t)T(s)$ and $T(0) = I$, the operator A, defined as the strong limit

$$A = \lim_{t \downarrow 0} \frac{T(t) - I}{t}.$$

infinitesimal transformation A group $\mathcal{G}$ of linear fractional transformations of $D = \{z : |z| < 1\}$ onto itself contains infinitesimal transformations if there exists a sequence $\{S_n\}$ of elements of $\mathcal{G}$ such that none of S_n is the identity mapping I in group $\mathcal{G}$ and $\lim_{n\to\infty} S_n(z) = z$ for all z in D.

infinitesimal wedge An open set $C \subseteq \mathbf{C}^n$ which is sitting in a set $U + i\Lambda$, where U is an open set in $\mathbf{R}^n$ and Λ is an open cone in the space $\mathbf{R}^n_y$ of the imaginary coordinates, having the property that for every *proper subcone* Λ' of Λ and for every $\epsilon > 0$ there exists $\delta > 0$ such that $\{x \in U : \text{dist}(x, \partial U) > \epsilon\} + i(\Lambda' \cap \{|y| < \delta\}) \subseteq C$. The set U is called the *edge* of the infinitesimal wedge C.

By a proper subcone of Λ we mean a subcone with the property that the intersection of its closure with the set $\{|y| = 1\}$ is contained in the interior of Λ.

infinity The symbols $+\infty$ and $-\infty$ by which one can extend the real number system $\mathbf{R}$. The extended real number system is obtained by adjoining these symbols to the set $\mathbf{R}$ and then extending the addition $+$ and multiplication $\cdot$ as in the following.

$$a + (+\infty) = (+\infty) + a = +\infty, \quad \text{for } a \in \mathbf{R};$$

$$a + (-\infty) = (-\infty) + a = -\infty, \quad \text{for } a \in \mathbf{R};$$

$$a \cdot (+\infty) = (+\infty) \cdot a = +\infty, \quad a > 0, a \in \mathbf{R};$$

$$a \cdot (-\infty) = (-\infty) \cdot a = -\infty, \quad a > 0, a \in \mathbf{R};$$

$$a \cdot (+\infty) = (+\infty) \cdot a = -\infty, \quad a < 0, a \in \mathbf{R};$$

$$a \cdot (-\infty) = (-\infty) \cdot a = +\infty, \quad a < 0, a \in \mathbf{R}.$$

This extension facilitates the study of behavior of real-valued functions of real variables in the intervals where the absolute value of either a variable or values of a function increase without bound.

However, any other extensions such as

$$(+\infty) - (+\infty); \quad \frac{+\infty}{+\infty}; \quad (+\infty)^0;$$

etc., would lead to some inconsistency in the system and hence should be left as an indeterminate form.

The extended complex plane is obtained by adjoining the symbol ∞ to the complex plane $\mathbf{C}$ and for every $\epsilon > 0$ the set $\{z : |z| > 1/\epsilon\}$ is considered as an open disk centered at infinity. This extension facilitates the study of behavior of functions of a complex variable in a region where the moduli of either a variable or values of a function increase without bound.

inflection point A function f defined on an open interval I containing a has an *inflection point* at a if f is a concave function on one of the intervals comprising $I\backslash\{a\}$ and convex on the other. If the second derivative f'' exists on $I\backslash\{a\}$, then a is a point of inflection of f if and only if $f''(x) > 0$ on one side of a while $f''(x) < 0$ on the other side.

infra-exponential growth The growth estimate

$$\forall \epsilon > 0 \; \exists M > 0 : \; |f(z)| \leq M e^{\epsilon |z|}$$

for a function $f : U \subseteq \mathbf{C} \longrightarrow \mathbf{C}$.

infrared divergence Total energy is finite but the total number of photons is not. This is a phenomenon occurring in radiation theory (quantum field theory).

inhomogeneous linear differential equation A linear differential equation that is not homogeneous. *See* linear ordinary differential equation.

initial function The non-zero function g in a linear differential equation of the form

$$p_1(x)\frac{d^n y}{dx^n} + p_{n-1}(x)\frac{d^{n-1}y}{dx^{n-1}} + \cdots + p_1(x)\frac{dy}{dx} + p_0(x) = g(x).$$

initial point of integration The initial point of the path (curve) γ associated with a line integral. That is, the point $\gamma(a)$, where $\gamma : [a, b] \to \mathbf{R}^2$. *See* line integral.

initial set The closed subspace M associated with a partial isometry $T : H_1 \to H_2$ such that $||Tx|| = ||x||$ for every x in M while $Tx = 0$ for x in $M^\perp$. *See* partially isometric operator.

initial value problem A differential equation together with a set of initial conditions.

An initial condition is a condition on the solution of a differential equation at one point (for example, $y(x_0) = y_0$, $y'(x_0) = y_1$, etc.), while a boundary condition is a condition at two or more points.

injectivity radius For a point p on a manifold M with Riemannian metric, the maximal radius $r_m(p)$ such that the disk $B_r(p)$ of radius r centered at p is a geodesic disk for all $r < r_m(p)$. Recall that $r_m(p)$, in general, is a function of p. Moreover, it is a positive number which, in the case of Euclidean n-space, equals $+\infty$.

inner area Let S be bounded set in the xy-plane contained in a closed rectangle $R = [a, b] \times [c, d]$. By a partition P imposed on R we mean a set of subrectangles $R_{ij} = [x_{i-1}, x_i] \times [y_{j-1}, y_j]$ obtained by taking two partitions $a = x_0 < x_1 < \ldots < x_m = b$; $c = y_0 < y_1 < \ldots < y_n = d$ of the intervals $[a, b]$; $[c, d]$, respectively, and drawing lines parallel to the coordinate axes through these partition points.

The partition P consists of mn subrectangles R_{ij}, $1 \le i \le m$; $1 \le j \le n$, such that some of these are subsets of S while others are not. Let $a(P)$ denote the total area of all $R_{ij} \subseteq S$. The *inner area* of S is denoted by $\underline{A}(S)$ and defined by

$$\underline{A}(S) = \sup_P a(P).$$

inner function An analytic function f on the open unit disc $U = \{z : |z| < 1\}$ for which $|f(z)| \le 1$ in U and $|f^*| = 1$, σ-a.e., on the unit circle $T = \{z : |z| = 1\}$. Here f^* is the radial limit function of f and σ denotes Lebesgue measure on T.

inner product A function $(\cdot, \cdot)$ from $\mathbf{V} \times \mathbf{V}$ to $\mathbf{F}$, where $\mathbf{V}$ is a vector space over a field $\mathbf{F}$ (= the real or complex numbers), that satisfies the following conditions for all vectors x, y, $z \in \mathbf{V}$ and for all scalars $\alpha, \beta \in \mathbf{F}$

(i) $(x, x) \ge 0$;
(ii) $(x, x) = 0$ if and only if $x = 0$;
(iii) $(x, y) = \overline{(y, x)}$;
(iv) $((\alpha x + \beta y), z) = \alpha(x, z) + \beta(y, z)$.

Also *scalar product, interior product.*

instantaneous rate of change The limit of the average rates of change of the values of a real-valued function $y = f(x)$ over all intervals in the domain which contain a specific point a, as their lengths tend to zero. In symbols,

$$\lim_{\Delta x \to 0} \frac{\Delta y}{\Delta x}, \quad \lim_{x \to a} \frac{f(x) - f(a)}{x - a},$$

or

$$\lim_{h \to 0} \frac{f(a + h) - f(a)}{h}.$$

integrability The property of being integrable.

integrable A function whose integral exists (finite or infinite). *See* Riemann integral, Lebesgue integral. *See also* integrable distribution.

integrable distribution A distribution $f_\lambda(t)$ which depends upon a parameter $\lambda \in [a, b]$ and is defined on some domain $\Omega \subseteq \mathbf{R}^n$

so that, for each testing function $\phi(t)$ whose support is contained in Ω, the ordinary function $g(\lambda)$ defined by $g(\lambda) = \int_\Omega f_\lambda(t)\phi(t)dt$ is integrable over the interval $[a, b]$.

integrable function *See* integrable.

integral (1.) A certain kind of linear functional, assigning a number to a function. *See* Riemann integral, Riemann-Stieltjes integral, Daniell integral, Lebesgue integral, line integral, surface integral.
(2.) Having an integer value, as in the *integral roots* of a polynomial.
(3.) The solution of a differential equation.
(4.) *See* integral function.

integral calculus The elementary study of the Riemann integral, its evaluation and its applications to geometrical and elementary physical problems.

integral character A character (the trace of a representation) on a complex group, which is holomorphic.

integral constant *See* constant of integration.

integral curvature A numerical invariant of a compact surface denoted by $\int\int K\alpha$, and equal to 2π times the Euler characteristic of the surface. Thus $\int\int K\alpha = 2\pi(v - e + t)$, where v, e, t, are the total number of vertices, the total number of edges, and the total number of triangles of a triangulation of the surface, respectively.

integral divisor *See* sheaf of divisors.

integral equation An equation involving an unknown function and its integral against a kernel or kernels. For example, the integral equation

$$\phi(x) = f(x) + \lambda \int_a^b k(x, y)\phi(y)\,dy$$

with the unknown function $\phi(x)$ and kernel $k(x, y)$.

See linear integral equation, Volterra integral equation of the first kind, Volterra integral equation of the second kind.

integral equation of Volterra type *See* Volterra integral equation of the first kind, Volterra integral equation of the second kind.

integral function An entire function (analytic in the complex plane **C**).

integral invariant An integral invariant (of order n), according to Poincairé, is an expression of the form

$$\int \cdots \int_D f(x_1, \ldots, x_n)\,dx_1 dx_2 \cdots dx_n,$$

where the integration is extended to any region D for which the expression has the property

$$\int \cdots \int_D f(x_1, \ldots, x_n)\,dx_1 dx_2 \cdots dx_n = \int \cdots \int_{D_t} f(x_1, \ldots, x_n) dx_1 dx_2 \cdots dx_n.$$

Here $D_t = g(D, t)$ is the region occupied at the instant t by the points which occupy the domain D at the instant $t = 0$.

The volume of an incompressible fluid is a simple dynamical interpretation of an integral invariant which was given, also, by Poincairé.

integral manifold Given a vector subbundle $\mathcal{V}$ of the tangent bundle TM of a manifold M, an *integral manifold* of $\mathcal{V}$ is a closed smooth submanifold N of M such that, for every point $p \in N$, $\mathcal{V}_p$ is equal to the tangent space to N at p.

integral of Cauchy type The integral

$$F(z) = \int_C \frac{f(s)}{s - z}\,ds,$$

where C is a contour and f is continuous, but not necessarily analytic, on C.

integral of function *See* integral.

integral operator An operator T acting on a suitable space of functions, measurable with respect to a measure space $(X, \mathcal{A}, \mu)$, defined by

$$(Tf)(x) = \int k(x, y) f(y)\, d\mu(y).$$

The function $k(x, y)$ is said to be the *kernel* of T.

An example is the operator A defined by

$$(Ax)(t) = \int_a^b k(t, s) x(s)\, ds,$$

where x is in the space $C[a, b]$ of real-valued continuous functions on $[a, b]$ and $k(t, s)$ is a real-valued continuous function on the square $a \le t \le b$, $a \le s \le b$.

integral transform An integral operator from a function space to itself, or to another function space. Usually an *integral transform* is viewed as a change of basis, and, if acting between Hilbert spaces, is often unitary. An example is the Laplace transform $L\{f(t)\} = \int_0^\infty e^{-st} f(t)\, dt$. Its kernel $k(s, t)$ is defined by $k(s, t) = 0$ for $t < 0$; $k(s, t) = e^{-st}$ for $t \ge 0$.

integrand The function being integrated. Hence, the function $f(\cdot)$ that appears between the integral sign $\int$ and $d(\cdot)$ in the symbolic expression of a definite or indefinite integral.

integration The operation of evaluating a definite or indefinite integral.

integration along a fiber Let X, Y be two real oriented differentiable manifolds of dimensions n and m, respectively, and $f : X \to Y$ be a submersion. For a given $x \in X$ and $y = f(x)$, let $T_y(Y)$ be the tangent space to Y at y, and $\bigwedge^k T_y(Y)$ $(k \le m)$ be a corresponding differential fiber bundle. Let ν be a C^∞ $(n - m + k)$-form on X with compact support. Giving each fiber $f^{-1}(y)$ the orientation induced by f from the orientation of Y, there exists a unique C^∞ k-form λ on Y such that for all $y \in Y$ and all k-vectors $z_y \in \bigwedge^k T_y(Y)$,

$$\langle \lambda(y), z_y \rangle = \int_{f^{-1}(y)} \eta_{z_y}(x),$$

where η_{z_y} is an $(n - m)$-form on $f^{-1}(y)$ which depends on ν and z_y. The k-form λ is called the integral of ν along the fibers of f.

integration by parts The process of evaluating an integral $\int f(x)\, dx$ or $\int_a^b f(x)\, dx$ where the integrand $f(x)$ has the form $f(x) = u(x)v'(x)$, by use of the identity

$$\int u(x)v'(x)\, dx = u(x)v(x) - \int v(x)u'(x)\, dx,$$

or

$$\int_a^b u(x)v'(x)\, dx = \big(u(b)v(b) - u(a)v(a)\big) - \int_a^b v(x)u'(x)\, dx,$$

respectively.

integration by substitution The process of evaluating an integral $\int f(x)\, dx$ by putting it into the form

$$\int f(u(x))u'(x)\, dx = \int f(u)\, du,$$

where $u = u(x)$ is a continuously differentiable function and f has an antiderivative on its range.

For a definite integral, the process of evaluating $\int_a^b f(x)\, dx$ by putting it into the form

$$\int_a^b f(u(x))u'(x)\, dx = \int_{u(a)}^{u(b)} f(u)\, du,$$

where u' is continuous on $[a, b]$ and the range of u is contained in the domain of f.

integrodifferential equation An equation which involves one (or more) unknown functions f, together with both derivatives and integrals. For example, the equation

$$L\frac{di}{dt} + Ri + \frac{1}{C}\int_0^t i(\tau)\,d\tau = e(t)$$

which involves the unknown function $i(t)$, and the equation

$$\frac{\partial f(\nu, x)}{\partial x} = -\mu f(\nu, x) + \int_0^\nu P(\nu, \tau) f(\tau, x)\,d\tau$$

which involves the unknown function $f(\nu, x)$.

integrodifferential equation of Fredholm type A generalization of the Fredholm integral equation, also involving derivatives. For example, the equation

$$P(\lambda)x''(\lambda) + Q(\lambda)x'(\lambda) + R(\lambda)x(\lambda) + \mu\int_a^b K(\lambda, t)x(t)\,dt = g(\lambda)$$

is a linear second order ordinary *integrodifferential equation of Fredholm type* with the unknown function $x(\lambda)$.

integrodifferential equation of Volterra type A generalization of the Volterra integral equation, also involving derivatives. For example, the equation

$$x'(\lambda) = f(\lambda, x(\lambda)) + \mu\int_a^\lambda K(\lambda, t, x(t))\,dt$$

is a nonlinear first order ordinary *integrodifferential equation of Volterra type* with the unknown function $x(\lambda)$.

intercept The point(s) where a curve or graph of a function in $\mathbf{R}^n$ crosses one of the axes. For the graph of $y = f(x)$ in $\mathbf{R}^2$, the *y-intercept* is the point $(0, f(0))$ and the *x-intercepts* are the points $(p, f(p))$ such that $f(p) = 0$.

interior of curve *See* Jordan Curve Theorem.

interior problem A geometric condition related to the famous Cauchy problem. This problem consists of finding a solution u of the equation

$$F(u) = \sum_{i,k} A_{ik}\frac{\partial^2 u}{\partial x_i \partial x_k} + \sum_i B_i \frac{\partial u}{\partial x_i} + Cu = f,$$

when the values of u and one of its first derivatives is given, at every point of a certain surface S. As the knowledge of any first derivative (u being assumed to be known all over S) is equivalent to the knowledge of any other one (provided the corresponding direction is not tangent to S), we shall assume that the derivative in question is the transversal one. When calculating the values of u in a certain region R of an m-dimensional space, it will be assumed that if, from any point a of R as vertex, we draw the characteristic conoid, one of its sheets will cut out a certain portion S_0 (finite in every direction) of S and together with S_0, be the boundary of a portion T of our space. This geometric condition is expressed by saying that we have to deal with the interior problem. Note that no interior problem exists for non-normal hyperbolic equations. Moreover the solution so obtained satisfies all required conditions. On the contrary, no exterior problem admits a solution for arbitrary regular data.

interior product (1.) The inner product on a vector space. *See* inner product.

(2.) Suppose F is a vector field on an oriented surface M with a Riemannian metric g given in each coordinate system (u_α, M_α) by (U_α, g_α). Then with respect to a positively oriented atlas $(u_\alpha, M_\alpha)_{\alpha\in A}$, the interior product of F with dM is the 1-form $i_F dM$ defined locally by

$$i_F dM = \omega = \sqrt{g_\alpha}(-\mu_\alpha^2 du_\alpha^1 + \mu_\alpha^1 du_\alpha^2),$$

where $F_\alpha = \sum_j \mu_\alpha^j e_j(u_\alpha)$, is well defined.

intermediate integral In solving a nonlinear partial differential equation of the second order

$$F(x, y, z, p, q, r, s, t) = 0,$$

it is often desirable to find a relation $u = \phi(v)$ in which ϕ is an arbitrary function and u, v are two functions of x, y, z, p, q such that the original differential equation can be solved by eliminating the arbitrary function ϕ. The existence of such a relation is not necessary; however, in case it exists, the relation

$$u = \phi(v) \quad \phi \text{ arbitrary},$$

is called an *intermediate integral* of the above equation.

It is known that equations of the second order for which intermediate integrals exist are of the general form

$$Rr + 2Ss + Tt + U(rt - s^2) = V,$$

where R, S, T, U, V, are functions of x, y, z, p, q. Moreover, among others, Monge's method is the one that can be used to construct an intermediate integral for the equation.

Intermediate Value Theorem If f is a continuous real-valued function on the interval $[a, b]$ and $f(a) \le k \le f(b)$, then there exists a number c, $a \le c \le b$ such that $f(c) = k$. In other words, the range of f is a connected set.

interpolating sequence A sequence $\{z_n\}$ of complex numbers in the open unit disk D with the property that, for any bounded sequence of complex numbers $\{w_m\}$, there exists a function f in the Banach algebra $H^\infty(D)$ of bounded analytic functions in D such that $f(z_m) = w_m$ for every m.

interpolation Extending the definition of a function by assigning a value *between* known values, as opposed to *outside* given values, which is *extrapolation*.

For example, if $f(1.51) = 2.1$ and $f(1.53) = 2.3$, one might interpolate $f(1.52) = 2.2$.

interpolation of operators *See* Riesz-Thorin Theorem.

interpolation problem The problem of selecting a function f from a given class of functions, such as polynomial functions, spline functions, trigonometric functions, etc., in such a way that the graph of $y = f(x)$ passes through a finite set of given points. That is, the problem of finding f such that $f(z_j) = w_j$, for finitely many given points $\{z_j\}$, $\{w_j\}$.

interval A set of real numbers of one of the following forms.
Open interval: $(a, b) = \{x : a < x < b\}$.
Closed interval: $[a, b] = \{x : a \le x \le b\}$.
Half-open interval: either $[a, b) = \{x : a \le x < b\}$ or $(a, b] = \{x : a < x \le b\}$.
Unbounded interval: $[a, \infty) = \{x : x \ge a\}$, $(a, \infty) = \{x : x > a\}$, $(-\infty, a] = \{x : x \le a\}$, or $(-\infty, a) = \{x : x < a\}$.

The infinite intervals $(-\infty, a]$ and $[a, \infty)$ [resp., $(-\infty, a)$, (a, ∞)] may also be called closed [resp., open] in agreement with the use of the words *closed* and *open* in topology.

Note that the symbol ∞, above, read as "infinity" does not denote any real number. *See* infinity.

invariance (**1.**) Invariance of Laplace's equation: Suppose $U = U(u, v)$ is a harmonic function in a domain G in the w-plane. Then U satisfies Laplace's equation $U_{uu} + U_{vv} = 0$, for $(u, v) \in G$. Now, if $w = f(z) = u(x, y) + iv(x, y)$ is a conformal mapping from a domain D in the z-plane onto G, then the composite function

$$V = U(f(x, y)) = U(u(x, y), v(x, y))$$

is harmonic in D and satisfies Laplace's equation $V_{xx} + V_{yy} = 0$, for $(x, y) \in D$.
(**2.**) Invariance of contour integrals: The contour integral of a continuous complex-valued function f along a contour C is defined using a parameterization $z = z(t)$, $a \le t \le b$, of C. However, the integral is invariant under certain changes in the parameterization. For

example, let $t = \psi(s)$, $c \leq s \leq d$, be a continuous real-valued function with the range $[a, b]$. Furthermore, let ψ have a continuous derivative ψ' with $\psi'(s) > 0$. Then, the new parameterization $z = z(\psi(s))$, $c \leq s \leq d$, will not change the result of the contour integral.

(3.) Invariance of arc lengths: The arc length of a smooth arc is also invariant under the same changes of parameterizations as for the contour integrals.

(4.) Translation-invariance: The property $\mu(E + \lambda) = \mu(E)$, for certain measures μ on a group, where $E + \lambda = \{x + \lambda : x \in E\}$ for $E \subseteq G$.

invariant subspace A closed linear subspace M of a Banach space X associated with a bounded linear operator $T : X \to X$ on X by $Tx \in M$ whenever x is in M.

Since, for any X and any $T : X \to X$, the subspaces $M = \{0\}$ and $M = X$ are invariant subspaces for T, they are called the *trivial* invariant subspaces.

inverse Fourier transform For a function $f \in L^1(\mathbf{R}^n)$, the function

$$\mathcal{F}^{-1}[f(t)](x) = \int_{\mathbf{R}^n} e^{-2\pi i(x,t)} f(t)\, dt,$$

$x \in \mathbf{R}^n$, where $x = (x_1, \ldots, x_n)$, $t = (t_1, \ldots, t_n)$, and $(x, t) = \sum_{i=1}^n x_i t_i$.

inverse function A relation $g = \{(y, x) : (x, y) \text{ is in } f\}$ associated with the injective function $f : X \to Y$ with the domain $D(f) = X$ and range $R(f) = Y$.

It is proved that such g is a function and is uniquely determined by f. The function g is denoted by f^{-1}.

The inverse function f^{-1} is characterized by $f^{-1}(f(x)) = x$ and $f(f^{-1}(y)) = y$.

inverse function element A point in the domain of the inverse of a function (or the range of the function).

inverse operator The linear operator T^{-1} associated with a one-to-one linear operator T from $D(T)$ onto $R(T)$ such that

$$T^{-1}Tx = x \text{ for } x \in D(T), \text{ and}$$
$$TT^{-1}y = y \text{ for } y \in R(T).$$

inverse relation The relation $S = \{(y, x) : (x, y) \in R\}$ associated with the relation $R \subseteq X \times Y = \{(x, y) : x \in, X, y \in Y\}$.

inverse transform The inverse operator of a transform. It is most advantageous when a specific formula (such as an integral formula) for the inverse transform can be found.

Example: Inverse Laplace transform: Suppose $F(s)$ is an analytic function throughout the complex s-plane and having poles only at $s_j = \alpha_j + \beta_j$, $j = 1, \ldots, n$. For $\alpha_0 > \alpha_j$, $j = 1, \ldots, n$, the function $f(t)$ defined by

$$f(t) = \frac{1}{2\pi i} \lim_{\beta \to \infty} \int_{\alpha_0 - i\beta}^{\alpha_0 + i\beta} e^{st} F(s)\, ds,$$

provided the limit exists, is the inverse Laplace transform of $F(s)$.

Recall that the Laplace transform $F(s)$ of $f(t)$ is defined by

$$F(s) = L\{f(t)\} = \int_0^\infty e^{-st} f(t)\, dt.$$

inverse trigonometric function An inverse function f^{-1}, where f is a trigonometric function, restricted to a subset S of its domain, so that it is one-to-one on S. For example, the function $\sin x$ is one-to-one on the interval $[\frac{\pi}{2}, \frac{3\pi}{2}]$ (onto $[-1, 1]$). Thus, $\sin x$ has an inverse function mapping $[-1, 1]$ to $[\frac{\pi}{2}, \frac{3\pi}{2}]$.

For each inverse trigonometric function f^{-1} there is an alternative notation which is used interchangeably with f^{-1} such as arcsin which is interchangeable with $\sin^{-1}$, arccos interchangeable with $\cos^{-1}$, etc.

inversion The transformation

$$w = \begin{cases} x/|x|^2 & \text{if } x \neq 0, \infty \\ \infty & \text{if } x = 0 \\ 0 & \text{if } x = \infty \end{cases}$$

of the topological space $\mathbf{R}^n \cup \{\infty\}$ (the one-point compactification of $\mathbf{R}^n$) relative to the unit sphere, $n \geq 2$.

This transformation takes a neighborhood of infinity onto a neighborhood of 0. However, it preserves harmonic functions only when $n = 2$. Note that when $n = 2$ this transformation is $w = \frac{1}{z}$ on $\mathbf{R}^2 = \mathbf{C}$.

inversion formula An algebraic expression for the inverse of a linear operator, usually an integral operator or other transform. *See* inverse transform.

For example, the formula for the Fourier series of an $L^2(0, 2\pi)$ function can be thought of as the *inversion formula* for the transform taking an L^2 function to its Fourier coefficients. The formula is

$$f(e^{it}) = \sum_{-\infty}^{\infty} c_n e^{int},$$

where

$$c_n = \frac{1}{2\pi} \int_{-\pi}^{\pi} f(e^{it}) e^{-int}\, dt,$$

$(n = 0, \pm 1, \pm 2, \dots)$, are the Fourier coefficients of $f \in L^2(0, 2\pi)$.

invertible jet An r-jet $u = j_x^r f$ from X into Y is invertible if there is an r-jet $v = j_y^r g$ from Y into X such that $u \circ v$ and $v \circ u$ are defined and equal to the jets of the identity maps I_X and I_Y, respectively.

Recall that if X, Y, Z are differential manifolds and $f : X \to Y$, $f : Y \to Z$ are two C^r maps, $r \geq 0$, with $y = f(x)$, then the jet $j_x^r(g \circ f)$ is called the composition of the two jets $j_x^r f$ and $j_y^r g$ and is denoted by $j_y^r g \circ j_x^r f$.

involute Suppose a line L moves tangent to a curve E from point Q_1 to point Q_2 on E. The locus of any fixed point P on L is an *involute* of E. The involute depends upon the point P, but all involutes are parallel.

An involute of a planar curve E is given by the parameterization

$$x(t) = \alpha(t) - \alpha'(t)\frac{S(t) + C}{S'(t)};$$

$$y(t) = \beta(t) - \beta'(t)\frac{S(t) + C}{S'(t)},$$

$a \leq t \leq b$, where C is a constant and E is parameterized by

$$\alpha = \alpha(t), \beta = \beta(t), a \leq t \leq b;$$

furthermore, it is assumed that $\alpha(t)$, $\beta(t)$ are differentiable functions and $S(t) = \int_a^t \sqrt{\alpha'^2(u) + \beta'^2(u)}\, du$.

See also involute evolute.

involute evolute The locus of centers of curvature of the points of a given curve is called the *evolute* of the curve. If a curve E is the evolute of a curve C, then C is an involute of E.

If C is a planar curve having parameterization

$$x = x(t), y = y(t), a \leq t \leq b,$$

where $x(t)$, $y(t)$ have continuous first- and second-order partial derivatives, then the evolute of C is given by

$$\begin{aligned} \alpha &= \alpha(t) \\ &\equiv x(t) - y'(t)\frac{x'^2(t) + y'^2(t)}{x'(t)y''(t) - x''(t)y'(t)}; \\ \beta &= \beta(t) \\ &\equiv y(t) - x'(t)\frac{x'^2(t) + y'^2(t)}{x'(t)y''(t) - x''(t)y'(t)}. \end{aligned}$$

involute of circle An involute of the circle $x = R\cos t$, $y = R\sin t$, $0 \leq t \leq 2\pi$ is given by

$$\begin{aligned} x(t) &= R(\cos t + t \sin t); \\ y(t) &= R(\sin t - t\cos t), 0 \leq t \leq 2\pi. \end{aligned}$$

involute of curve *See* involute, involute of circle.

irreducible component Each member of the family $\{X_\nu\}_{\nu \in M}$ of irreducible analytic subsets of a reduced complex space X having the properties

(i) $\{X_\nu\}$ is a locally finite covering of X;
(ii) $X_\nu \subseteq X_\mu$ for all $\nu \neq \mu$ in M.

Recall that X is a reduced complex space if it is reduced at every $x \in X$. Moreover, $A \subseteq X$ is an irreducible subset if it is an irreducible complex space furnished with its canonical reduced complex structure.

irreducible germ of analytic set Let Y be an analytic set in the analytic space X. The germ Y_y of Y at $y \in Y$ is irreducible if whenever Y_y is represented by the union of two germs Y'_y, Y''_y at y, then $Y_y = Y'_y$ or $Y_y = Y''_y$.

irregular boundary point *See* regular boundary point.

irregular singular point *See* regular singular point.

island An (apparently) off-the-edge subset of the Mandelbrot set, the set interior to all bounded orbits of 0 in the complex z-plane associated with the quadratic polynomials of the form $f(z) = z^2 + c$.

Despite its appearance, it is proved that those islands are actually connected to the whole set by thin "filaments."

isobaric polynomial The polynomial

$$P = P(x_0, \ldots, x_m, y_0, \ldots, y_n)$$

is *isobaric* in the variables x_i and y_j, of weight mn, whenever any term $cx_0^{i_0}x_1^{i_1} \cdots x_m^{i_m}y_0^{j_0} \cdots y_n^{j_n}$ of P satisfies the equation $\sum_{r=0}^{m} ri_r + \sum_{s=0}^{n} sj_s = mn$.

isolated point (1.) A point p of a set S in a topological space which has a neighborhood disjoint from $S \backslash \{p\}$.
(2.) A point p of a set S in a metric space lying at a positive distance from $S \backslash \{p\}$.

isolated singularity A point z_0 in the complex plane **C** at which a function f fails to be analytic but such that there is a deleted neighborhood $0 < |z - z_0| < \epsilon$ of z_0 in which f is analytic.

isometric *See* isometric mapping.

isometric mapping A distance preserving mapping; hence, a mapping $J : X_1 \rightarrow X_2$ between two metric spaces, such that $d(x, y) = d(Jx, Jy)$ for all $x, y \in X_1$.

isometric operator *See* isometry.

isometry A linear operator T between normed vector spaces, satisfying $\|Tx\| = \|x\|$, for all x.

isomorphism For sets bearing any type of algebraic structure, an isomorphism of that type of structure is a 1-to-1, and onto map that preserves the structure.
(1.) An isomorphism of vector spaces must be 1-to-1, onto and *linear;*
(2.) An isomorphism of metric spaces must be 1-to-1, onto and *isometric;*
(3.) An isomorphism of Banach spaces must be 1-to-1, onto, linear, and *norm-preserving.*

A Banach space isomorphism between two Hilbert spaces is automatically a Hilbert space isomorphism (i.e., it is inner-product preserving) by the polarization identity. *See* polarization identity.

isoperimetric A general term which is assigned to any optimization problem in which a class of competing curves must satisfy certain integral conditions of the form $\int_0^1 F(x, y, x', y')\,dt = l$ or $\int_0^L G(x, y, x', y')\,dt = H$.

isoperimetric problem To determine a curve, among all curves $\gamma(t) = (x(t), y(t))$ joining two given points a and b, that extremizes the integral $E = \int_{t_0}^{t_1} G(x, y, x', y')\,dt$ for which the integrals $\int_{t_0}^{t_1} F(x, y, x', y')\,dt$ are given a prescribed value.

Example 1: Find a simple closed contour among all simple closed contours C of a fixed

length in the (x, y)-plane that encloses the largest area. Recall that the arc length and the enclosed area of a simple closed contour C with parameterization $x = x(t)$, $y = y(t)$, $a \le t \le b$, can be expressed by
$\int_a^b \sqrt{x'^2 + y'^2}\, dt$ and $\frac{1}{2}\int_a^b (xy' - x'y)\, dt$, respectively.

Example 2: Determine the shape of a perfectly flexible rope of uniform density that hangs at rest with its end points fixed.

isothermal coordinates If $u = (u_1, u_2)$ are local coordinates on a manifold M such that u_1, u_2 are conjugate harmonic functions, then $u = (u_1, u_2)$ are called *isothermal coordinates*. Note that the line element in isothermal coordinates looks like

$$ds^2 = \frac{(du_1)^2 + (du_2)^2}{\lambda(u_1, u_2)}, \ \lambda(u_1, u_2) \ne 0.$$

isothermal parameters If M is a manifold with a Riemannian metric and carrying a structure Φ of class $\mathbf{C}^1$, then to every mapping $\phi \in \Phi$ with domain V there correspond three real-valued functions F, G, and H in $\phi(V)$, satisfying the conditions $F > 0$, $FH - G^2 > 0$. An *isothermal parameter system* is a mapping $\phi \in \Phi$ whose associated coefficients F, G, H satisfy $F = H$, $G = 0$.

iterated integral An integral of the form

$$\int_{a_1}^{b_1}\left\{\int_{a_2}^{b_2}\left\{\cdots\left\{\int_{a_n}^{b_n} f(x_1, x_2, \ldots, x_n)\, dx_1\right\} dx_2\right\}\cdots\right\} dx_n$$

which consists of n, $n > 1$, successive single-variable integrals. Fubini's Theorem gives conditions under which an n-dimensional integral can be written in this form.

iterated kernel **(1.)** The general term $k_{n+1}(x, t)$ of the iterated kernel associated with the solution of Volterra integral equation of the second kind, $\phi(x) = f(x) + \lambda \int_a^x k(x, y)\phi(y)\, dy$, is given by

$$k_{n+1}(x, t) = \int_t^x k(x, y) k_n(y, t)\, dy$$

$(n = 1, 2, \ldots)$, and $k_1(x, y) \equiv k(x, y)$.

(2.) The general term $k_{n+1}(x, t)$ of the iterated kernel associated with the approximation for the solution of Fredholm integral equation of the second kind, $\phi(x) = f(x) + \lambda \int_a^b k(x, y)\phi(y)\, dy$, is given by

$$k_{n+1}(x, t) = \int_a^b k(x, y) k_n(y, t)\, dy$$

$(n = 1, 2, \ldots)$, and $k_1(x, y) \equiv k(x, y)$.

iterative method of solving integral equations The method in which one starts with an approximation for the solution, say $\phi_0(x)$. Then substitutes $\phi_0(x)$ in the integral equation to get the next approximation, $\phi_1(x)$. If one continues this process successively, then under appropriate conditions the sequence $\{\phi_n(x)\}$ obtained would converge to a solution $\phi(x)$.

For example, in the case of the Volterra integral equation of the second kind $\phi(x) = f(x) + \int_0^x k(x, y)\phi(y)\, dy$, one can show that if $f(x)$ is continuous on an interval $0 \le x \le a$ and $k(x, y)$ is continuous on the square $0 \le x \le a$, $0 \le y \le a$, then the sequence $\phi_n(x) = \int_0^x k(x, y)\phi_{n-1}(y)\, dy$, $n \ge 1$, would converge to the solution $\phi(x)$.

J

Jackson polynomials For a continuous function $f(x)$ on $[0, 2\pi]$, the polynomials

$$J_N * f = c_N \int_0^{2\pi} \frac{\sin\frac{1}{2}N't}{\sin\frac{1}{2}t} f(t-x)dt,$$

where $N' = [\frac{N}{2}]+1$ and c_N is chosen so that the kernel J_N has L^1-norm 1.

Jacobi field Let $c = c(t)$, $0 \le t \le a$, be a geodesic, parameterized by arc length, $|\dot{c}(t)| = 1$, on a manifold M with Riemannian metric. A vector field $Y(t)$ along c is a *Jacobi field* if Y is orthogonal to c and

$$\frac{\nabla^2 Y}{dt^2}(t) + K \circ c(t)Y(t) = 0,$$

where K is the Gauss curvature.

Jacobi polynomial *See* orthogonal polynomials.

Jacobi's condition The necessary condition for a maximum or minimum of the second variation $\delta^2 J$, $J = \displaystyle\int_{x_0}^{x_1} F(x, y, y')\,dx$, which is that

$$r_1(x)r_2(x_0) - r_2(x)r_1(x_0) \neq 0$$

throughout the interval (x_0, x_1), where r_1, r_2 are two particular linearly independent integrals of the Jacobi's differential equation associated with J.

Jacobi's elliptic functions Recall that if ω_1, ω_3 are two complex numbers such that 0, ω_1, ω_3 are noncolinear, then the Weierstrass $\wp$ function $\wp(z)$, with fundamental periods $2\omega_1$ and $2\omega_3$, is an elliptic function of order 2. *See* Weierstrass $\wp$ function. Moreover, associated with $\omega_1, \omega_2 = \omega_1 + \omega_3$, and ω_3 there are constants μ_j, $j = 1, 2, 3$, satisfying the Legendre's relations: $2\omega_1\mu_2 - 2\omega_2\mu_1 = -\pi i$, $2\omega_2\mu_3 - 2\omega_3\mu_2 = -\pi i$, $2\omega_3\mu_1 - 2\omega_1\mu_3 = \pi i$.

Using the Weierstrass sigma function $\sigma(z)$ with fundamental periods $2\omega_1$ and $2\omega_3$, (*See* Weierstrass σ function) and the associated sigma functions $\sigma_j(z) = \frac{\sigma(z+\omega_j)}{\sigma(\omega_j)} e^{-\mu_j z}$, $j = 1, 2, 3$, the Jacobi's elliptic functions snz, cnz, and dnz are defined by

$$\text{sn}z = \sqrt{\wp(\omega_1) - \wp(\omega_3)}\, \frac{\sigma\left(\frac{z}{\sqrt{\wp(\omega_1)-\wp(\omega_3)}}\right)}{\sigma_3\left(\frac{z}{\sqrt{\wp(\omega_1)-\wp(\omega_3)}}\right)},$$

$$cnz = \frac{\sigma_1\left(\frac{z}{\sqrt{\wp(\omega_1)-\wp(\omega_3)}}\right)}{\sigma_3\left(\frac{z}{\sqrt{\wp(\omega_1)-\wp(\omega_3)}}\right)},$$

$$dnz = \frac{\sigma_2\left(\frac{z}{\sqrt{\wp(\omega_1)-\wp(\omega_3)}}\right)}{\sigma_3\left(\frac{z}{\sqrt{\wp(\omega_1)-\wp(\omega_3)}}\right)}.$$

The Jacobi's elliptic functions snz, cnz, and dnz are all elliptic functions of order 2 with simple poles $2m\mathcal{K} + (2n-1)i\mathcal{L}$, where

$$\mathcal{K} = \omega_1\sqrt{\wp(\omega_1) - \wp(\omega_3)}$$

and

$$\mathcal{L} = \omega_3\sqrt{\wp(\omega_1) - \wp(\omega_3)}.$$

Jacobi's imaginary transformation In the notation for Jacobi's elliptic functions sn(u), cn(u) and dn(u), indicate the modulus of the lattice, by writing $f(u, k)$, for a function corresponding to the normal lattice $2\Omega^*$, with modulus k. *See* Jacobi's elliptic functions. Recall that the lattice $2i\Omega^*$ is normal with modulus k'. Jacobi's *imaginary transformation* is given by

$$\text{sn}(iu, k) = i\,\text{sc}(u, k')$$
$$\text{sn}(u, k') = -i\,\text{sc}(iu, k)$$
$$\text{cn}(iu, k) = i\,\text{nc}(u, k')$$
$$\text{cn}(u, k') = -i\,\text{nc}(iu, k)$$
$$\text{dn}(iu, k) = i\,\text{dc}(u, k')$$
$$\text{dn}(u, k') = -i\,\text{dc}(iu, k).$$

0-8493-0320-6/00/$0.00+$.50

Jacobi's real transformation The transformation

$$\mathrm{sn}(u, \frac{1}{k}) = k\,\mathrm{sn}(\frac{u}{k}, k),$$
$$\mathrm{cn}(u, \frac{1}{k}) = \mathrm{dn}(\frac{u}{k}, k),$$
$$\mathrm{dn}(u, \frac{1}{k}) = \mathrm{cn}(\frac{u}{k}, k).$$

See Jacobi's imaginary transformation.

Jacobian determinant The determinant of the Jacobian matrix of a function $f : \mathbf{R}^n \to \mathbf{R}^n$, having coordinate functions $f_1, \ldots, f_n$. It is denoted by either $\frac{\partial(f_1, \ldots, f_n)}{\partial(x_1, \ldots, x_n)}$, or $\frac{\partial(y_1, \ldots, y_n)}{\partial(x_1, \ldots, x_n)}$.

See Jacobian matrix.

Jacobian matrix The matrix associated with a function $f : \mathbf{R}^n \to \mathbf{R}^m$, having coordinate functions $f_1, \ldots, f_m$, with the (i, j)-entry $\frac{\partial f_i}{\partial x_j}(x_0)$, the first-order partial derivative of f_i with respect to x_j, $1 \leq i \leq m$; $1 \leq j \leq n$, at a point x_0 in the domain of f:

$$\begin{bmatrix} \frac{\partial f_1}{\partial x_1}(x_0) & \frac{\partial f_1}{\partial x_2}(x_0) & \cdots & \frac{\partial f_1}{\partial x_n}(x_0) \\ \frac{\partial f_2}{\partial x_1}(x_0) & \frac{\partial f_2}{\partial x_2}(x_0) & \cdots & \frac{\partial f_2}{\partial x_n}(x_0) \\ \cdot & \cdot & \cdot & \\ \frac{\partial f_m}{\partial x_1}(x_0) & \frac{\partial f_m}{\partial x_2}(x_0) & \cdots & \frac{\partial f_m}{\partial x_n}(x_0) \end{bmatrix}.$$

Jacobian variety Let X be a compact Riemann surface of genus $g > 0$, marked by a canonical homology basis $a_1, \ldots, a_g : b_1, \ldots, b_g$ and let $\{dw_1, \ldots, dw_g\}$ be a basis for the space $\Omega(X)$ of holomorphic differentials on X dual to the above canonical homology basis. We then have

$$\int_{a_k} dw_j = \delta_{jk}, \quad j, k = 1, 2, \ldots, g.$$

Also, the entries of the B-period matrix τ are given by

$$\int_{b_k} dw_j = \tau_{jk}, \quad j, k = 1, 2, \ldots, g.$$

The Jacobian variety of X, denoted by $Jac(X)$, is defined by

$$Jac(X) = \mathbf{C}^g/(\mathbf{Z}^g + \tau\mathbf{Z}^g),$$

where $\mathbf{Z}^g + \tau\mathbf{Z}^g$ is the period lattice (over $\mathbf{Z}$). Here the elements $\underline{z} \in \mathbf{C}^g$ and $\underline{n} \in \mathbf{Z}^g$ are considered as column vectors.

Jensen's formula Let f be analytic on a closed disk $|z| \leq r$, $r > 0$, with $f(0) \neq 0$. If f does not vanish on the circle $|z| = r$ and has zeros $z_1, z_2, \ldots, z_n$ in the open disk $|z| < r$, repeated according to multiplicities, then

$$\log \frac{r^n}{|z_1 z_2 \cdots z_n|} = \frac{1}{2\pi} \int_0^{2\pi} \log \frac{|f(re^{i\theta})|}{|f(0)|}\, d\theta.$$

In particular, if f is non-vanishing on $|z| \leq r$, then

$$\log |f(0)| = \left(\frac{1}{2\pi} \int_0^{2\pi} \log |f(re^{i\theta})|\, d\theta \right).$$

jet Let X and Y be two $\mathbf{C}^r$ manifolds, $0 \leq r < \infty$. An *r-jet* from X into Y is an equivalence class $[x, f, U]_r$ of triples (x, f, U), where $U \subseteq X$ is an open set, $x \in U$, and $f : U \to Y$ is a C^r map. Here the equivalence relation is $[x, f, U]_r = [x', g, V]_r$ if $x = x'$ and in some (and hence any) pair of charts applied to f at x, f and g have the same derivatives up to order r. The point x is called the *source*, and $y = f(x)$ the *target* of $[x, f, U]_r$. When f is specified, $[x, f, U]_r$ is called the *r-jet* of f at x and is denoted by $j_x^r f$ or $j^r f(x)$.

John-Nirenberg space *See* bounded mean oscillation.

joint resolvent set *See* joint spectrum.

joint spectrum An attempt to generalize the notion of spectrum of an operator to two or more commuting operators. *See* spectrum.

One definition is as follows. Let $T_1, \dots, T_n$ be commuting operators on a normed linear space X, over the complex numbers. We say a point $(\lambda_1, \dots, \lambda_n)$ belongs to the *joint resolvent set* $\rho(T_1, \dots, T_n)$ if there exist bounded operators $S_1, \dots, S_n$, commuting with the operators $T_1, \dots, T_n$, such that

$$\sum_{j=1}^{n}(T_j - \lambda_j I)S_j = I,$$

where I is the identity on X. The *joint spectrum* is the complement, in the complex plane, of the joint resolvent set.

joint variation *See* covariance.

Jordan arc The image γ of a continuous mapping $z = (x(t), y(t))$ in the xy-plane defined on an interval $[a, b]$ such that $z(t_1) \neq z(t_2)$ when $t_1 \neq t_2$. The mapping $z = (x(t), y(t))$, $a \leq t \leq b$, is called a parameterization of γ. γ may also be referred to as *simple arc*.

Jordan content *See* set of Jordan content 0.

Jordan curve The image γ of a continuous mapping $z = (x(t), y(t))$ in the xy-plane defined on an interval $[a, b]$ such that $z(t_1) = z(t_2)$ (for $z_1 \neq z_2$) if and only if $\{z_1, z_2\} = \{a, b\}$. The term *simple closed curve* may also be used.

Jordan Curve Theorem The complement of any simple closed curve (Jordan curve) γ in the xy-plane consists of two disjoint planar domains D_1, D_2 such that D_1 is bounded and D_2 unbounded having the common boundary γ. The domain D_1 is said to be *interior* to γ and D_2 *exterior*.

Jordan Decomposition For a real measure μ, the decomposition

$$\mu = \mu^+ - \mu^-,$$

where $\mu^\pm$ are mutually singular positive measures. The decomposition is obtained by setting

$$\mu^+ = \frac{|\mu| + \mu}{2} \quad \mu^- = \frac{|\mu| - \mu}{2},$$

where $|\mu|$ is the total variation of μ. *See* total variation.

Jordan's test Let f be a complex-valued function in $L^1(\mathbf{R})$. If f is of bounded variation on some neighborhood of a point a, then

$$\lim_{N\to\infty} \frac{1}{2\pi} \sum_{|n|\leq N} \left(e^{ina} \int_{\mathbf{R}} f(x)e^{-inx}\, dx \right)$$
$$= \frac{1}{2}\left[\lim_{x\to a^+} f(x) + \lim_{x\to a^-} f(x) \right].$$

Julia direction A ray $Arg\, z = \theta_0$ in the complex plane associated with a function $f : U \subseteq \mathbf{C} \to \mathbf{C}$ such that, for any $\epsilon > 0$, the function f takes every finite value w, with at most n_0 possible exceptions, in the sector: $\theta_0 - \epsilon < Arg\, z < \theta_0 + \epsilon$, infinitely often. Such a function f is said to have a *Julia direction*.

For example, if f has an essential singularity at $z = 0$, then there exists a ray L emanating from the origin, called a Julia direction, such that f takes every finite value w, with one possible exception, infinitely often near L in every neighborhood of $z = 0$. Note that Picard's Theorem does not address the distribution of w-points of f in a neighborhood of an essential singularity.

Julia exceptional function Any meromorphic function with no Julia direction.

Evidently, Julia exceptional functions are among those meromorphic functions which fail to have asymptotic values. As an example, the function

$$f(z) = \frac{\Pi_{n=0}^{\infty}(1 - \frac{z}{2^n})}{\Pi_{m=0}^{\infty}(1 + \frac{z}{2^m})}$$

is a *Julia exceptional function*.

jump A discontinuity x_0 of a function $f(x)$ of a real variable, such that

$$\lim_{x \to x_0-} f(x) \quad \text{and} \quad \lim_{x \to x_0+} f(x)$$

exist and are unequal. For instance, a monotonic function has a *jump* at each point of its discontinuity. *See also* discontinuity point of the first kind.

K

K D V equation *See* Korteweg-deVries equation.

K-complete analytic space A complex space X in which, for every point $x_0 \in X$, there exists a finite number of holomorphic functions $f_1, \ldots, f_n$ on X such that x_0 is an isolated point of the set $\cap_{j=1}^{n}\{x \in X : f_j(x) = f_j(x_0)\}$.

k-valued algebroid function An analytic function f in a domain G satisfying an irreducible algebraic equation

$$A_0(z)f^k + A_1(z)f^{k-1} + \cdots + A_k(z) = 0,$$

where all $A_j(z)$ are single-valued meromorphic functions in the domain G in the complex z-plane.

Kahler manifold A complex manifold with a Kahler metric. *See* Kahler metric.

Kahler metric A Hermitian metric on an almost complex manifold such that the fundamental 2-form is closed.

Kelvin transformation A transformation by reciprocal radii. That is, if E is a given subset of $\mathbf{R}^n \setminus \{0\}$, we transform its elements x, in $(x_1, \ldots, x_n)$- coordinates, into elements x^*, in $(x_1^*, \ldots, x_n^*)$-coordinates, by means of the transformation

$$\mathrm{x}^* = \frac{\mathrm{x}}{|\mathrm{x}|^2}.$$

The *Kelvin transformation* transforms a neighborhood of the point at infinity into a finite domain near the origin. It plays a role in harmonic function theory. For example, Lord Kelvin observed that if U is a function harmonic in a neighborhood of infinity then its Kelvin transform $\mathcal{K}[U]$ defined by $\mathcal{K}[U](\mathrm{x}^*) = |\mathrm{x}|^{2-n}U(\mathrm{x})$ will be harmonic in the image domain near the origin of the $(x_1^*, \ldots, x_n^*)$- coordinates.

kernel (1.) The set $\ker T$ associated with a linear transformation $T : X \to Y$, between vector spaces, and defined by

$\ker T = \{x \in X : Tx = 0\}$.

(2.) A function $k(x, y)$ of two variables, used to define an integral operator

$$f \to \int k(x, y)f(y)d\mu(y)$$

on one of the spaces $L^p(\mu)$.

Many "kernels" bearing mathematicians names fall into this category. A number are delineated in the following list.

(3.) Aronszajn-Bergman reproducing kernel: the function $k(a, b) : A \times A \to \mathbf{C}$ associated with a Hilbert space X of complex-valued functions on a set A with the inner product (f, g) with the property that for any fixed b, $k(a, b) \in X$ is a function of a and $f(b) = (f(a), K(a, b))$ for all $f \in X$.

(4.) Cauchy kernel: the function $k(s, z) = \frac{1/2\pi i}{s-z}$ in the Cauchy representation of a function analytic in a region with boundary contour C:

$$f(z) = \int_C k(s, z)f(s)\,ds = \frac{1}{2\pi i}\int_C \frac{f(s)\,ds}{s-z}.$$

(5.) Dirichlet kernel: the function $D_n(t)$, $n = 0, 1, \ldots$, given by

$$D_n(t) = \sum_{k=-n}^{n} e^{ikt} = \frac{\sin[(n+1/2)t]}{\sin(t/2)}$$

$(t \in [-\pi, \pi])$.

(6.) Dirichlet conjugate kernel:

$$\frac{1}{2}\tilde{D}_n(t) = \sum_{k=1}^{n} \sin(kt) = \frac{\cos(\frac{1}{2}t) - \cos(n+\frac{1}{2})t}{2\sin(\frac{1}{2}t)}$$

0-8493-0320-6/00/$0.00+$.50

$t \in [-\pi, \pi]$).

(7.) Fejér kernel: the function $F_n(t)$, $n = 0, 1, \ldots$, given by

$$F_n(t) = \sum_{k=-n}^{n} (1 - \frac{|k|}{n+1}) e^{ikt} = \frac{1}{n+1} \left[\frac{\sin[(n+1/2)t]}{\sin(t/2)} \right]^2$$

($t \in [-\pi, \pi]$).

(8.) Hilbert kernel: the function $k(z, \alpha) = 2/[(x-\alpha)^2 + y^2]$, $z = x + iy$, in the solution to the Dirichlet problem in the upper half plane with an unknown harmonic function $f(z)$ having boundary values $f(\alpha)$ continuous on the real axis with $\lim_{|\alpha| \to \infty} f(\alpha) = f(\infty)$: $f(z) = \frac{1}{2\pi} \int_{-\infty}^{\infty} k(z, \alpha) f(\alpha)\, d\alpha$.

(9.) Kernel of an integral operator: the function $k : (X, \mathcal{A}, \mu) \times (X, \mathcal{A}, \mu) \to \mathbf{F}$ in the formula of an integral operator $T : L^2(X, \mathcal{A}, \mu) \to L^2(X, \mathcal{A}, \mu)$ given by

$$(Tf)(x) = \int k(x, y) f(y)\, d\mu(y).$$

Here $\mathbf{F}$ denotes either the real field, $\mathbf{R}$, or the complex field, $\mathbf{C}$.

(10.) Poisson kernel: the function P_r, $0 \le r < 1$, given by

$$P_r(x) = \frac{1 - r^2}{1 - 2r \cos x + r^2}$$

($x \in \mathbf{R}$).

kernel differential The *multiplicative Cauchy kernel differential* $\lambda_{q-q^*}(p)$ on a closed marked Riemann surface X is defined by $\lambda_{q-q^*}(p) = A(p, q)\, dz(p)$, where

(i.) For $q \neq q^*$ fixed $A(p, q)\, dz(p)$ is a meromorphic differential on X with the divisor $\mathcal{D} \ge -p - q^*$ and residue $+1$ at q;

(ii.) For $p \neq q^*$ fixed in local coordinates at p, the function $A_p(q) = A(p, q)$ has the form of an Abelian integral in its dependence on q with

$$\int_{a_j} d_q A(p, q)\, dz(p) = 0;$$

$$\int_{b_j} d_q A(p, q)\, dz(p) = 2\pi i \zeta_j,$$

where $\{\zeta_j\}$, $j = 1, \ldots, g$, is the basis for the space $\Omega(X)$ of all holomorphic differentials dual to the marking.

kernel function *See* kernel.

kernel representation Let Ω_j, $j = 1, 2$, denote simply connected domains in the z_j-plane and C_j, $j = 1, 2$, denote rectifiable simple curves, beginning at $t = -1$ and ending at $t = 1$, in the closed unit disk centered at the origin. For $\delta > 0$ the sets $N(C_j, \delta)$, $j = 1, 2$, are defined by $N(C_j, \delta) = \{t \in \mathbf{C} : \exists t^* \in C_j \text{s.t.} |t - t^*| < \delta\}$. For $\mu_j, \nu_j \in \mathbf{C}$, $j = 1, 2$, the sets $\Omega_j^* \subseteq \mathbf{C}$ are domains containing $\lambda_j = \mu_j + \frac{1}{2}(z_j - \nu_j)(1 - t^2)$ for all z_j in Ω_j and t in $N(C_j, \delta)$. Also, $D_j = \partial/\partial z_j$, $j = 1, 2$, $D_t = \partial/\partial t$, $L = D_1 D_2 + a_1(z) D_1 + a_2(z) D_2 + a_0(z)$, $L_j = D_1 D_2 + b_j(z) D_{(3-j)} + c_j(z)$, $j = 1, 2$. Moreover, for $j = 1, 2$, the functions $e_j(z)$ defined on $\Omega = \Omega_1 \times \Omega_2$ are associated with the given elements $s_{(3-j)}$ in $\Omega_{(3-j)}$ and η_j in the set $C^\omega(\Omega_j)$ of all holomorphic functions on Ω_j:

$$e_j(z) = \exp\left[\eta_j(z_j) - \int_{s_{(3-j)}}^{z_{(3-j)}} a_j(z^*) \Big|_{z_j^* = z_j} dz^*_{(3-j)} \right].$$

Theorem. Considering the differential equation $Lw = D_1 D_2 w + a_1(z) D_1 w + a_2(z) D_2 w + a_0(z) w = 0$, $a_0, a_1, a_2 \in \mathbf{C}^\omega(\Omega)$, let $s = (s_1, s_2) \in \Omega = \Omega_1 \times \Omega_2$ be given and let Ω_0 be a simply connected subdomain of Ω. For a given small $\delta > 0$ and $\mu_j, \nu_j \in \mathbf{C}$ if k_j is not an identically vanishing $\mathbf{C}^\omega(\Omega_0 \times N(C_j, \delta))$-solution of the equation

$$\left[(1 - t^2) D_{(3-j)} D_t - t^{-1} D_{(3-j)} + 2(z_j - \nu_j) t L_j \right] k_j = 0$$

satisfying the condition

$$(z_j - \nu_j)^{-1} t^{-1} D_{(3-j)} k_j \in \mathbf{C}^0\left(\Omega_0 \times N(C_j, \delta)\right),$$

then $k_j^* = e_j k_j$ is a Bergman kernel for the above differential equation on Ω_0.

Recall that, for $j = 1, 2$, the integral operator T_j from the vector space $V(\Omega_j^*)$ of functions $f_j \in \mathbf{C}^\omega(\Omega_j^*)$ to the vector space $V(\Omega_0)$ of $w \in \mathbf{C}^\omega(\Omega_0)$ defined by

$$T_j f_j(z) = \int_{C_j} k_j^*(z, t) f_j(\lambda_j)(1-t^2)^{-1/2}\, dt$$

is a Bergman operator for the above differential equation, on Ω_0, with the Bergman kernel k_j^*, if $T_j \neq 0$, and if for every $f_j \in \mathbf{C}^\omega(\Omega_j^*)$ the image $w_j = T_j f_j$ is a solution of the above equation on Ω_0.

Kernel Theorem Let X, Y be open sets in $\mathbf{R}^m$ and $\mathbf{R}^n$, respectively. An operator T from $L^2(X)$ to $L^2(Y)$ is a Hilbert-Schmidt operator if and only if it is an integral operator associated with a kernel $k \in L^2(X \times Y)$. This means that k is a square integrable function on $X \times Y$ and

$$Tf(y) = \int_X k(x, y) f(x)\, dx.$$

Recall that a bounded linear operator T from a separable Hilbert space H_1 to a separable Hilbert space H_2 is called a Hilbert-Schmidt operator if

$$\|T\|_2 = \Big[\sum_{i=1}^{\infty} \|Te_i\|_{H_2}^2\Big]^{1/2} < \infty,$$

where $\{e_i\}$ is an orthonormal basis for H_1.

Kleene's hierarchy Projective set theory is regarded as the theory of the $\mathbf{N}^N$ analytic hierarchy of Kleene, where $\mathbf{N}^N$ is the set of all number-theoretic functions with one argument. It was S.C. Kleene who, by using the theory of recursive functions, established a theory of hierarchies that generalizes the classical descriptive set theory.

Klein's line coordinates All lines in a 3-dimensional projective space $\mathbf{P}^3$ form a quadric space in a 5-dimensional projective space $\mathbf{P}^5$. When $\mathbf{P}^3$ is a complex projective space, every line in $\mathbf{P}^3$ can be represented by homogeneous coordinates $(\xi_0, \xi_1, \xi_2, \xi_3, \xi_4, \xi_5)$ of $\mathbf{P}^5$ satisfying $\xi_0^2 + \xi_1^2 + \xi_2^2 + \xi_3^2 + \xi_4^2 + \xi_5^2 = 0$. They are called *Klein's line coordinates*. Note that homogeneous coordinates are projective coordinates, and if $(\xi_0, \xi_1, \cdots, \xi_5) \neq (0, 0, \cdots, 0)$, and $\lambda \neq 0$, then $(\xi_0, \xi_1, \cdots, \xi_5)$ and $(\lambda\xi_0, \lambda\xi_1, \cdots, \lambda\xi_5)$ represent the same point in $\mathbf{P}^5$.

Kloosterman sum The function

$$K(u, v, q) = \sum_{x (\mathrm{mod}\ q), (x,q)=1} \exp\left(\frac{2\pi i}{q}(ux + \frac{v}{x})\right),$$

where u, v, q are integers.

H.D. Kloosterman studied the above sum when he tried to improve the Fourier coefficient a_n of the Eisenstein series. The Kloosterman sum is also related to the arithmetic of the quadratic forms.

Kobayashi pseudodistance In the theory of analytic functions of several variables, the largest pseudodistance on a complex analytic space X among all pseudodistances δ_X on X for which all holomorphic mappings $(X, \delta_X) \to (D, \rho)$ are distance-decreasing, where ρ is the Poincaré distance on the unit disk $D = \{z : |z| < 1\}$. *See* pseudodistance.

Kodaira's Theorem A Hodge manifold has a biholomorphic embedding into a projective space.

Kodaira's Vanishing Theorem Let L be the sheaf associated with a positive line bundle (i.e., its first Chern class is represented by a Hodge metric) on a compact complex manifold X. Then $H^i(X, K_X \otimes L) = 0$ for any $i > 0$, where K_X is the canonical sheaf, i.e., the sheaf of holomorphic n-forms, $n = \dim X$, and $H^i(X, K_X \otimes L)$ is the cohomology group with coefficient sheaf $K_X \otimes L$.

Koebe function The function $f(z) = z(1-z)^{-2}$. *See* Branges' Theorem.

Koebe's Theorem Write $B_r(P)$ for an open ball centered at P with radius r, and $S_r(P)$ for its surface. Suppose that u is continuous in a domain D and at every point $P \in D$, there is a sequence $\{r_k\}$ decreasing to zero such that the mean value of u over $B_{r_k}(P)$ or $S_{r_k}(P)$ is equal to $u(P)$ for each k. Then u is a harmonic function in D.

Kondô's Uniformization Theorem Every coanalytic set is uniformizable by a coanalytic set.

This is the most important result in descriptive set theory. Also called the Novikov-Kondô-Addison Theorem.

Korteweg-deVries equation The partial differential equation $u_t + uu_x + u_{xxx} = 0$.

Köthe space *See* echelon space.

Kronecker set Let G be a locally compact Abelian group. A subset K in G is said to be a Kronecker set if for every continuous function f on K of absolute value 1 and every $\epsilon > 0$ there exists a $g \in \widehat{G}$ such that $|f(x) - g(x)| < \epsilon$, for all $x \in K$, where $\widehat{G}$ is the character group of G.

Kuramochi compactification Let R be an open Riemann surface. For a function f on R, $(R)\frac{\partial f}{\partial n} = 0$ means that there exists a relatively compact subregion R_f such that f is of class $\mathbf{C}^\infty$ on R outside R_f and the Dirichlet integral of f over $R - R_f$ is not greater than those of functions on $R - R_f$ that coincide with f on the boundary of R_f. The *Kuramochi compactification* R_K^* is a compactification determined by the family K of bounded continuous functions f on R satisfying $(R)\frac{\partial f}{\partial n} = 0$.

Kuramochi kernel Let R be an open Riemann surface, and R_K^* be the Kuramochi compactification determined by the family K of bounded continuous functions f on R satisfying $(R)\frac{\partial f}{\partial n} = 0$. The continuous function $k(p, q)$ on R satisfying $(R)\frac{\partial f}{\partial n} = 0$ vanishes in a fixed parametric disk R_0 in R and is harmonic in $R \backslash \overline{R_0}$ except for a positive logarithmic singularity at a point q which can be extended continuously to $R \times R_K^*$, which is called the *Kuramochi kernel*. By using this kernel, R_K^* is metrizable.

L

L–space A space X of arbitrary elements for which, associated with certain sequences $\{a_n\}_{n=1}^{\infty}$ in the space, there is an $a \in X$, called the limit of the sequence and denoted $a = \lim_{n\to\infty} a_n$, such that:
1. If $\lim_{n\to\infty} a_n = a$ and $\{a_{n_k}\}_{k=1}^{\infty}$ is a subsequence of $\{a_n\}$, then $\lim_{k\to\infty} a_{n_k} = a$, and
2. If $a_n = a$, for all n, then $\lim_{n\to\infty} a_n = a$.
$\{a_n\}$ is called a convergent sequence and is said to converge to a.

L^*–space A space X which is an L-space and also satisfies: If $\{a_n\}_{n=1}^{\infty}$ does not converge to a, then there is a subsequence $\{a_{n_k}\}_{k=1}^{\infty}$ for which no further subsequence converges to a. *See* L–space.

L'Hôpital's Rule A rule applied to compute the limit of an indeterminate form.

Suppose that the functions f and g are real-valued and differentiable in a deleted neighborhood of the point c, for example, in $(c-\delta, c) \cup (c, c+\delta)$, where δ is a positive number. Suppose $g'(x) \neq 0$ in that neighborhood, and $\lim_{x\to c} f(x) = 0 = \lim_{x\to c} g(x)$. Then

$$\lim_{x\to c} \frac{f(x)}{g(x)} = \lim_{x\to c} \frac{f'(x)}{g'(x)}$$

provided that the limit on the right either exists (as a real number), or is $\pm\infty$.

This rule extends to the case where $\lim_{x\to c} f(x) = \pm\infty = \lim_{x\to c} g(x)$.

Also spelled L'Hospital's Rule. *See also* limit of an indeterminate form.

lacunary power series A power series $\sum a_j z^{\lambda_j}$ such that the sequence $\{\lambda_j\}$ is lacunary. *See* lacunary sequence. Also called *Hadamard lacunary*.

lacunary sequence A sequence of positive integers $\{\lambda_j\}$ such that there exists $q > 1$ such that $\lambda_{j+1} > q\lambda_j$, for all j. Or, a sequence $\{a_j\}$ of real or complex numbers, such that $a_j = 0$ for all j except for a lacunary sequence.

ladder method A method by which the main part of a differential equation is decomposed, in two different ways, into two factors, each of which has order one, and from that decomposition is obtained a recurrence formula involving the parameter. This method, also called the *factorization method*, is helpful in solving the hypergeometric (Gaussian) differential equation.

As an example, consider Legendre's differential equation with a parameter n

$$\frac{d}{dx}[(1-x^2)\frac{dy}{dx}] + n(n+1)y = 0.$$

Let $B_n = (1-x^2)\frac{d}{dx} + nx$, $F_n = (1-x^2)\frac{d}{dx} - nx$, and $L_n[y] = (1-x^2)(\frac{d}{dx}[(1-x^2)\frac{dy}{dx}] + n(n+1)y)$. Then L_n is decomposed in two different ways, $L_n \equiv F_n \cdot B_n + n^2$, or $L_n \equiv B_{n+1} \cdot F_{n+1} + (n+1)^2$. If y_n is a solution of $L_n[y] = 0$, then multiplying both sides of $F_n \cdot B_n[y_n] + n^2 y_n = 0$ by B_n yields $B_n \cdot F_n \cdot (B_n[y_n]) + n^2 B_n[y_n] = 0$. So $B_n[y_n]$ is a solution of $L_{n-1}[y] = 0$. Similarly, multiplying both sides of $B_{n+1} \cdot F_{n+1}[y_n] + (n+1)^2 y_n = 0$, we see that $F_{n+1}[y_n]$ is a solution of $L_{n+1}[y] = 0$. Hence we obtain a recurrence formula involving the parameter n.

Lagrange identity Consider the linear homogeneous ordinary differential equation

$$F(y) = p_0 y^{(n)} + p_1 y^{(n-1)} + \cdots + p_n y = 0.$$

Integrating $\int \bar{z} F(y) dx$ by parts yields the *Lagrange identity*

$$\bar{z}F(y) - y\overline{G(z)} = \frac{d}{dx} R(y, z),$$

where
$G(z) = (-1)^n ((\overline{p_0} z)^{(n)} - (\overline{p_1} z)^{(n-1)} + \cdots + (-1)^n \overline{p_n} z)$,

0-8493-0320-6/00/$0.00+$.50

and

$$R(y,z)=\sum_{k=1}^{n}\sum_{h=0}^{k-1}(-1)^h y^{(k-h-1)}(p_{n-k}\overline{z})^{(h)}.$$

Lagrange multiplier Consider the simplest constrained maximum-minimum problem in the calculus of variations: Find two functions $y(x)$ and $z(x)$ so that the functional

$$J(y,z)=\int_{x_0}^{x_1} F(x,y,z,y',z')dx$$

attains its extrema subject to the condition

$$G(x,y,z)=0$$

and the fixed boundary conditions

$$y(x_0)=y_0, \qquad z(x_0)=z_0,$$
$$y(x_1)=y_1, \qquad z(x_1)=z_1.$$

The *Lagrange multiplier method* is to introduce an auxiliary function $F^*=F+\lambda(x)G$, where $\lambda(x)$ is to be determined, and convert the above constrained extrema problem into the constraint-free extrema problem of the following functional

$$J^*=\int_{x_0}^{x_1} F^*(x,y,z,y',z')dx.$$

The corresponding Euler-Lagrange equation is

$$F_y^*-\frac{d}{dx}F_{y'}^*=0,$$
$$F_z^*-\frac{d}{dx}F_{z'}^*=0;$$

or

$$F_y+\lambda(x)G_y-\frac{d}{dx}F_{y'}=0,$$
$$F_z+\lambda(x)G_z-\frac{d}{dx}F_{z'}=0.$$

Together with the constraint equation $G(x,y,z)=0$ we can eliminate $\lambda(x)$ and one unknown function (e.g., z), and obtain a differential equation for $y(x)$ of second order. Then we follow the routine procedure to solve for $y(x)$, and hence $z(x)$. The associated $\lambda(x)$ is called the *Lagrange multiplier.*

See also Lagrange's method of indeterminate coefficients.

Lagrange's bracket Let $x=(x_1,x_2,\cdots,x_n)$, $p=(p_1,p_2,\cdots,p_n)$. For two C^∞ functions $F(x,u,p)$ and $G(x,u,p)$ of x,u,p, the *Lagrange's bracket* is defined by

$$[F,G]=\sum_{\nu=1}^{n}\Big(\frac{\partial F}{\partial p_\nu}\Big(\frac{\partial G}{\partial x_\nu}+p_\nu\frac{\partial G}{\partial u}\Big)-\frac{\partial G}{\partial p_\nu}\Big(\frac{\partial F}{\partial x_\nu}+p_\nu\frac{\partial F}{\partial u}\Big)\Big).$$

This bracket has the following properties:

(i.) $[F,G]=-[G,F]$,

(ii.) $[F,\varphi(G_1,\cdots,G_k)]=\sum_{i=1}^{k}\frac{\partial\varphi}{\partial G_i}[F,G_i]$,

(iii.) $[[F,G],H]+[[G,H],F]+[[H,F],G]=\frac{\partial F}{\partial u}[G,H]+\frac{\partial G}{\partial u}[H,F]+\frac{\partial H}{\partial u}[F,G]$.

Lagrange's method of indeterminate coefficients A method of finding the maximum and minimum of a given real-valued function $f(x_1,\cdots,x_n)$, of n variables, subject to $m(m<n)$ constraints or side conditions

$$g_k(x_1,\cdots,x_n)=0,\ k=1,\cdots,m.$$

Lagrange's method of indeterminate coefficients, or the *method of Lagrange multipliers* is to introduce the function

$$F(x_1,\cdots,x_n)=f(x_1,\cdots,x_n)+\sum_{k=1}^{m}\lambda_k g_k(x_1,\cdots,x_n)$$

where $\lambda_1,\cdots,\lambda_m$ are indeterminate coefficients called the *Lagrange multipliers,* then take partial derivatives of F with respect to $x_1,\cdots,x_n$ and $\lambda_1,\cdots,\lambda_m$. The critical points will follow by solving the following equations:

$$\frac{\partial F}{\partial x_i}=0, \quad i=1,\cdots,n$$
$$g_k=0, \quad k=1,\cdots,m.$$

Lagrange's method of variation of constants A method to find a particular solution of the nonhomogeneous linear differential equation

$$y^{(n)} + p_{n-1}(x)y^{(n-1)} + \cdots + p_1(x)y' + p_0(x)y = f(x),$$

provided that we already know the general solution

$$y(x) = \sum_{i=1}^{n} C_i y_i(x)$$

of the associated homogeneous equation

$$y^{(n)} + p_{n-1}(x)y^{(n-1)} + \cdots + p_1(x)y' + p_0(x)y = 0.$$

We consider $C_1, C_2, \cdots, C_n$, not as constants, but as functions of x, and determine them by the following equations

$$\begin{aligned} &y_1(x)C_1'(x) + y_2(x)C_2'(x) + \cdots + y_n(x)C_n'(x) = 0, \\ &y_1'(x)C_1'(x) + y_2'(x)C_2'(x) + \cdots + y_n'(x)C_n'(x) = 0, \\ &\cdots \quad \cdots \quad \cdots \quad \cdots \\ &y_1^{(n-1)}(x)C_1'(x) + \cdots + y_n^{(n-1)}(x)C_n'(x) = f(x). \end{aligned}$$

Then $y(x) = \sum_{i=1}^{n} C_i(x)y_i(x)$ will be a particular solution of the nonhomogeneous equation. This is always possible because the *Wronskian determinant* $W(y_1(x), y_2(x), \cdots, y_n(x))$ of the n linearly independent solutions of the homogeneous equation is never zero, and the solution of $C_1'(x), C_2'(x), \cdots, C_n'(x)$ can be obtained by the *Cramer rule.*

Lagrange's problem This is an extremal problem of a functional imposed with a finiteness condition. In the calculus of variations, the extremal problems imposed with some conditions are typical. Around the year 1760, L.J. Lagrange introduced a general method of solving variational problems connected with mechanics.

Laguerre polynomial For $\alpha > -1$ and $n = 0, 1, 2, \ldots$, the polynomials

$$\begin{aligned} L_n^{(\alpha)}(x) &\equiv \frac{e^x x^{-\alpha}}{n!}\frac{d^n}{dx^n}(e^{-x}x^{n+\alpha}) \\ &= \sum_{j=0}^{n} \binom{n+\alpha}{n-j} \frac{(-x)^j}{j!}. \end{aligned}$$

The *associated Laguerre polynomials*, or *Sonine polynomials* are

$$S_n^{(\alpha)}(x) \equiv \frac{(-1)^n}{\Gamma(\alpha+n+1)} L_n^{(\alpha)}(x),$$

which appear in the solutions of the Schrödinger equation. *See* orthogonal polynomials.

lambda-function (λ-function) The function defined by

$$\lambda(z) = \frac{\wp(\frac{1+z}{2}) - \wp(\frac{z}{2})}{\wp(\frac{1}{2}) - \wp(\frac{z}{2})}$$

where $\wp$ is the Weierstrass $\wp$-function with the fundamental period $(1, z)$. The λ-function is a modular function of level 2, and is employed to construct an analytic isomorphism of a Riemann surface to a Riemann sphere. *See* Weierstrass $\wp$ function.

Lamé function of the first species In Lamé's differential equation

$$4\Delta_\lambda \frac{d}{d\lambda}(\Delta_\lambda \frac{d\Lambda}{d\lambda}) = (K\lambda + C)\Lambda,$$

let $K = n(n+1)$, for $n = 0, 1, 2, \ldots$. Then, for a suitable value of C, Lamé's differential equation has a solution that is a polynomial in λ or a polynomial multiplied by one, two, or three of the following $\sqrt{a^2+\lambda}$, $\sqrt{b^2+\lambda}$, and $\sqrt{c^2+\lambda}$.

Among the solutions there are $2n+1$ linearly independent solutions. Denote them by $f_n^m(\lambda)$, $m = 1, 2, \cdots, 2n+1$. They are classified into the following four families:

If n is an even number, $n = 2p$, then $p + 1$ solutions among the $2n+1$ $(= 4p+1)$ solutions are polynomials in λ of degree p

$$f_n^m(\lambda) = (\lambda - \theta_1)(\lambda - \theta_2)\cdots(\lambda - \theta_{\frac{n}{2}}),$$

which are called *Lamé's functions of the first species.*

The other $3p$ solutions are polynomials in λ of degree $p-1$ multiplied by one of $\sqrt{(b^2+\lambda)(c^2+\lambda)}$, $\sqrt{(c^2+\lambda)(a^2+\lambda)}$, or $\sqrt{(a^2+\lambda)(b^2+\lambda)}$ which are called *Lamé's functions of the third species.*

If n is an odd number, $n=2p+1$, then among the $2n+1$ $(=4p+3)$ solutions $3(p+1)$ are of the type

$$f_n^m(\lambda)=\left\{\begin{matrix}\sqrt{a^2+\lambda}\\ \sqrt{b^2+\lambda}\\ \sqrt{c^2+\lambda}\end{matrix}\right\}\times(\lambda-\theta_1)(\lambda-\theta_2)\cdots(\lambda-\theta_{\frac{n-1}{2}})$$

which are called *Lamé's functions of the second species.*

The other p solutions are of the type

$$f_n^m(\lambda)=\sqrt{(a^2+\lambda)(b^2+\lambda)(c^2+\lambda)}\times(\lambda-\theta_1)(\lambda-\theta_2)\cdots(\lambda-\theta_{\frac{n-3}{2}})$$

which are called *Lamé's functions of the fourth species.*

Lamé function of the fourth species *See* Lamé function of the first species.

Lamé function of the second species *See* Lamé function of the first species.

Lamé function of the third species *See* Lamé function of the first species.

Lamé's differential equation Let a, b, c be constants $(a>b>c>0)$. For any given $(x, y, z)\in\mathbf{R}^3$, the cubic equation in θ

$$\frac{x^2}{a^2+\theta}+\frac{y^2}{b^2+\theta}+\frac{z^2}{c^2+\theta}-1=0$$

has three real roots λ, μ, ν satisfying

$$\lambda>-c^2>\mu>-b^2>\nu>-a^2.$$

Write $\Delta_\lambda=\sqrt{(a^2+\lambda)(b^2+\lambda)(c^2+\lambda)}$. Then the ordinary differential equation

$$4\Delta_\lambda\frac{d}{d\lambda}(\Delta_\lambda\frac{d\Lambda}{d\lambda})=(K\lambda+C)\Lambda$$

is called *Lamé's differential equation,* where K and C are separation constants. The solution $\Lambda(\lambda)$ is also a factor of the function $\Psi(\lambda,\mu,\nu)=\Lambda(\lambda)M(\mu)N(\nu)$ which is a solution of Laplace's equation $\Delta\Psi=0$, and Ψ can be obtained by the method of separation of variables. Lamé's differential equation is also satisfied by $M(\mu)$ and $N(\nu)$ if we replace λ by μ and ν.

Landau equation Equations emerging from the Feynman integrals. Let G be a Feynman graph. A well known result of Landau, Nakanishi, and Bjorken (1959) asserts that the singularities of a Feynman amplitude $f_G(p)$ are confined to the subset $L^+(G)$, which is called a positive-α *Landau-Nakanishi variety,* or *Landau variety,* of the set

$$M=\left\{p\in\mathbf{R}^{4n}:\sum_{j,r}[j:r]p_r=0\right\},$$

defined by a set of equations called *Landau-Nakanishi equations,* or *Landau equation.* The notation $[j:k]$ is the *incidence number* which is $+1$, -1, or 0.

The following are the Landau-Nakanishi equations:

$$u_r=\sum_{j=l}^{n'}[j:r]w_j\quad(r=1,\cdots,n),$$

$$\sum_{r=1}^{n}[j:r]p_r+\sum_{l=1}^{N}[j:l]k_l=0\quad(j=1,\cdots,n'),$$

$$\sum_{j=1}^{n'}[j:l]w_j=\alpha_l k_l\quad(l=1,\cdots,N),$$

$$\alpha_l(k_l^2-m_l^2)=0\quad(l=1,\cdots,N),$$

$$\alpha_l\geq 0,$$

with some $\alpha_l\neq 0$.

In the above equations, u_r, w_j, and k_l are real four-vectors and α_l is a real number, all of which are to be eliminated to define relations among the p_r.

Landau variety *See* Landau equation.

Landau's symbols The symbols O and o, indicating the word "order," and due to E.G.H. Landau. Let f and g be two real- or complex-valued functions. If $\left|\frac{f(x)}{g(x)}\right|$ is bounded as $x \to a$, then f is called at most of the order of g as $x \to a$, written $f(x) = O(g(x))$ as $x \to a$. If $\frac{f(x)}{g(x)} \to 0$, as $x \to a$, then f is said to be of lower order than g as $x \to a$, written $f(x) = o(g(x))$ as $x \to a$.

Landau's Theorem If the complex valued function $f(z) = a_0 + a_1 z + \cdots (a_1 \neq 0)$ is holomorphic in $|z| < R$ and $f(z) \neq 0, 1$ in the same domain $|z| < R$, then there exists a constant $L(a_0, a_1)$ depending only on a_0 and a_1, such that $R \leq L(a_0, a_1)$.

Landau-Nakanishi equations *See* Landau equation.

Landau-Nakanishi variety *See* Landau equation.

Landen's transformation The elliptic integral of the first kind in Legendre-Jacobi standard form is

$$F(k, \varphi) = \int_0^{\sin\varphi} \frac{dx}{\sqrt{(1-x^2)(1-k^2x^2)}} = \int_0^{\varphi} \frac{d\psi}{\sqrt{(1-k^2\sin^2\psi)}},$$

where the constant k is called the *modulus*.

The *complete* elliptic integral of the first kind is defined by

$$K = K(k) = F(k, \frac{\pi}{2}) = \int_0^1 \frac{dx}{\sqrt{(1-x^2)(1-k^2x^2)}} = \int_0^{\frac{\pi}{2}} \frac{d\psi}{\sqrt{(1-k^2\sin^2\psi)}}.$$

Define the *complementary modulus* $k' = \sqrt{1-k^2}$, and $K' = K(k')$. Let $\sin\varphi_1 = \frac{(1+k')\sin\varphi\cos\varphi}{\sqrt{1-k^2\sin^2\varphi}}$, and $k_1 = \frac{1-k'}{1+k'}$. Then we have

$$F(k, \varphi) = (1 + k_1)F(k_1, \varphi_1)/2,$$

which is called *Landen's transformation.* Since $k_1 < k$ when $0 < k < 1$, this transformation reduces the calculation of the elliptic integral to those k_1 with smaller value than k.

Laplace equation The linear second order partial differential equation

$$\sum_{i=1}^{n} \frac{\partial^2 u}{\partial x_i^2} = 0.$$

The operator $\Delta \equiv \sum_{i=1}^{n} \frac{\partial^2}{\partial x_i^2}$ is called the *Laplace operator,* or the *Laplacian.*

Laplace operator *See* Laplace equation.

Laplace transform *See* Laplace-Stieltjes transform.

Laplace's method A method for obtaining asymptotic expansions of a function represented by an integral with a large parameter.

An asymptotic sequence $\{\varphi_n(x)\}_{n=1}^{\infty}$ is a sequence of functions defined in $\mathbf{R}^+$ such that $\varphi_{n+1}(x) = o(\varphi_n(x))$ as $x \to \infty$ for each n. The corresponding formal series $\sum a_n\varphi_n(x)$ is called an *asymptotic series.* A function $f(x)$ defined on $\mathbf{R}^+$ is said to have the *asymptotic expansion,* denoted by

$$f(x) \sim \sum_{n=0}^{\infty} a_n\varphi_n(x)$$

as $x \to \infty$, if $f(x)$ and the above asymptotic series satisfy, for each integer $n \geq 0$,

$$f(x) - \sum_{i=0}^{n} a_i\varphi_i(x) = O(\varphi_{n+1}(x))$$

as $x \to \infty$, where the symbols o, O are Landau's symbols.

Note that the coefficients a_n, $n = 1, 2, \cdots$, are uniquely determined.

Laplace's method works as follows:

Let $h(x)$ be a smooth real-valued function in $[a-\delta, a+\delta], \delta > 0$, satisfying $h(a) < h(x)$ for all $x \in [a-\delta, a) \cup (a, a+\delta]$, and $h''(a) > 0$, and let $g(x) \in \mathbf{C}^\infty([a-\delta, a+\delta])$. Then $I(x) = \int_{a-\delta}^{a+\delta} e^{-xh(t)} g(t)dt$ has an asymptotic expansion $I(x) \sim e^{-xh(a)} \frac{\sqrt{2\pi}}{\sqrt{x}} \{g(a)\frac{1}{\sqrt{h''(a)}} + c_1 x^{-1} + \cdots + c_n x^{-n} + \cdots\}$ for $x \to \infty$, where the c_n are determined by derivatives of $h(x)$ at $x = a$ of order $\leq 2n+2$ and those of $g(x)$ at $x = a$ of order $\leq 2n$.

As an example, applying Laplace's method to the gamma function $\Gamma(x) = \int_0^\infty t^{x-1} e^{-t} dt$ we can obtain its asymptotic expansion

$$\int_{a-\delta}^{a+\delta} e^{-x(t-\log t)} t^{-1} dt \sim e^{-x} \frac{\sqrt{2\pi}}{\sqrt{x}} (1 + c_1 x^{-1} + \cdots).$$

This implies Stirling's formula

$$\Gamma(x) = \sqrt{2\pi} x^{x-\frac{1}{2}} e^{-x} (1 + O(x^{-x})).$$

Laplace-Beltrami operator An operator emerging from the theory of harmonic integrals.

Let X be an oriented n-dimensional differential manifold of class $\mathbf{C}^\infty$ with a Riemannian metric ds^2 of class $\mathbf{C}^\infty$. We define a linear mapping $*$ that transforms ($\mathbf{C}^\infty$) p-forms to $(n-p)$-forms. Using $*$ we can define an inner product for p-forms. Let d be the exterior derivative, and let $\delta = (-1)^{np+n+1} * d *$ which operates on p-forms. Then d and δ are adjoint to each other with respect to the inner product. The *Laplace-Beltrami operator* is defined by

$$\Delta = d\delta + \delta d,$$

which is a self-adjoint elliptic differential operator. Note that Δ operates on p-form, and $*\Delta = \Delta *$.

Laplace-Stieltjes transform The transform $\alpha \to L\alpha$, defined by

$$(L\alpha)(s) = \int_0^\infty e^{-st} d\alpha(t) = \lim_{A\to\infty} \int_0^A e^{-st} d\alpha(t)$$

whenever the limit exists, where $\alpha(t)$ is a function of bounded variation in the interval $[0, A]$ for every positive number A.

If $\alpha(t) = \int_0^t \varphi(u)du$, where $\varphi(u)$ is Lebesgue integrable in the interval $[0, A]$ for every positive number A, then

$$(L\varphi)(s) = \int_0^\infty e^{-st} \varphi(t)dt$$

is called the *Laplace transform* of $\varphi(t)$.

Note that, in the case of the Laplace-Stieltjes transform, if the integral converges for some complex number s_0, then it converges for all s with $\Re s > \Re s_0$. This transform is used to solve some differential equations.

Laplacian *See* Laplace equation.

lattice-point formula If a function $f(x)$ is invariant under the Hankel transform, that is,

$$f(x) = \int_0^\infty \sqrt{xy} J_\nu(xy) f(y)dy,$$

where J_ν is the Bessel function, then $f(x)$ is called a *self-reciprocal* function (with respect to the Hankel transform).

Applying a self-reciprocal function we can get the *lattice-point formula* in number theory as follows.

Let $r(n)$ be the number of possible ways in which a nonnegative integer n can be represented as the sum of two square numbers. Put

$$\overline{p}(x) = \sum_{0\leq n\leq x}{}' r(n) - \pi x,$$

where $\sum'$ means that if x is an integer, we take $\frac{r(n)}{2}$ instead of $r(n)$. Then $f(x) = x^{-\frac{3}{2}}(\overline{p}(\frac{x^2}{2\pi}) - 1)$ is self-reciprocal with respect to the Hankel transform with $\nu = 2$.

Using this, G.H. Hardy proved (1925) the formula

$$\overline{p}(x) = \sqrt{x}\sum_{n=1}^{\infty}\frac{r(n)}{\sqrt{n}}J_1(2\pi\sqrt{nx}).$$

lattice-point problem In the Euclidean plane, let O be the origin, and P, Q be two points not colinear with O. Passing through P and Q, draw lines parallel to OQ and OP, respectively, and obtain a parallelogram $OPRQ$. Then extend its sides indefinitely, and draw the two systems of equidistant parallel lines, of which OP, QR and OQ, PR are consecutive pairs, thus dividing the plane into infinitely many congruent (equal) parallelograms. Such a figure of lines is called a *lattice.* The points of intersection of the lines are called *lattice points.*

We choose the lattice whose vertices are the points in the (x, y)-plane with integer coordinates. We denote by **D** the closed region in the upper right half quadrant contained between the axes and the hyperbola $xy = n$, that is, the closed region defined by $xy \leq n, x \geq 1$ and $y \geq 1$. Let $D(n)$ be the number of lattice points lying in the closed region **D**. Dirichlet proved (1849) that

$$D(n) = n\log n + (2\gamma - 1)n + O(\sqrt{n}),$$

where γ is the Euler's constant.

Note that $D(n)$ has another interpretation. Denote by $d(n)$ the number of divisors of n, including 1 and n, then $D(n) = d(1)+d(2)+\cdots+d(n)$.

Write $D(n) = n\log n + (2\gamma - 1)n + O(n^\beta)$. The *Dirichlet divisor problem* is to find better bounds for the error in the approximation. Suppose θ is the lower bound of numbers β. Dirichlet actually showed that $\theta \leq \frac{1}{2}$. G. Voronöi proved (1903) $\theta \leq \frac{1}{3}$, and van der Corput in 1922 that $\theta < \frac{33}{100}$, and these numbers have been improved further by later mathematicians. On the other hand, Hardy and Landau independently in 1915 proved that $\theta \geq \frac{1}{4}$. The true value of θ is still unknown.

There are many extensions of this problem.

latus rectum The chord that passes through the focus and is perpendicular to the major axis of an ellipse, or the chord that passes through the focus and is perpendicular to the principal axis of a hyperbola.

For the ellipse, $\frac{x^2}{a^2}+\frac{y^2}{b^2} = 1$, where $a > b$, and the hyperbola $\frac{x^2}{a^2}-\frac{y^2}{b^2} = 1$, the half length of the *latus rectum* is equal to $l = \frac{b^2}{a}$.

Laurent expansion An expansion of the form

$$f(z) = \sum_{n=0}^{\infty}a_n(z-z_0)^n + \sum_{n=1}^{\infty}\frac{b_n}{(z-z_0)^n},$$

for a complex-valued function $f(z)$, analytic throughout an annular domain $R_1 < |z - z_0| < R_2$, where $0 \leq R_1 < R_2 \leq \infty$. Here

$$a_n = \frac{1}{2\pi i}\int_C\frac{f(z)dz}{(z-z_0)^{n+1}}, n = 0, 1, 2, \cdots,$$

$$b_n = \frac{1}{2\pi i}\int_C\frac{f(z)dz}{(z-z_0)^{-n+1}}, n = 1, 2, \cdots,$$

and C is any positively oriented simple closed contour.

The above *Laurent series* is often written as

$$f(z) = \sum_{n=-\infty}^{\infty}c_n(z-z_0)^n,$$

where $R_1 < |z - z_0| < R_2$, and

$$c_n = \frac{1}{2\pi i}\int_C\frac{f(z)dz}{(z-z_0)^{n+1}},$$
$$n = 0, \pm 1, \pm 2, \cdots.$$

Lax-Halmos Theorem *See* Beurling's Theorem.

least upper bound Let **S** be a set, and the relation $\leq$ a partial order on **S**. Suppose **T** is a subset of **S**. An *upper bound* of **T** is an element $b \in$ **S** such that $x \leq b$ for all $x \in$ **T**. A *least upper bound* of **T** in **S** is an upper bound b of T such that if c is another upper bound of T, then $b \leq c$. The notation is $b = l.u.b.\mathrm{T}$, or $b = \sup \mathrm{T}$.

Lebesgue area Let A be a plane domain, and T be a continuous mapping of A into the Euclidean space $\mathbf{R}^3$. The pair (T, A) is called a *surface.* For surfaces (T, A) and (T', A'), and a set $B \subseteq A \cap A'$, define $d(T, T', B) = \sup_{p \in B} |T(p) - T'(p)|$. Let (T, A), (T_1, A_1), $(T_2, A_2), \cdots$ be given. If $A_n \uparrow A$, and $d(T, T_n, A_n) \to 0$, then (T_n, A_n) (or simply T_n) is said to converge to (T, A) (or T), and we write $T_n \to T$.

If A consists of a finite number of triangles and T is linear on each triangle, i.e., the image of A under T consists of triangles, then we replace the notation (T, A) by (P, F), and denote the area $T(A)$ by $a(P, F)$.

Given a surface (T, A), let Φ be the set of all sequences $\{(P_n, F_n)\}$ converging to (T, A). Then the *Lebesgue area* of (T, A) is defined to be

$$L(T, A) = \sup_{\Phi}\{\liminf_{n} a(P_n, F_n)\}.$$

Lebesgue decomposition If μ is a positive, σ-finite measure and λ a complex measure on the same σ-algebra, the *Lebesgue decomposition* of λ with respect to μ is

$$\lambda = \lambda_a + \lambda_s,$$

where λ_a is absolutely continuous with respect to μ and λ_s is purely singular with respect to μ.

The *Lebesgue Decomposition Theorem* asserts the existence and uniqueness of the decomposition, under the above hypotheses. When the measure μ is not specified, it is assumed to be Lebesgue measure. *See* absolute continuity, singular measure.

Lebesgue integral A real-valued Lebesgue measurable function ϕ that assumes only a finite number of values, $a_1, \cdots, a_n$, is called a *simple function.*

If ϕ vanishes outside a set of finite Lebesgue measure, the integral of ϕ is defined by

$$\int \phi(x)dx = \sum_{i=1}^{n} a_i m(A_i),$$

where A_i is the set on which ϕ assumes the value a_i, $A_i = \{x : \phi(x) = a_i\}$, and $m(A_i)$ is the Lebesgue measure of the set A_i, $i = 1, \cdots, n$.

If E is a Lebesgue measurable set, we define the integral of ϕ on E by

$$\int_E \phi(x)dx = \int \phi(x) \cdot \chi_E(x)dx,$$

where χ_E is the characteristic function of E, i.e., $\chi_E(x) = 1$, if $x \in E$; 0, elsewhere.

If f is a bounded Lebesgue measurable function defined on a Lebesgue measurable set E with $m(E)$ finite, we define the *Lebesgue integral of f over E* by

$$\int_E f(x)dx = \inf \int_E \phi(x)dx$$

where the infimum is taken for all simple functions $\phi \geq f$.

The Lebesgue integral was introduced by Henri Lebesgue (1902). It generalized the Riemann integral and overcame some of its shortcomings. For example, the Lebesgue integral gives more powerful and useful convergence theorems than the Riemann integral and it makes the spaces L^p complete. Roughly speaking, the Riemann integral is calculated by dividing the area under the graph of a function into vertical rectangles, while Lebesgue's theory is to divide the area horizontally.

Lebesgue measurable function *See* Lebesgue measure.

Lebesgue measurable set *See* Lebesgue measure.

Lebesgue measure Denote the length of an open interval I of $\mathbf{R}^1$ by $l(I)$. For each set E of real numbers we define the *Lebesgue* *outer measure* $m^*(E)$ of E as

$$m^*(E) = \inf_{E \subseteq \bigcup I_n} \sum l(I),$$

where $\{I_n\}$ is any countable collection of open intervals which cover E, that is, $E \subseteq \bigcup I_n$, and $\sum l(I)$ is the sum of lengths of the intervals in the collection. The infimum is taken on all such sums.

A set $E \subseteq \mathbf{R}^1$ is said to be *Lebesgue measurable* (due to Carathéodory) if, for each set F of $\mathbf{R}^1$ we have $m^*(F) = m^*(F \bigcap E) + m^*(F \bigcap E^c)$, where E^c is the complement of E in $\mathbf{R}^1$.

If E is a Lebesgue measurable set, we define the *Lebesgue measure* $m(E)$ to be the Lebesgue outer measure of E, i.e., $m(E) = m^*(E)$.

A *Lebesgue measurable function* is a function f with values in the extended real numbers, $\mathbf{R} \cup \{-\infty\} \cup \{\infty|$ whose domain is a Lebesgue measurable set and such that, for each real number C, the set $\{x : f(x) < C\}$ is Lebesgue measurable.

The Lebesgue measure of an interval is its length, the Lebesgue measure of any geometric figure in $\mathbf{R}^2$ is its area, and the Lebesgue measure of any geometric solid in $\mathbf{R}^3$ is its volume.

Lebesgue outer measure *See* outer measure.

Lebesgue space Let (Ω, μ) be a measure space, and $0 < p < \infty$. We say that a function $f(x)$ on Ω is *of class* $L^p(\Omega)$ if f is measurable and the integral of $|f(x)|^p$ is finite. We consider

two measurable functions to be equivalent if they are equal almost everywhere, and define the space $L^p(\Omega)$ to be the set of equivalence classes. We introduce the norm $\|f\|_p$, defined by

$$\|f\|_p = \left(\int_\Omega |f(x)|^p \, d\mu(x)\right)^{1/p}.$$

Then, for $1 \leq p < \infty$, $L^p(\Omega)$ is a Banach space, with the norm $\|f\|_p$ and for $0 < p < 1$, $L^p(\Omega)$ is a metric space with the distance function

$$d(f, g) = \int_\Omega |f - g|^p d\mu.$$

For any $0 < p < \infty$, $L^p(\Omega)$ is called a *Lebesgue space*.

Lebesgue's Convergence Theorem Suppose that E is a Lebesgue measurable set of real numbers and the function g is Lebesgue integrable over E. Let $\{f_n\}$ be a sequence of Lebesgue measurable functions such that on E

$$|f_n(x)| \leq g(x)$$

and such that $f_n(x)$ converges to $f(x)$ almost everywhere on E. Then

$$\int_E f(x)dx = \lim_{n\to\infty} \int_E f_n(x)dx.$$

The theorem is sometimes called the *Dominated Convergence Theorem.*

Another convergence theorem of Lebesgue is the *Monotone Convergence Theorem,* in which it is assumed that $\{f_n\}$ are Lebesgue integrable functions, with $f_1(x) \leq f_2(x) \leq \ldots$ on E. The conclusion is

$$\int \lim_{n\to\infty} f_n(x)dx = \lim_{n\to\infty} \int f(x)dx \leq \infty.$$

Lebesgue's density function The function $D(E, x)$, defined by the following limit, if it exists,

$$D(E, x) = \lim_{r\to 0} \frac{m(E \bigcap B(x, r))}{m(B(x, r))},$$

where E is a subset of n-dimensional Euclidian space $\mathbf{R}^n$, $B(x, r)$ denotes the open ball (with radius r and center x) and $m(\cdot)$ is Lebesgue measure.

Lebesgue's Density Theorem states that, if $E \subseteq \mathbf{R}^n$ is a Lebesgue measurable set, we have

$D(E, x) = 1$ for almost all $x \in E$, and
$D(E, x) = 0$ for almost all $x \notin E$.

A point x in E with $D(E, x) = 1$ is called a *point of density* for E. Hence the Lebesgue

Density Theorem asserts that almost all the points of E are points of density.

A set E is said to be *density open* if E is Lebesgue measurable and $D(E, x) = 1$ for all $x \in E$. The density open sets form a topology. So the Lebesgue Density Theorem implies that almost every point of a Lebesgue measurable set is interior to it (in the density open topology).

These notions are used to describe *approximate continuity* and the *approximate derivative.*

Lebesgue's Density Theorem *See* Lebesgue's density function.

Lebesgue's test For a function $f(x)$ which is Lebesgue integrable in $(-\pi, \pi)$, denote by $\mathcal{F}(f)$ its Fourier series. Let $\varphi_x(t) = f(x + t) + f(x - t) - 2f(x)$. *Lebesgue's test* asserts that if

$$\int_0^h |\varphi_x(t)|dt = o(h)$$

and

$$\lim_{\eta \to 0+} \int_\eta^\pi \frac{|\varphi_x(t) - \varphi_x(t + \eta)|}{t} dt = 0,$$

then $\mathcal{F}(f)$ converges at x to $f(x)$.

Lebesgue-Stieltjes integral Suppose that $\alpha(x)$ is a monotone increasing and right continuous function on an interval $[a, b]$. For any interval $I = (x_1, x_2] \subset [a, b]$, we define an interval function $U(I) = \alpha(x_2) - \alpha(x_1)$. It is nonnegative and countably additive. Using $U(I)$ we can construct a corresponding outer measure and hence a completely additive measure. The *Lebesgue integral* with respect to this measure is called the *Lebesgue-Stieltjes integral,* and is denoted by $\int_a^b f(x)d\alpha(x)$.

Lebesgue-Stieltjes measure Let $f(x)$ be a right continuous function of bounded variation on a closed interval $[a, b]$. Then $f(x)$ admits the Jordan decomposition $f(x) = \pi(x) - \nu(x)$, where $\pi(x)$ and $\nu(x)$ are monotone increasing right continuous functions. It follows that $\pi(x)$ and $\nu(x)$ induce bounded measures $d\pi(x)$ and $d\nu(x)$ on $[a, b]$ respectively. *See* Lebesgue-Stieltjes integral. The difference $d\pi - d\nu$ of these two measures is a completely additive set function on $[a, b]$ which is called the *(signed) Lebesgue-Stieltjes measure* induced by f.

left adjoint linear mapping Let F be a field (not necessarily commutative) with an antiautomorphism J. For left linear spaces V, W over F, a mapping $\Phi : V \times W \to F$ is called a (right) *sesquilinear form relative to J* if the following four conditions are satisfied:
(i.) $\Phi(x + x', y) = \Phi(x, y) + \Phi(x', y)$;
(ii.) $\Phi(x, y + y') = \Phi(x, y) + \Phi(x, y')$;
(iii.) $\Phi(\alpha x, y) = \alpha\Phi(x, y)$;
(iv.) $\Phi(x, \alpha y) = \Phi(x, y)\alpha^J$, for $x, x' \in V$, $y, y' \in W$, and $\alpha \in F$.

Let $\Phi' : V' \times W' \to F$ be another sesquilinear form relative to J. Then for any linear mapping $u : V \to V'$, there exists a unique linear mapping $u^* : W' \to W$ such that

$$\Phi'(u(x), y') = \Phi(x, u^*(y'))$$

for $x \in V$ and $y' \in W'$.

This linear mapping u is called the *left adjoint linear mapping* of u.

left derivative If the left (or right) limit

$$\lim_{h \to 0-, x+h \in I} \frac{f(x + h) - f(x)}{h}$$

(or

$$\lim_{h \to 0+, x+h \in I} \frac{f(x + h) - f(x)}{h})$$

exists then f is called *left* (or *right*) *differentiable at* x, and the corresponding left (or right) limit is denoted by $f'_-(x)$ (or $f'_+(x)$) which is called the *left derivative* (or *right derivative*) of f at x.

A function f is differentiable at x if and only if f is both left and right differentiable at x.

left differentiable *See* left derivative.

left linear space A vector space over a noncommutative field, in which scalar multiplication is permitted on the left only. *See* vector space.

Legendre polynomial *See* orthogonal polynomials.

Legendre relation The Weierstrass $\wp$-function has the fundamental periods $2\omega_1$ and $2\omega_3$. Weierstrass's zeta function $\zeta(u)$ and Weierstrass's sigma function $\sigma(u)$ have quasiperiodicity in the sense that

$$\zeta(u+2\omega_i)=\zeta(u)+2\eta_i,$$
$$\sigma(u+2\omega_i)=-e^{2\eta_i(u+\omega_i)}\sigma(u),$$
$$\eta_1+\eta_2+\eta_3=0,\ \eta_i=\zeta(\omega_i),\ i=1,2,3,$$

where $\omega_1+\omega_2+\omega_3=0$.

By taking the integral $\int\zeta(u)du$ once around the boundary of a fundamental period parallelogram, we have the following (*Legendre*) relations

$$\eta_1\omega_3-\eta_3\omega_1=\pm\frac{\pi}{2}i,$$
$$\eta_2\omega_1-\eta_1\omega_2=\pm\frac{\pi}{2}i,$$
$$\eta_3\omega_2-\eta_2\omega_3=\pm\frac{\pi}{2}i,$$

for $\Im(\frac{\omega_3}{\omega_1})\neq 0$.

Legendre's associated differential equation The equation

$$(1-z^2)\frac{d^2w}{dz^2}-2z\frac{dw}{dz}$$
$$+(n(n+1)-\frac{m^2}{1-z^2})w=0,$$

where n is a positive integer.

When $m=0$ and n is replaced by a complex number ζ, we obtain *Legendre's differential equation*

$$(1-z^2)\frac{d^2w}{dz^2}-2z\frac{dw}{dz}+\zeta(\zeta+1)w=0.$$

Legendre's transformation A transformation of $2n+1$ variables $z, x_j, p_j, j=1,\cdots,n$,

$$Z=Z(z,x_1,\cdots,x_n,p_1,\cdots,p_n),$$
$$X_j=X_j(z,x_1,\cdots,x_n,p_1,\cdots,p_n),$$
$$j=1,\cdots,n,$$
$$P_j=P_j(z,x_1,\cdots,x_n,p_1,\cdots,p_n),$$
$$j=1,\cdots,n,$$

is said to be a *contact transformation* in the Euclidean space $\mathbf{R}^{n+1}$ with the coordinate system $(z,x_1,\cdots,x_n)$ if the total differential equation

$$dz-p_1dx_1-p_2dx_2-\cdots-p_ndx_n=0$$

is invariant under the transformation.

Legendre's transformation

$$X=-p,\quad Y=xp-y,\quad P=-x,$$

is a contact transformation which is derived from the relation between poles and polar lines with respect to the parabola $x^2+2y=0$ in a plane.

Legendre-Jacobi standard form Any elliptic integral can be expressed by an appropriate change of variables as a sum of elementary functions and elliptic integrals of the following three kinds

$$\int\frac{dx}{\sqrt{(1-x^2)(1-k^2z^2)}},$$
$$\int\sqrt{\frac{1-k^2x^2}{1-x^2}}dx,$$

and

$$\int\frac{dx}{(1-a^2x^2)\sqrt{(1-x^2)(1-k^2x^2)}}.$$

These three kinds of integrals are called the elliptic integrals of the *first, second,* and *third kind,* respectively, in *Legendre-Jacobi standard form.*

Leibniz's formula A formula giving the nth derivative of the product of two functions

f and g:

$$(fg)^{(n)} = \sum_{k=0}^{n} \binom{n}{k} f^{(n-k)} g^{(k)}$$
$$= f^{(n)} g + \binom{n}{1} f^{(n-1)} g'$$
$$+ \binom{n}{2} f^{(n-2)} g'' + \cdots + f g^{(n)},$$

where $\binom{n}{k} = \frac{n!}{k!(n-k)!}$ are the binomial coefficients.

Leibniz's series The series

$$\sum_{n=1}^{\infty} \frac{(-1)^{n+1}}{(2n-1)} = 1 - \frac{1}{3} + \frac{1}{5} - \frac{1}{7} + \cdots$$

which converges (by Leibniz's test) and has the sum $\frac{\pi}{4}$.

Leibniz's Theorem (Leibniz's Test) Let $\sum(-1)^n a_n$ be an alternating infinite series where $a_n > 0$ for all n. If $a_n \geq a_{n+1}$ for all n and $a_n \to 0$ as $n \to \infty$, then $\sum(-1)^n a_n$ converges.

lemniscate A *lemniscate* or *Bernoulli's lemniscate* is a two-leaf shaped plane curve. In a rectangular coordinate system, a typical equation of a lemniscate is

$$(x^2 + y^2)^2 = 2a^2(x^2 - y^2),$$

where a is a constant.

length A fundamental notion in mathematics, of which many different versions exist. For example, the length of an interval of real numbers $[a, b]$ is $b - a$, the length of a vector $x = (x_1, \cdots, x_n)$ in $\mathbf{R}^n$ is $\sqrt{\sum_{i=1}^{n} x_i^2}$, etc.. *See also* arc length, length of curve.

length of curve Given a curve $\gamma : [a, b] \to \mathbf{R}^k$ and $P = \{x_0 = a < x_1 < \ldots < x_n = b\}$ a partition of $[a, b]$, associate the number

$$\Lambda(P, \gamma) = \sum_{n=1}^{n} |\gamma(x_i) - \gamma(x_{i-1})| .$$

The ith term in this sum is the distance (in $\mathbf{R}^k$) between the points $\gamma(x_{i-1})$ and $\gamma(x_i)$ of the curve γ, and the sum is the length of the polygonal path with vertices at $\gamma(x_0), \gamma(x_1), \cdots, \gamma(x_n)$, in this order.

The *length of the curve* γ is defined as

$$\Lambda(\gamma) = \sup_P \Lambda(P, \gamma),$$

where the supremum is taken over all partitions of $[a, b]$. If $\Lambda(\gamma) < \infty$, then γ is called *rectifiable*.

If γ is continuously differentiable, i.e., γ' is continuous on $[a, b]$, then γ is rectifiable and

$$\Lambda(\gamma) = \int_a^b \|\gamma'(t)\| \, dt,$$

where $\|\gamma'(t)\|$ is the Euclidean norm of $\gamma'(t)$ in $\mathbf{R}^k$.

Leray-Schauder Fixed-Point Theorem Let D be a bounded open set of a Banach space X containing the origin O. Let $F(x, t) : \overline{D} \times [0, 1] \to X$ be a compact mapping such that $F(x, 0) \equiv 0$. Suppose that $F(x, t) \neq x$ for any $x \in \partial D$ and $t \in [0, 1]$. Then the compact mapping $F(x, 1)$ has a fixed point x in D. (Here, $\overline{D}$ is the closure of D.)

level surface Let u be a harmonic function. The set $\{p : u(p) = \text{constant}\}$ is called a *level surface* of u.

Levi problem The problem of deciding whether every pseudoconvex domain is a domain of holomorphy.

E.E. Levi proposed this problem. It was affirmatively solved by Oka (1942, for $n = 2$, and 1953, for manifolds spread over $\mathbf{C}^n$ for $n \geq 2$), H. Bremermann, and F. Norguet. The problem was also solved by H. Grauert in a more general form in 1958 by using results on linear topological spaces, and by L. Hörmander in 1965 by using methods of the theory of partial differential equations.

See pseudoconvex domain.

Levi pseudoconvex domain A domain $\Omega \subset \mathbf{C}^n$ with $\mathbf{C}^2$ boundary and defining function ρ such that the (Levi) form

$$\sum_{j,k=1}^{n} \frac{\partial^2 \rho}{\partial z_j \partial \bar{z}_k}(P) w_j \bar{w}_k$$

is nonnegative semidefinite on the complex tangent space at P (i.e., $\sum_{j=1}^{n}(\partial\rho/\partial z_j)(P)w_j = 0$), for every $P \in \partial\Omega$.

Levi-Civita parallelism In the theory of surfaces in differential geometry, a vector field $\lambda^\alpha(t)\mathbf{x}_\alpha$ defined along a curve $u^\alpha = u^\alpha(t)$ on a surface is said to be *parallel in the sense of Levi-Civita* along the curve if its covariant derivative along the curve vanishes, that is,

$$\frac{\delta\lambda^\alpha}{dt} = \frac{d\lambda^\alpha}{dt} + \left\{ \begin{matrix} \alpha \\ \beta\gamma \end{matrix} \right\} \lambda^\beta \frac{du^\gamma}{dt} = 0.$$

Lewy-Mizhota equations The equations

$$\left(\frac{\partial}{\partial x_{r+2s+l}} + \sqrt{-1} x_{r+2s+l} \frac{\partial}{\partial x_n} \right) f = 0,$$

$l = 1, \cdots, a;$

$$\left(\frac{\partial}{\partial x_{r+2s+l}} - \sqrt{-1} x_{r+2s+l} \frac{\partial}{\partial x_n} \right) f = 0,$$

$l = a + 1, \cdots, a + b.$

These equations are employed in the structure theorems of the system of microdifferential equations in microlocal analysis.

Liapunov function *See* Lyapunov function.

Lie derivative For a $\mathbf{C}^\infty$-manifold M, K a tensor field and X a vector field (both of class $\mathbf{C}^\infty$) on M, the *Lie derivative* of K with respect to the vector field X is a tensor field $L_X K$, defined by

$$(L_X K)_p = \lim_{t\to 0}(K_p - (\tilde{\varphi}_t K)_p)/t,$$

where φ_t denotes the local one-parameter group of local transformations around p generated by X.

Lifting Theorem Let T be a contraction operator on a Hilbert space H and U a unitary dilation of T, acting on $K \supset H$. *See* unitary dilation. The *Lifting Theorem* asserts that if a bounded operator S commutes with T, then there exists a dilation $V : K \to K$ of S such that $\|V\| = \|S\|$ and V commutes with U.

limaçon of Pascal Suppose $r = f_1(\theta)$ and $r = f_2(\theta)$ are equations of curves C_1 and C_2 in polar coordinates. A curve C having the equation $r = \lambda_1 f_1(\theta) + \lambda_2 f_2(\theta)$, where λ_1, λ_2 are constants, and C_1 is a circle with center at the origin O, is called a *conchoidal curve* of C_2 with respect to O. Moreover, if C_2 is a circle and O is on C_2, the conchoidal curve of C_2 with respect to O is called a *limaçon*, or *limaçon of Pascal*.

The equation of a limaçon C in a polar coordinate system having the diameter of a circle passing through O as its initial line is

$$r = a\cos\theta \pm b,$$

while the equation of C with respect to the Cartesian coordinate system is

$$(x^2 + y^2 - ax)^2 = b^2(x^2 + y^2).$$

If $a > b$, O is a node of the curve, if $a = b$, O is a cusp. When $a = b$, the curve is also called a *cardioid*.

limit A sequence $\{s_n\}$ of real numbers is said to be *convergent to a limit* s, written $\lim_{n\to\infty} s_n = s$, or $s_n \to s$ as $n \to \infty$, if for each positive number ε there exists a positive integer N such that whenever $n > N$, we have $|s_n - s| < \varepsilon$. The number s is called the *limit* of the sequence $\{s_n\}$.

A sequence $\{p_n\}$ of points in a topological space is said to *converge* to a point p if for each neighborhood U of p there exists a positive integer N such that $p_n \in U$ whenever $n > N$. The point p is called *a limit* (or *a limit point*) of the sequence $\{p_n\}$. Note that in a Hausdorff topological space if a sequence has a limit, then this limit is unique.

For a real-valued function $f(x)$ of a real variable x defined on a neighborhood of a point c except possibly the point c itself, if L is a real number, and for each positive number ε there exists a positive number δ such that $|f(x) - L| < \varepsilon$ holds whenever $0 < |x - c| < \delta$, then we say that the function $f(x)$ tends to the *limit* L as x approaches c. We write $\lim_{x \to c} f(x) = L$ or $f(x) \to L$ as $x \to c$.

In the above paragraph replacing $0 < |x - c| < \delta$ by $c < x < c+\delta$ (or $c-\delta < x < c$), we define $f(x) \to L$ as $x \to c+0$ (or as $x \to c-0$). Also write $\lim_{x \to c+} f(x) = L$ (or $\lim_{x \to c-} f(x) = L$), and call L the *limt on the right (on the left.)*

Another notion of convergence is *Cèsaro convergence.* For a given sequence $\{s_n\}$ of real numbers, we define the *Cèsaro means* of this sequence to be the numbers

$$\sigma_1 = s_1,$$
$$\sigma_2 = \frac{s_1 + s_2}{2},$$
$$\sigma_3 = \frac{s_1 + s_2 + s_3}{3}, \ldots,$$
$$\sigma_n = \frac{s_1 + \cdots + s_n}{n}, \ldots.$$

If this new sequence $\{\sigma_n\}$ converges to, say, σ, then the original sequence $\{s_n\}$ is *Cèsaro convergent* to σ, or *convergent in Cèsaro means* to σ. The number σ is the *limit in Cèsaro means.* Note that a sequence $\{s_n\}$ converges (in the original sense) to L, then $\{s_n\}$ also converges to σ in Cèsaro means, but not vice versa.

limit in the mean A sequence $\{f_n\}$ in $L^p(\Omega)$ $(1 \le p < \infty)$ is said to converge to f *in the mean,* if $\lim_{n \to \infty} \|f_n - f\|_p = 0$. The function f is called the *limit in the mean,* sometimes written as $f = l.i.m.f_n$.

limit of an indeterminate form The limit of an algebraic expression (as $x \to a$), such that, substituting $x = a$ in the expression is not defined. Thus, the expression may assume the form $0/0, \infty/\infty, 0^\infty$, etc., when $x = a$.

For example, let $f(x) = \cos x$ and $g(x) = x - \frac{\pi}{2}$, then $\lim_{x \to \frac{\pi}{2}} f(x) = \lim_{x \to \frac{\pi}{2}} g(x) = 0$, and the limit of the quotient $\frac{f(x)}{g(x)}$ is of the form $0/0$, when $x = \frac{\pi}{2}$, and $\frac{f(x)}{g(x)}$ has an *indeterminate form.*

Other indeterminate forms are $0 \cdot \infty$, $\infty - \infty$, 0^0, ∞^0, and 1^∞.

The limit of an indeterminate form can sometimes be found by a special algebraic or geometric manipulation. However, if the functions involved are appropriately differentiable, L'Hôpital's Rule is more conveniently applied to solve the limit. In the above example $f'(x) = -\sin x, g'(x) = 1$, and L'Hôpital's Rule implies that $\lim_{x \to \frac{\pi}{2}} \frac{f(x)}{g(x)} = \lim_{x \to \frac{\pi}{2}} \frac{f'(x)}{g'(x)} = -1$.

See also L'Hôpital's Rule.

limit on the left *See* limit.

limit on the right *See* limit.

Lindelöf's Asymptotic Value Theorem Let $f(z)$ be a complex-valued function that is holomorphic and bounded in a closed angular domain $\overline{D} = \{z : \alpha \le \arg z \le \beta\}$, except for the point at infinity. Suppose that $f(z) \to a$ as $z \to \infty$ along one side of the angle and that $f(z) \to b$ as $z \to \infty$ along the other side of it. Then $a = b$ and $f(z) \to a$ uniformly as $z \to \infty$ in $\overline{D}$.

See also maximum principle, Schwarz's Lemma, Phragmén-Lindelöf Theorem.

Lindelöf's Theorem Let D be a domain in the complex plane. Suppose that there exists an arc of angular measure α that is on a circle of radius R centered at a point z_0 of D and not contained in D. Let C denote the intersection of the boundary of D with the disk $|z - z_0| < R$. If $f(z)$ is a single-valued holomorphic function that satisfies $|f(z)| \le M$, and if $\limsup_{z \to \zeta} |f(z)| \le M$ for every $\zeta \in C$, then the inequality $|f(z_0)| \le M^{1-1/n} m^{1/n}$ holds for every positive integer n satisfying $2\pi/n \le \alpha$.

line In Euclidian space $\mathbf{R}^3$, the locus of the equation $\gamma(t) = (x_0 + at, y_0 + bt, z_0 + ct)$ is a *straight line* which passes through the point (x_0, y_0, z_0) and parallel to the vector (a, b, c).

More generally, the term *line* may refer to a *curve*, a continuous mapping γ of an interval $[a, b]$ into $\mathbf{R}^n$.

Historically, Euclid distinguished straight lines and curves. A straight line in space is determined by any two points on it. However, nowadays, lines in the sense of Euclid are called curves, and a straight line is considered a curve. C. Jordan was the first mathematician to give an exact definition of a curve by using analytic method in his Cours d' analyse I (1893).

Older treatises may refer to *right lines* instead of straight lines.

line element Let M be a 1-dimensional differentiable manifold and $f : M \to \mathbf{R}^n$ an immersion. Let x be a point in $\mathbf{R}^n$ and $\mathbf{e}_1, \cdots \mathbf{e}_n$ orthonormal vectors in $\mathbf{R}^n$. Denote the tangent space of M at x by M_x. For $X \in M_x$, the form ds defined by $ds(X) = (df(X), \mathbf{e}_1)$ is called the *line element*, which is used in the theory of curves in differential geometry.

line integral (1.) Let γ be a smooth curve in a region $\Omega \subset \mathbf{R}^2$, with a parameterization $x = \phi(t), y = \psi(t)(a \le t \le b)$ and $\omega = A(x, y)dx + B(x, y)dy$ a 1-form in Ω. The *line integral* of ω over γ is defined by

$$\int_\gamma \omega = \int_a^b [A(\gamma(t))\phi'(t) + B(\gamma(t))\psi'(t)]dt.$$

(2.) For a continuous function $w = f(z)$ in a region Ω in the complex plane $\mathbf{C}$ and γ a curve in Ω, Let $z_0, z_1, \ldots z_n$ be a subdivision of γ and, for each j, let $\zeta_j \in \gamma$, between z_{j-1} and z_j, and form the sum

$$I_n = \sum_{j=1}^{n} f(\zeta_j)(z_j - z_{j-1}).$$

Then f is integrable over γ if there is a number I such that, for every $\epsilon > 0$ there is $\delta > 0$ such that the sum I_n is closer to I than ϵ whenever $\max |z_j - z_{j-1}| < \delta$. Then I is called the *line integral* of f over γ, denoted

$$I = \int_\gamma f(z)dz.$$

line of curvature In the theory of curves in differential geometry, a curve C on a surface such that the tangent line at each point of the curve coincides with a principal direction at the point.

line of regression In the theory of curves and surfaces in differential geometry, at each point of a space curve C we define the *osculating plane*, and the family of osculating planes of C envelops a developable surface S and coincides with the locus of tangent lines to C. The surface S is called the *tangent surface* of C, and C itself is called the *line of regression* of S.

line of swiftest descent *See* cycloid, which has the property that a particle slides down along the cycloid in the gravitational field from one point to another point in minimal elapsed time.

linear Similar in some way to a straight line. Linear problems are usually more tractible and their solutions can often be used as approximations to solutions of more general ones. *See* linear system, linear operator, linear combination.

linear approximation A linear function approximating a given function. For $f : V \to \mathbf{R}^m$ (V a domain in $\mathbf{R}^n$), the differencial of f at $x \in V$ is a linear approximation, in a small neighborhood of x. *See* differential.

linear combination An element of the form $x = \sum_{i=1}^{n} \alpha_i x_i$, where $x_1, \ldots, x_n$ come from a vector space $\mathbf{V}$ and $\alpha_1, \ldots, \alpha_n$ from its scalar field $\mathbf{F}$.

linear coordinates Let P be an affine space of dimension n. A frame f in P is a list $f_0, \ldots, f_n$ of $n+1$ points of P such that the vectors $f_1 - f_0, \ldots, f_n - f_0$ form a basis of $P^\#$ the vector space of all transformations of P. Relative to a frame f, each point can be written as

$$p = [(f_1 - f_0)\xi_1 + \cdots + (f_n - f_0)\xi_n] + f_0$$

for a unique list α of n scalars. These scalars are called the *linear coordinates* of the point p relative to the frame f.

linear dependence The property of vectors $x_1, x_2, \cdots, x_n$ in a vector space **V** over a field **F** that there exist scalars $\alpha_1, \alpha_2, \cdots, \alpha_n$, not all equal 0, such that $\alpha_1 x_1 + \alpha_2 x_2 + \cdots + \alpha_n x_n = 0$.

Vectors $x_1, x_2, \cdots, x_n$ of **V** are called *linearly independent* if they are not linearly dependent. That is, $\alpha_1 x_1 + \alpha_2 x_2 + \cdots + \alpha_n x_n = 0$, only if $\alpha_1 = \alpha_2 = \cdots = \alpha_n = 0$.

linear discriminant function The function $t = \mathbf{a}^T\mathbf{X}$, where $\mathbf{X}$ is the matrix of observation, $\mathbf{a}$ is a vector and $\mathbf{a}^T$ is the transpose of $\mathbf{a}$ which is to be determined so that if there are k observation populations $\mathbf{X}_i, i = 1, \cdots, k$, then the vector $\mathbf{a}$ can reveal the differences of these k populations in a sense that this is the best possible one. This notion appears in the multivariate analysis of statistics.

linear form A linear transformation L from **V** to **F**, where **V** is a linear (vector) space over the field **F**. That is, L satisfies

$$L(\alpha x + \beta y) = \alpha L(x) + \beta L(y)$$

for all vectors x and y in **V** and all scalars α and β in **F**.

linear fractional transformation *See* linear function.

linear function (1.) A linear operator between vector spaces. *See* linear operator.
(2.) A function $f : \mathbf{R} \to \mathbf{R}$ (or $f : \mathbf{C} \to \mathbf{C}$), of the form $f(x) = ax + b$, where a and b are constants.
(3.) A function $f : \mathbf{C} \to \mathbf{C}$ of the form $f(z) = (az+b)/(cz+d)$, where $ad - bc \neq 0$. Also called *linear fractional transformation* or *bilinear transformation*.

linear functional A linear operator from a vector space into its field of scalars. *See* linear operator.

linear independence *See* linear dependence.

linear integral equation An integral equation in an unknown function y which is linear in y, that is, of the form $Ly = g$, where L satisfies $L(a_1 f_1 + a_2 f_2) = a_1 L f_1 + a_2 L f_2$. *See* integral equation.

linear operator A mapping $f : \mathbf{V} \to \mathbf{W}$, where **V** and **W** are vector spaces over the same field **F**, such that

$$f(\alpha x + \beta y) = \alpha f(x) + \beta f(y)$$

for all vectors x and y in **V** and all scalars α and β in **F**.

linear ordinary differential equation An ordinary differential equation in an unknown function y, which is linear in y. That is, an equation of the form

$$p_n(x)y^{(n)} + p_{n-1}(x)y^{(n-1)} + \cdots + p_1(x)y' + p_0(x)y = f(x)$$

where $p_0(x), p_1(x), \cdots, p_n(x)$ and $f(x)$ are given functions of a real (or complex) variable x and $y' = \frac{dy}{dx}$.

The equation is said to have *order n*, if $f(x) \equiv 0$, it is said to be *homogeneous*.

linear space *See* vector space.

linear subspace A non-empty subset of a vector space **V** over a field **F**, closed under the addition operation of **V** and under scalar multiplications by elements of **F**. *See* vector space.

linear system A simultaneous set of n linear ordinary differential equations in m unknown functions

$$\frac{dy_1}{dt} = p_{11}(t)y_1 + p_{12}(t)y_2 + \cdots + p_{1m}(t)y_m + f_1(t)$$
$$\frac{dy_2}{dt} = p_{21}(t)y_1 + p_{22}(t)y_2 + \cdots + p_{2m}(t)y_m + f_2(t)$$
$$\cdots$$
$$\frac{dy_n}{dt} = p_{n1}(t)y_1 + p_{n2}(t)y_1 + \cdots + p_{mm}(t)y_m + f_n(t)$$

where $y_1, y_2, \cdots, y_m$ are unknowns and $p_{ij}(t)$ and $f_i(t)$ are given functions. If the functions $f_1, f_2, \cdots, f_n$ are all identically zero, then the system is *homogeneous*, otherwise, it is *nonhomogeneous*.

Frequently, it is convenient to write column vectors $\mathbf{y} = (y_1, \ldots, y_m)$ and $\mathbf{f} = (f_1, \ldots, f_n)$ and a matrix $P = (p_{ij})$, so that the above system has the simple form

$$\mathbf{y}' = P\mathbf{y} + \mathbf{f}.$$

Analogously, *linear system* may refer to a linear set of partial differential equations.

linear topological space *See* topological vector space.

linear transformation *See* linear operator.

linearized operator In *global analysis*, consider a nonlinear differential operator on a finite-dimensional manifold. By using functional analytical techniques it often happens that its domain is neither a linear space nor an open subset, but an infinite-dimensional manifold, and that such a nonlinear operator can be regarded as a differentiable mapping between infinite-dimensional manifolds. The differential at a point in that source manifold is called a *linearized operator*, to which one can apply various theorems of linear functional analysis.

Liouville's First Theorem An elliptic function with no poles in a period parallelogram is constant. *See* elliptic function, period parallelogram.

Liouville's formula The formula

$$W(y_1(x), y_2(x), \cdots, y_n(x)) = W(y_1(x_0), y_2(x_0), \cdots, y_n(x_0)) \times \exp\left(-\int_1^x p_{n-1}(t)dt\right),$$

for the Wronskian $W(y_1, y_2, \ldots, y_n)$ of the solutions of the homogeneous linear ordinary differential equation of order n

$$y^{(n)} + p_{n-1}(x)y^{(n-1)} + \cdots + p_1(x)y' + p_0(x)y = 0,$$

in an interval D, where $x_0 \in D$.

Example: $y'' + \frac{3}{x}y' + \frac{1}{x^2}y = 0$, $x_0 = 1$.

Solutions $y_1(x) = \frac{1}{x}, y_2(x) = \frac{1}{x}\ln x$ are linearly independent. Then

$$W(y_1(x), y_2(x)) = \frac{1}{x^3} = W(y_1(1), y_2(1))\exp\left(-\int_1^x \frac{3}{t}dt\right).$$

Liouville's Fourth Theorem The sum of the zeros minus the sum of poles of an elliptic function, in a period parallelogram, is itself a period. *See* elliptic function, period parallelogram.

Liouville's Second Theorem The sum of the residues of an elliptic function at its poles in any period parallelogram is zero.

An immediate corollary of Liouville's Second Theorem is that there exists no elliptic function of order 1. *See* elliptic function, period parallelogram.

Liouville's Theorem A function $f : \mathbf{C} \to \mathbf{C}$ which is bounded and entire (analytic in all of $\mathbf{C}$ and $|f(z)| \leq M$ there) is constant.

Liouville's Third Theorem An elliptic function of order n assumes every value n

times in a period parallelogram. *See* elliptic function, period parallelogram.

Liouville-Green approximation The approximate solution

$$w = Af^{-\frac{1}{4}} \exp[\int f^{\frac{1}{2}} dx] + Bf^{-\frac{1}{4}} \exp[-\int f^{\frac{1}{2}} dx]$$

of the equation

$$\frac{d^2 w}{dx^2} = f(x)w.$$

Theoretical physicists may refer to this as the *WKB* or *BKW approximation.*

Lipschitz condition The condition $|f(s) - f(t)| \leq M|s-t|^{\alpha}$, (for $x, y \in I$), for a real or complex valued function f, on an interval I. More precisely called a *Lipschitz condition of order* α.

A real function satisfying the Lipschitz condition is *absolutely continuous,* and hence differentiable almost everywhere in the interval I.

Lipschitz space The set $Lip_{\alpha}(I)$, of real-valued functions f satisfying a *Lipschitz condition of order* α on I ($0 < \alpha < 1$, and I an interval of real numbers). That is, $f \in Lip_{\alpha}(I)$ if there exists a constant c such that $|f(x) - f(y)| \leq c|x-y|^{\alpha}$ for all x, y in I. The smallest such constant c is called the Lip α norm of f and is denoted by $\|f\|_{Lip_{\alpha}}$. Endowed with this norm $Lip_{\alpha}(I)$ becomes a Banach space provided we identify functions which differ by a constant almost everywhere in I.

Littlewood-Paley theory Write the dyadic blocks of the Fourier series of the function f by

$$\Delta_j f(x) = \sum_{2^j \leq |k| < 2^{j+1}} c_k e^{ikx}, \; j \in \mathbf{N}.$$

The fundamental result of the *Littlewood-Paley theory* is that for $1 < p < \infty$, the two norms $\|f\|_p$ (L^p-norm) and $|c_0| + \|(\sum_0^{\infty} |\Delta_j f(x)|^2)^{1/2}\|_p$ are equivalent. This theory is used in harmonic analysis and wavelet theory. Some of the deepest results on $L^p(\mathbf{R}^n)$, $1 < p < \infty$, are often derived from this theory. This theory is due to J.E. Littlewood and R.E.A.C. Paley, A. Zygmund, and E.M. Stein.

lituus A plane curve having the equation $r^2\theta = a$ in polar coordinates.

local cohomology group Let Ω be an open subset of $\mathbf{R}^n$. The kth *local cohomology group* with support in Ω is the kth *derived functor* of $\mathcal{F} \mapsto \Gamma_{\Omega}(\mathbf{C}^n, \mathcal{F})$, where $\mathcal{F}$ is a *sheaf* on $\mathbf{C}^n$, and $\Gamma_{\Omega}(\mathbf{C}^n, \mathcal{F})$ is the totality of sections of $\mathcal{F}$ defined on a neighborhood of Ω and with support in Ω and is calculated as the kth cohomology group of the complex $\Gamma_{\Omega}(\mathbf{C}^n, \mathcal{L})$, where $\mathcal{L}$ denotes any *flabby* resolution (i.e., resolution by flabby sheaves) of $\mathcal{F}$.

This notion is used in the theory of distributions and hyperfunctions.

local concept A concept that can be defined in an arbitrarily small neighborhood of a point of a given figure or a space, such as the concept of a tangent line.

local coordinate system Let X and Y be topological spaces. A homeomorphism $\varphi : U \to V$ of an open set U in X to an open set V in Y is called a *local coordinate system* of Y with respect to X. For two local coordinate systems $\varphi_1 : U_1 \to V_1$ and $\varphi_2 : U_2 \to V_2$, the homeomorphism $\varphi_2^{-1} \circ \varphi_1 : \varphi_1^{-1}(V_1 \cap V_2) \to \varphi_2^{-1}(V_1 \cap V_2)$ is called a *transformation of local coordinates.*

local coordinates *See* local coordinate system.

local dimension If an *analytic set* A is irreducible at a point z^0, then there exists a system of local coordinates $(z_1, \cdots, z_n)$ centered at z^0 and a pair of positive numbers $d \leq n$ and k such that, in a neighborhood of

z^0, A is a k-sheeted *ramified covering space* with covering mapping φ: $(z_1, \cdots, z_n) \to (z_1, \cdots, z_d)$. The number d is called the *local dimension* of A at z^0 and is denoted by $\dim_{z^0} A$. This concept is introduced in the theory of analytic spaces.

local maximum The greatest value a function f takes on in an open interval.

local minimum The least value a function f takes on in an open interval.

local one parameter group Let V be a $\mathbf{C}^k$ manifold of dimension n with $k \geq 3$ having a countable basis. Let $U \subset V$ be open, $\epsilon > 0$, and $I := \{t \in R : |t| < \epsilon\}$. Let $g : I \times U \to V$ be a C^r mapping and $g_t : U \to V$, where $g_t(x) = g(t, x)$. Then g is a local one parameter group of $\mathbf{C}^r$ transformations of U into V if, for each $t \in I$, g_t is a $\mathbf{C}^r$ diffeomorphism of U onto an open subset of V; g_0 is the identity; and whenever $s, t, s + t \in I$ and $x, g_t(x) \in U$ we have $g_{s+t} = g_s \circ g_t(x)$.

local operator A mapping T from the $\mathbf{C}^\infty$ functions with compact support into the $\mathbf{C}^\infty$ functions such that whenever a test function φ vanishes on an open set O, then $T\varphi$ also vanishes on O. For example, differential operators are local operators.

local parameter Let G be a Lie group and X be a $\mathbf{C}^\infty$ paracompact manifold of dimension n. Consider a transformation $x \mapsto y = f(a, x)$ where $x = (x^1, \ldots, x^n)$ are coordinates in X and $y = (y^1, \ldots, y^n)$ are coordinates in Y and $(a^1, \ldots, a^n)$ are coordinates for a neighborhood of e in G. The $(a^1, \ldots, a^n)$ are called the *local parameters* of the transformation.

local transformation An atlas defining a structure on a topological manifold. *See* atlas.

locally Cartan pseudoconvex domain Let G be a domain in $\mathbf{C}^n$ and $z^0 \in \partial G$. If there exists an open neighborhood U of z^0 such that every connected component of $G \cap U$ is a domain of holomorphy, then G is said to be *Cartan pseudoconvex* at z^0.

locally convex space A topological vector space $\mathbf{X}$ in which the topology has a base consisting of convex sets. The topology with which $\mathbf{X}$ becomes a locally convex space is called a *locally convex topology.*

locally convex topology *See* locally convex space.

locally Levi pseudoconvex domain Let G be a domain in $\mathbf{C}^n$ and $z^0 \in \partial G$. If every 1-dimensional analytic set that has z^0 as an ordinary point contains points not belonging to $G \cup \{z^0\}$ in the neighborhoods of z^0, then G is called *Levi pseudoconvex* at z^0.

locally symmetric space A topological vector space L such that for each neighborhood U of the origin O of L there exists a neighborhood $V \subset U$ such that for each $v \in V$ the line segment joining v and O is contained in V and $v \in V$ implies $-v \in V$.

locus A set of points satisfying some given conditions.

Example: The locus of points in the plane that are equidistant from a fixed point form a circle.

log-log coordinates *See* logarithmic coordinates.

logarithm function Any function of the form $y = \log_a(x)$ where $a^y = x$ and a is a positive constant (the base of the logarithm). *See* exponential function.

logarithmic coordinates Coordinates that use a logarithmic scale. That is, in the $\log_a(\cdot)$–scale, the logarithmic coordinate (c, d) corresponds to the rectangular Cartesian coordinate (a^c, a^d). Also *log-log coordinates.*

logarithmic curve The set of points (x, y) in the rectangular Cartesian plane satisfying $y = \log_a x, a > 1$.

logarithmic function *See* logarithm function.

logarithmic function of base a *See* logarithm function.

logarithmic integral The analytic function $li(z)$ defined, for $|\arg z| < \pi$, $|\arg(1 - z)| < \pi$, by $li(z) = \int_0^z \frac{dt}{\log t}$, for $z \neq 1$, where the integral is over any path in the plane which includes the real intervals $(-\infty, 0]$ and $[1, \infty)$.

logarithmic series The Taylor series for $\log_e(1 + x)$; that is, the series

$$\sum_{n=1}^{\infty}(-1)^{n+1}\frac{x^n}{n}, \quad -1 < x \leq 1.$$

logarithmic singularity The function $f(z)$ has a logarithmic singularity at $z = w$ in the complex plane if

$$f(z) = c \log|z - w| + h(z)$$

for some constant $c \in R$ and some harmonic function $h(z)$ in a neighborhood of $z = w$.

logarithmic spiral The set of points (r, θ) in the polar plane satisfying, $\log r = a\theta, a > 0$.

logarithmically convex domain A Reinhardt domain G such that $\{(\log|z_1|, \ldots, \log|z_n|) : z \in G \cap \mathbf{C}^{*n}\}$ is a convex subset of $\mathbf{R}^n$.

long line Let ω_1 denote the first uncountable ordinal and $[0, \omega_1)$ be the set of all ordinals from 0 to ω_1. The long line is the set $\mathbf{R} \times [0, \omega_1)$ with the dictionary (or lexicographic) order topology.

Lorentz space Let $p \in [1, \infty)$ and $q \in [1, \infty]$. The Lorentz space $L(p, q)$ of measurable functions on $\mathbf{R}^n$ is given by

$$L(p, q) := \{f : \|f\|_{(p,q)} < \infty\}$$

where

$$\|f\|_{(p,q)} := \int_0^{\infty}\left[(t^{1/p} f^{**}(t))^q \frac{dt}{t}\right]^{1/q},$$

for $p, q \in [1, \infty)$, while if $p \in [1, \infty]$, and $q = \infty$, define

$$\|f\|_{(p,q)} := \sup_{t>0} t^{1/p} f^{**}(t),$$

where

$$f^{**}(t) = \frac{1}{t}\int_0^t f^*(r)\, dr,$$

$$f^*(t) = \inf\{s > 0 : |E_s^f| \leq t\},$$

and

$$E_s^f = \{x : |f(x)| > s\}.$$

lower bound A number that is less than or equal to all numbers in a given nonempty set S of the real number set. That is, if S is a nonempty subset of the real numbers and if $a \leq s$, for all $s \in S$, then a is a lower bound of S.

lower limit (1.) For a sequence $S = \{a_1, a_2, \ldots\}$, the *limit inferior* of the sequence. It can be described as the infimum of all limit points of S, in which case, it is denoted $\underline{\lim}_{n\to\infty} a_n$, and can be found by $\underline{\lim}_{n\to\infty} a_n = \lim_{n\to\infty} \inf_{k\geq n} a_k$.
(2.) For a sequence of sets $A_1, A_2, \ldots$, it is not the same as the limit inferior, but rather the lower limit for this sequence of sets, denoted $Li_{n\to\infty} A_n$, is the set of points, p, such that every neighborhood of p intersects all A_n, for n sufficiently large.
(3.) The left endpoint of an interval over which an integrable function is being integrated. For example, in $\int_a^b f(x)dx$, the lower limit (of integration) is a.

lower limit function The lower limit of a function (also called the *lower envelope*)

$f : A \to R$, where A is a nonempty subset of R, is the function $g : A \to R \cup \{-\infty, +\infty\}$ defined by

$$g(a) = \sup_{\delta>0} \inf_{0<|x-a|<\delta} f(x), \ a \in A.$$

lower semicontinuity A real-valued function f defined on a set S is *lower semicontinuous* at a point $s_0 \in S$, if for every $\epsilon > 0$, there is a neighborhood N of s_0, such that $f(s) > f(s_0) - \epsilon$, for all $s \in N$. If f is lower semicontinuous at each point $s \in S$, then f is lower semicontinuous on S.

lower semicontinuous function at point x_0 *See* lower semicontinuity.

lower semicontinuous function on a set *See* lower semicontinuity.

loxodromic spiral A solid curve C ($C : [a, b] \to \mathbf{R}^3$ continuous), on a surface of revolution, such that C intersects the meridians at a constant angle.

Example: If S is the cylinder formed by rotating $y = 1$ about the x-axis, for $0 \leq x \leq 2\pi$, the curve $C : [0, 2\pi] \to R^3$ given by $C(x) = (x, \cos x, \sin x)$ is a Loxodromic spiral with respect to S.

lune A spherical figure formed by two great semicircles which have the same endpoints. *See also* spherical wedge.

Luzin space An uncountable topological space for which every nowhere dense subset is countable.

Luzin's Unicity Theorem Let $E \times F$ be a product space where E is analytic, and let H be an $(E \times F)$−bianalytic region from $E \times F$. Then $\{y \in F : H(y)$ is a singleton$\}$ is F−coanalytic.

Lyapunov function Consider an autonomous system $y' = f(y)$, where we assume $f(y)$ is continuous in an open set containing 0 and $f(0) = 0$ (so that $y(t) = 0$ is a solution). A function $V(y)$, defined in a neighborhood of $y = 0$ is called a *Lyapunov function* for the above problem if (i.) V has continuous partial derivatives; (ii.) $V(y) \geq 0$ with $V(y) > 0$, if $y \neq 0$; (iii.) the trajectory derivative $\dot{V}(y)$, along a solution, is nonpositive. The *trajectory derivative along* y of (iii.) is $\dot{V} = \frac{dV(y(t))}{dt}$.

Existence of a Lyapunov function implies stability of the solution $y(t) = 0$.

Also *Liapunov function.*

Lyapunov stability *See* asymptotic stability.

Lyapunov-Schmidt procedure A method for finding the solution of a nonlinear bifurcation problem in infinite dimensions by reducing the problem to finding solutions of a finite number of nonlinear equations in a finite number of real or complex variables.

M

M. Riesz's Convexity Theorem *See* Riesz-Thorin Theorem, which Riesz proved in the case in which both measure spaces are finite sets, with counting measure and $1 \le p \le q \le \infty$.

MacLaurin series A Taylor series centered at, or expanded about, $x = 0$. *See* Taylor series.

major axis of an ellipse The longer of the two perpendicular line segments about which an ellipse is symmetric that connects the foci of the ellipse.

major function *See* Perron integrable function.

manifold An n-dimensional manifold is a connected Hausdorff space such that each point has a neighborhood which is homeomorphic to an open ball in $\mathbf{R}^n$. *See also* atlas.

Mannheim curve The curve given by

$$\tau^2(s) = \kappa(s)\frac{[1 - a\kappa(s)]}{a}$$

where a is a constant, τ is the torsion, κ is the curvature, and s is the arc length.

mantissa The mantissa of a real number is the decimal part of its decimal expansion. For example, the real number 3.14159 has mantissa .14159.

mapping A function (i.e., a rule or an assignment) of exactly one element in a set to each object from another set. *See* function.

mapping of class $\mathbf{C}^r$ A function which is r times continuously differentiable where $r \in \mathbf{Z} \cup \{\infty\}$.

Martin boundary Let (E, ρ) be a locally compact metric space with a countable basis, and let δ be a metric on E, compatible with the topology of E, such that the Cauchy completion (E, δ) coincides with the Alexandroff compactification. Let r be a positive m-integrable function (m the measure on E), such that the potential $\hat{G}r$ is bounded, continuous and strictly positive. Let $\{f_n\}_{n=1}^{\infty}$ be dense in the space of functions of compact support on E. Let $\hat{K}f_n = (\hat{G}f_n)/\hat{G}r$, and for $x, y \in E$, define the pseudometric $d(x, y) = \sum_{n=1}^{\infty} W_n|\hat{K}f_n(x) - \hat{K}f_n(y)|$, where the $W_n > 0$ are chosen such that $\sum_{n=1}^{\infty} W_n|\hat{K}f_n| < \infty$. E^*, the Cauchy completion of the metric space $(E, \delta + d)$ is compact and is called the *Martin compactification* for the random variable X started with distribution rm. The set $M = E^* \backslash E$ is called the *Martin boundary* or *Martin exit boundary* for X started with distribution rm. Martin boundary and Martin compactification play an important role in Markov chains.

Martin compactification *See* Martin boundary.

Martin kernel Every harmonic function $h(z)$ has the integral representation

$$h(z) = \int_{\Delta M} K(z, \xi)\, d\mu(\xi), \quad z \in R,$$

where μ is a measure on ΔM (the Martin boundary of the given domain). $K(z, \xi)$ is the Martin kernel given by

$$K(z, \xi) = \frac{G(z, \xi)}{G(z, \xi_0)}, \quad \text{if } z \neq \xi_0,$$

while $K(z, \xi) = 0$ for $z = \xi_0$, $\xi \neq \xi_0$, and $K(z, \xi) = 1$ for $z = \xi = \xi_0$ where $z, \xi \in R$, $\xi_0 \in R$ is fixed, and G is the generalized Green's function for the domain. (That is, for fixed z, $G(z, \xi)$ is a positive harmonic in ξ except at $\xi = z$ where it has a logarithmic singularity, is symmetric in (z, ξ), and approaches zero at every regular boundary point of the domain and ∞ if the domain is unbounded.) *See* Martin boundary.

0-8493-0320-6/00/$0.00+$.50

Maslov index The Maslov index of a curve on a Lagrangian surface is the intersection index with the cycle of singularities of the Lagrangian projection.

Mathieu function of the first kind Consider Mathieu's differential equation written in the form

$$y'' + (\lambda - 2q\cos 2x)y = 0,$$

where λ, q are parameters. The *Mathieu functions of the first kind*, denoted ce_n and se_n, are solutions of the differential equation (for different values of λ), given by

$$ce_{2n}(x,q) = \sum_{k=0}^{\infty} A_{2k}^{(2n)} \cos 2kx,$$

$$ce_{2n+1}(x,q) = \sum_{k=0}^{\infty} A_{2k+1}^{(2n+1)} \cos(2k+1)x,$$

$$se_{2n}(x,q) = \sum_{k=0}^{\infty} B_{2k}^{(2n)} \sin 2kx,$$

and

$$se_{2n+1}(x,q) = \sum_{k=0}^{\infty} B_{2k+1}^{(2n+1)} \sin(2k+1)x,$$

where the As and Bs are known constants that depend on the parameter q.

Mathieu function of the second kind The general solution of Mathieu's differential equation written in the form

$$y'' + (\lambda - 2q\cos 2x)y = 0,$$

can be expressed either as

$$y(x) = Ace_n(x,q) + Bfe_n(x,q),$$

or

$$y(x) = Cse_n(x,q) + Dge_n(x,q),$$

where A, B, C, D are arbitrary constants and n is an integer, and where the functions fe_n and ge_n are known as *Mathieu functions of the second kind* and are obtained from the Mathieu functions of the first kind, ce_n and se_n, respectively, by the method of variation of parameters. *See* Mathieu function of the first kind, method of variation of parameters.

Mathieu's differential equation Any differential equation of the form

$$y'' + (a + b\cos(2x))y = 0.$$

The general solution of *Mathieu's differential equation* is given by

$$y = Ae^{kx} f(x) + Be^{-kx} f(-x)$$

where k is a constant and f is a 2π-periodic function.

matrix of a bilinear form Let V be a vector space of dimension n and F a bilinear form on V. Let $B = (e_1, \ldots, e_n)$ be a fixed, ordered basis of V and

$$F(e_i, e_j) = a_{ij}, \; i, j = 1, 2, \ldots, n.$$

Then F determines the matrix $A = (a_{ij})$ which is the matrix of the bilinear form F relative to the basis B.

maximal accretive operator An accretive operator with no proper accretive extension. *See* accretive operator.

maximal dissipative operator A dissipative operator with no proper dissipative extension. *See* dissipative operator.

maximally almost periodic group An amenable group G, which is injectable into a compact group, is called a *maximally almost periodic group*. *See* amenable group.

maximum principle If f is analytic and nonconstant in the interior of a connected region R, then $|f(z)|$ has no maximum value in the interior of R.

maximum value of a function Let A be a nonempty set, and let $f : A \to \mathbf{R}$. $f(a)$ is the maximum value of f, if $f(x) \le f(a)$, for all $x \in A$. *See* local maximum.

mean convergence The sequence of functions $\{f_k\}$ converges in the mean (of order n) to f on the set S provided

$$\lim_{k\to\infty} \int_S |f_k(x) - f(x)|^n \, dx = 0.$$

Mean convergence of order n is synonymous with *convergence in the mean of order n* or *convergence in L^n*.

mean curvature The *mean curvature*, K_m, of a surface is the average of the principal curvatures of the surface. That is, $K_m = (\rho_1 + \rho_2)/2$ where ρ_1, ρ_2 are the principal curvatures of the surface.

mean value of a function For the integrable function f on $[a, b]$, the quantity given by

$$\frac{1}{b-a} \int_a^b f(x) \, dx.$$

Also *average value of f on* $[a, b]$.

Mean Value Theorem of differential calculus Let f be a continuous function on $[a, b]$ and differentiable on (a, b). Then there is a number $c \in (a, b)$ such that

$$f'(c) = \frac{f(b) - f(a)}{b - a}.$$

Mean Value Theorem of integral calculus If f is continuous on $[a, b]$, then there is a number $c \in [a, b]$ such that

$$f(c) = \frac{1}{b-a} \int_a^b f(x) \, dx.$$

measure Let S be a set and let A be an algebra of subsets of S. An extended, nonnegative, real-valued function, μ, defined on A is called a measure if

1. $\mu(\emptyset) = 0$,
2. $\mu(U \cup V) = \mu(U) + \mu(V)$, where $U, V \in A$ and $U \cap V = \emptyset$, and
3. $\mu(\cup_{n=1}^{\infty} U_n) = \sum_{n=1}^{\infty} \mu(U_n)$ for every countable subcollection $\{U_n\}_{n=1}^{\infty}$ of pairwise disjoint subsets of A for which $\cup_{n=1}^{\infty} U_n \in A$.

measurable function A real-valued function $f(x)$ such that $\{x : f(x) > \alpha\}$ is a measurable set, for all real α. Or, a complex linear combination of measurable functions. *See* real-valued function.

measurable set For a measure on a σ-algebra M, a set belonging to M.

Mellin transform The Mellin transform F, of a real-valued function f, defined on the positive real numbers, is a complex-valued function of a complex variable, z, defined by $F(z) = \int_0^{\infty} f(x) x^{z-1} dx$.

Mercer's Theorem If $K(x, y)$ is a continuous, nonnegative definite Hermitian kernel on the square $a \leq x, y \leq b$, with positive eigenvalues $\{\lambda_i\}_{i=1}^{\infty}$ and corresponding eigenfunctions $\{\phi_i\}_{i=1}^{\infty}$, then

$$K(x, y) = \sum_{i=1}^{\infty} \frac{\phi_i(x)\bar{\phi}_i(y)}{\lambda_i}$$

uniformly and absolutely on the square $a \leq x, y \leq b$.

meridian A meridian of a surface is a section of the surface of revolution containing the axis of revolution.

meromorphic curve *See* meromorphic function.

meromorphic function A function which is defined and analytic except at isolated points, all of which are poles.

meromorphic mapping *See* meromorphic function.

method of moving frames The method of moving frames associates with each point on a curve (where the curvature $\kappa \neq 0$) the orthonormal set of vectors $\{T, N, B\}$ where T is the tangent vector field, N is the normal vector field ($N(s) = \frac{T'(s)}{\kappa(s)}$), and B is the binormal vector field ($B(s) = T(s) \times N(s)$).

method of Picard iterations *See* method of successive approximations.

method of successive approximations The process of approximating a solution, $y(x)$, of the first order differential equation, $y'(x) = f(x, y)$, satisfying the initial condition $y(x_0) = y_0$, by a sequence $\{y_n(x)\}_{n=1}^{\infty}$ defined by $y_0(x) = y_0$ and, for $n \geq 1$,

$$y_n(x) = y_0 + \int_{x_0}^{x} f(s, y_{n-1}(s))\, ds.$$

method of successive iterations *See* method of successive approximations.

method of variation of constants *See* method of variation of parameters.

method of variation of parameters Let $X(t)$ be a fundamental matrix solution of $X' = A(t)X$, where $A(t)$ is an $n \times n$ matrix function which is continuous on an interval I. Given a continuous $n \times 1$ vector function, $f(t)$, defined on I, the method of variation of parameters seeks a solution of $x' = A(t)x + f(t)$ of the form $z(t) = X(t)v(t)$ for some $v(t)$. A function $v(t)$ determined by this method is given by

$$v(t) = \int_{t_0}^{t} X^{-1}(s) f(s)\, ds,$$

where t_0 is a fixed point in I. A particular solution of $x' = A(t)x + f(t)$ thus produced is given by

$$z(t) = \int_{t_0}^{t} X(t)X^{-1}(s) f(s)\, ds.$$

This formula for $z(t)$ is called the *formula for the variation of parameters*. Also *variation of constants*.

Metric Comparison Theorem On a semi-Riemannian manifold M there is a unique connection D such that

(1.) $[V, W] = D_V W - D_W V$, and

(2.) $X(V, W) = (D_X V, W) + (V, D_X W)$

for all $X, V, W \in \chi(M)$. D is called the Levi–Civita connection of M, and is characterized by the Koszul formula

$$\begin{aligned} 2(D_V W, X) &= V(W, X) + W(X, V) \\ &\quad - X(V, W) - (V, [W, X]) \\ &\quad + (W, [X, V]) + (X, [V, W]). \end{aligned}$$

$D_V W$ is the covariant derivative of W with respect to V for the connection D; $[V, W]$ is the bracket of V and W; $\chi(M)$ is the set of all smooth vector fields on M, and $(\ ,\ )$ denotes metric tensor.

The above theorem has been called the miracle of semi-Riemannian geometry.

metric connection Let N and M be manifolds and $\mu : N \to M$ be a $\mathbf{C}^\infty$ mapping. A *connection*, D, on μ assigns to each $t \in N_n$ an operator D_t which maps vector fields over μ into $M_{\mu n}$ and satisfies the following conditions for $t, \nu \in N_n$, X, Y vector fields over μ, C^∞ functions $f : N \to R$, $a, b \in R$, and C^∞ vector fields Z on N:

1. $D_{at+b\nu} X = a D_t X + b D_\nu X$,
2. $D_t(aX + bY) = a D_t X + b D_t Y$,
3. $D_t(fX) = (tf)X(n) + (fn) D_t X$,
4. The vector fields $D_Z X$ over μ defined by $(D_Z X)(n) = D_{Z(n)} X$ is C^∞.

A connection D on a map $\mu : N \to M$ is said to be *compatible* with the metric $\langle\ ,\ \rangle$ on μ or is a *metric connection* if parallel translation along curves in N preserves inner products.

metric space A pair (S, d), where S is a set and $d : S \times S \to [0, \infty)$ is a metric. That is, for each $x, y, z \in S$,

1. $d(x, y) = 0$ if and only if $x = y$,
2. $d(x, y) = d(y, x)$, and
3. $d(x, y) \leq d(x, z) + d(z, y)$.

For example, the set of real numbers, $\mathbf{R}$, with the absolute value metric, $d(x, y) = |x - y|$, forms a metric space.

metric vector space A pair (X, G), of a vector space X and an associated metric tensor $G : X \times X \to \mathbf{R}$.

Meusnier's Theorem The normal curvature of a curve on a surface equals the product of its curvature by the cosine of the angle θ between the principal normal of the curve and the normal of the surface.

micro-analytic hyperfunction A hyperfunction is a linear functional on the space of analytic functions. If M is a real analytic oriented manifold, let $S^*M = (T^*M\backslash\{0\})/(0, \infty)$ be the cotangent bundle. For a point $(x_0, \sqrt{-1}\alpha_0\infty) \in \sqrt{-1}S^*M$, a hyperfunction f defined on a neighborhood of x_0 is called a *micro-analytic hyperfunction* at $(x_0, \sqrt{-1}\alpha_0\infty)$ if f can be expressed as a sum of boundary values, $f = \sum_i b_{\Gamma_i}(F_i)$ in a neighborhood of x_0 with the inner product $\langle\beta, \alpha_0\rangle < 0$, for all $\beta \in \cup_i\Gamma_i$.

microdifferential equation An equation of the form

$$\sum_{k=-\infty}^{m} P_k u = 0,$$

where $\sum_k P_k$ is a microdifferential operator on an open subset U of P^*X, u is a microfunction on U, and P^*X is the cotangent projective bundle of a complex analytic manifold X. *See* microdifferential operator, microfunction.

microdifferential operator Given a complex analytic manifold X, let P^*X be its cotangent projective bundle, and let $\pi : T^*X\backslash\{0\} \to P^*X$ be the canonical projection. For an open subset U of P^*X and an integer m, a *microdifferential operator* on U is a formal expression

$$\sum_{k=-\infty}^{m} P_k(z, \xi)$$

such that (a.) for each $k \leq m$, P_k is a holomorphic function in $(z, \xi) \in \pi^{-1}(U)$, which for fixed $z \in X$ is homogeneous of degree k in $\xi \in T^*X$, and (b.) for every compact subset K of $\pi^{-1}(U)$, there exists a constant $C_K > 0$ such that on K,

$$|P_k(z, \xi)| \leq C_K^{|k|}|k|!, \ k \leq m.$$

microfunction Given a real analytic oriented manifold M, with $B(M)$ its space of hyperfunctions and S^*M the cotangent bundle, and given U an open subset of M, a function u is a microfunction on U, if $u \in B(m)/\{f \in B(M) : SS(f) \cap U = \emptyset\}$, where $SS(f)$ is the set of points in $\sqrt{-1}S^*M$ for which f is not micro-analytic. *See* micro-analytic hyperfunction.

microlocal analysis The study (via microfunctions), of the local structure of solutions of partial differential equations on analytic manifolds by means of the cotangent bundle. *See* microfunction.

midpoint of a line segment The point M on the line segment $\overline{AB}$ which is equidistant from both A and B. In analytic geometry, given the points $P_1(x_1, y_1)$, $P_2(x_2, y_2)$, the midpoint, M, of $\overline{P_1P_2}$ is given by $M = (\frac{x_1+x_2}{2}, \frac{y_1+y_2}{2})$.

minimal operator Let $p, q \in [1, \infty]$, $D = \frac{d}{dt}$, and consider the ordinary differential expression

$$\tau = \sum_{k=0}^{n} a_k D^k,$$

where $a_k \in \mathbf{C}^k(I)$, $0 \leq k \leq n$, $a_n(t) \neq 0$, $t \in I$ (arbitrary interval). The minimal operator corresponding to (τ, p, q), denoted by $T_{0,\tau,p,q}$, is defined to be the minimal closed extension of $T^R_{\tau,p,q}$ when $1 \leq p, q < \infty$. When $1 < p, q \leq \infty$, the minimal operator is defined to be $T'_{\tau^*,q',p'}$. The operator $T^R_{\tau,p,q}$ is defined to be the restriction of $T_{r,p,q}$ to those functions

$$f \in D(T_{r,p,q}) \\ = \{f : f \in A_n(I) \cap L_p(I), \tau f \in L_p(I)\},$$

where $A_0(I) = C(I)$, and where, for each positive integer n, $A_n(I)$ denotes the set of complex-valued functions f on I for which $f^{(k-1)} = D^{k-1}f$ exists and is absolutely continuous on every compact subinterval of

1. For $g \in A_n(I)$, $\tau^* g = \sum_{k=0}^{n} b_k g^{(k)}$ a.e., where

$$b_k = \sum_{j=k}^{n} (-1)^j \binom{j}{k} D^{j-k} a_j, \quad 0 \le k \le n.$$

Finally,

$$T'_{\tau^*, q', p'} = \overline{T}^R_{\tau, p, q}, \quad 1 < p, q < \infty.$$

minimal surface A surface whose mean curvature is 0. *See* mean curvature.

minimizing sequence If A is a bounded, positive, self-adjoint operator and $Ax^* = y$, a minimizing sequence $\{x_n\}$ is a sequence determined by the method of steepest descent such that

$$||x_{n+1} - x^*|| \le ||x_n - x^*||$$

and

$$Q(x_n) - Q(x^*) = (A(x_n - x^*), x_n - x^*),$$

where $(A(x_n - x^*), x_n - x^*) = O(1/n)$ and Q is the quadratic functional such that (in some sense) $Ax - y$ is the derivative of $Q(x)$.

minimum value of a function Let A be a nonempty set, and let $f : A \to \mathbf{R}$. $f(a)$ is the minimum value of f, if $f(x) \ge f(a)$, for all $x \in A$. *See* local minimum.

Minkowski function *See* gauge function.

minor function *See* Perron integrable function.

modified Mathieu function of the first kind The functions given by

$$Ce_k(x, q) = ce_k(ix, q),$$

and

$$Se_k(x, q) = -i\, se_k(ix, q),$$

where $i = \sqrt{-1}$ and ce, se are the Mathieu functions of the first kind. *See* Mathieu function of the first kind.

modular form Any automorphic form on a modular group. *See* modular group.

modular function An automorphic function for a modular group. *See* modular group.

modular group The transformation group consisting of $SL_2(\mathbf{Z})$ or any of its subgroups of finite index are known as modular groups, where

$$SL_2(\mathbf{Z}) = \left\{ \begin{bmatrix} a & b \\ c & d \end{bmatrix} \in SL_2(\mathbf{R}) \right\}$$

for $a, b, c, d \in \mathbf{Z}$, and, in general,

$$SL_n(\mathbf{R}) = \{\alpha \in M_n(\mathbf{R}) \mid \det \alpha = 1\}$$

and

$$M_n(S) = \{n \times n \text{ matrices over } S\}.$$

module Let R be a ring. A module A over R (also called an R-module) is an additive Abelian group such that $ra = ar \in A$ for all $r \in R$ and $a \in A$ and that satisfies the following properties for each $r_1, r_2, r_3 \in R$ and $a_1, a_2, a_3 \in A$.
1. $r(a_1 + a_2) = ra_1 + ra_2$,
2. $(r_1 + r_2)a = r_1 a + r_2 a$, and
3. $r_1(r_2 a) = (r_1 r_2)a$.

modulus The modulus of a complex number $z = a + bi$ is given by

$$|z| = \sqrt{a^2 + b^2}.$$

For a real number, the modulus is just the absolute value.

See also periodicity modulus.

modulus of continuity Let $f : A \to B$, where $A, B \subset \mathbf{R}$. If there exists a function $m_f : (0, \infty) \to (0, \infty)$ such that, given $\epsilon > 0$ and $x, y \in A$, $|x - y| < m_f(\epsilon)$ implies $|f(x) - f(y)| < \epsilon$, then m_f is called the modulus of continuity for f. Such a function m_f exists for each uniformly continuous f.

moment generating function For a random variable X, with density $f(x)$, the moment generating function is the expected value of e^{tX} given by

$$E(e^{tX}) = \int_{-\infty}^{\infty} e^{tx} f(x)dx,$$

if X is continuous, and given by

$$E(e^{tX}) = \sum_x e^{tx} f(x),$$

if X is discrete.

Monge-Ampere equation The partial differential equation

$$\det \left(\frac{\partial^2 u}{\partial x_j \partial x_k} \right)_{j,k=1}^n = 0,$$

for u in a domain $D \subset R^n$. Also *Monge-Ampere differential equation.*

monodromy group The matrix group generated by the monodromy matrices $T_1, \ldots, T_m$. *See* monodromy matrix.

monodromy matrix Let $q_1, \ldots, q_n$ be holomorphic functions in a simply connected domain D on the Riemann sphere excluding the points (possibly even poles of the q_i) $a_1, \ldots, a_m$. Fix a_0 distinct from $a_1, \ldots, a_m$, let U be a simply connected neighborhood of a_0, and consider the differential equation

$$w^{(n)} + q_1(z)w^{(n-1)} + \cdots + q_n(z)w = 0.$$

In U there is a holomorphic fundamental system of solutions of the differential equation which is given by

$$w(z) = \begin{bmatrix} w_1(z) \\ w_2(z) \\ \vdots \\ w_n(z) \end{bmatrix}.$$

Let γ_j be a simple, closed curve starting and ending at a_0 which traverses around a_j in a positive orientation. Continuing $w(z)$ analytically along γ_j we get the fundamental system of solutions

$$\gamma_j w(z) = w^j(z) = w(z)T_j$$

where T_j is a constant, nonsingular $n \times n$ matrix which is called a *monodromy matrix.*

Monodromy Theorem Suppose $f(z)$ is analytic in a finite domain D and that D is bounded by a simple, closed curve C. Suppose further that f is continuous in $D \cup C$, and f is single-valued for each z on C. Then f is single-valued for every $z \in D$.

monogenic function If

$$\lim_{z \to z_0} \frac{f(z) - f(z_0)}{z - z_0}$$

is the same for all paths in the complex plane, then $f(z)$ is said to be *monogenic* at z_0. Monogenic, therefore, essentially means having a single derivative at a point. Functions are either monogenic or have infinitely many derivatives (in which case they are called polygenic); intermediate cases are not possible.

monoidal transformation Let $p : \mathbf{C}^{n+1} \times P^n(C) \to C^{n+1}$ denote the projective mapping and $\tilde{\mathbf{C}}^{n+1} = G([\rho])$ where $G([\rho])$ denotes the irreducible component of X containing the Hopf fibering ρ where $X := \{((z^j), [w^k]) \in \mathbf{C}^{n+1} \times P^n(\mathbf{C}) : z^j w^k - z^k w^j = 0, \ \forall j, k\}$ and $P^n(\mathbf{C})$ is the projective space on the complex plane $\mathbf{C}$. The holomorphic mapping

$$\tilde{p} : \tilde{\mathbf{C}}^{n+1} \to \mathbf{C}^{n+1}, \quad \tilde{p} := P|_{\tilde{C}^{n+1}}$$

is called the *monoidal transformation* or *blowing up.*

monotone operator Let B be a Banach space and K an ordered cone in B with induced order $\leq$. An operator $T : B \to B$ is *monotone increasing* if for any $x, y \in B$, with $x < y$ (that is, $x \leq y$ and $x \neq y$), then $Tx \leq Ty$. T is *monotone decreasing* if for any $x, y \in B$, with $x < y$, then $Ty \leq Tx$.

monotonic function A function which is either monotonically decreasing or monotonically increasing. See *monotonically de-*

creasing function, monotonically increasing function.

monotonic sequence A sequence which is either monotonically decreasing or monotonically increasing. See *monotonically decreasing sequence, monotonically increasing sequence.*

monotonically decreasing function A function, $f : S \to \mathbf{R}$, where S is a nonempty subset of $\mathbf{R}$, such that, for each $s_1, s_2 \in S$, $s_1 < s_2$ implies $f(s_1) \geq f(s_2)$.

monotonically decreasing sequence A sequence $\{s_n\}_{n=1}^{\infty}$ such that $s_i \geq s_{i+1}$ for each $1 \leq i < \infty$.

monotonically increasing function A function, $f : S \to R$, where S is a nonempty subset of R such that, for each $s_1, s_2 \in S$, $s_1 < s_2$ implies $f(s_1) \leq f(s_2)$.

monotonically increasing sequence A sequence $\{s_n\}_{n=1}^{\infty}$ such that $s_i \leq s_{i+1}$ for each $1 \leq i < \infty$.

Morera's Theorem If $f(z)$ is defined and continuous in a region R and if $\int_\gamma f\, dz = 0$ for all closed curves γ in R, then $f(z)$ is analytic in R.

Morse's Lemma Suppose that f is twice continuously differentiable and that $f(0) = f'(0) = 0$, but $f''(0) \neq 0$. Then, in a neighborhood of $x = 0$, $f(x) = x^2 g(x)$, where g is continuous and $2g(0) = f''(0)$.

moving coordinate system *See* method of moving frames.

moving coordinates *See* method of moving frames.

moving frame The moving trihedral formed by the tangent, normal and binormal lines to a curve in three-dimensional space.

multi-index Numbering for the coefficients or terms in a sequence or series, based on the Cartesian product of $n(\geq 2)$ copies of the nonnegative integers. For example, the power series expansion of a function $f(x, y)$ of two variables

$$f(x, y) = \sum_{i,j=0}^{\infty} a_{ij} x^i y^j,$$

involves the multi-index $(i, j) \in \mathbf{Z}_+ \times \mathbf{Z}_+$.

multilinear form A real- or complex-valued function of several vector space variables which is linear in each variable separately.

multilinear mapping A mapping, from a cartesian product of vector spaces to a vector space, which is linear in each variable separately.

multiplanar coordinates Multiple Cartesian coordinates.

multiple Fourier series Fourier series for a function of more than one variable. Hence, for $f(s, t)$, periodic, of period 2π in s and t, the series expansion

$$f(s, t) \sim \sum_{j,k=-\infty}^{\infty} a_{jk} e^{i(js+kt)},$$

where

$$a_{jk} = \frac{1}{(2\pi)^2} \int_0^{2\pi} f(s, t) e^{-i(js+kt)} ds dt.$$

multiple integral An expression of the form

$$\int\int \cdots \int f(x_1, x_2, \ldots, x_k) dx_1 dx_2 \cdots dx_k$$

where f is a function of several variables. A multiple integral requires integrating over each variable separately. *See also* double integral, triple integral.

multiple point (1.) A point in the domain of a given function $f(x_1, \ldots, x_k)$ at which

all of the first $n-1$ partial derivatives of f equal zero, while at least one of the nth partial derivatives is non-zero; in this case, the point is a multiple point of order n.
(2.) A point where two or more branches of a curve intersect each other.

multiple-valued function A function which assigns more than one value to certain elements of the domain. A k-valued function assigns k distinct values to almost every element of the domain; for example, $f : \mathbf{C} \to \mathbf{C}$ defined by $f(z) = \sqrt[k]{z}$ is k-valued.

Technically, the definition of a function does not permit such multiple-valued behavior. However, it is convenient to use such an abuse of terminology in dealing with mappings such as the above example.

multiplication operator A linear operator on a vector space whose elements are functions, which is defined as multiplication by a specified associated function; for example, given the function ϕ, the associated multiplication operator M_ϕ is defined by

$$M_\phi(f)(x) = \phi(x)f(x).$$

multiplicative function A function which distributes over multiplication; a function f such that

$$f(xy) = f(x)f(y).$$

multiplicity function A cardinal number-valued function associated with a specific linear operator T which assigns to each complex number λ the dimension of the kernel of $T - \lambda I$; hence, the function M_T defined by

$$M_T(\lambda) = \dim(\ker(T - \lambda I)).$$

multiplicity of the eigenvalue The dimension of the corresponding eigenspace, the subspace spanned by the corresponding eigenvectors of a linear operator; hence, for a linear operator T and an eigenvalue λ, the number dim ker$(T - \lambda I)$.

multiply connected region A region which is not simply connected; a region in which there exists some simple closed curve, which cannot be shrunk continuously to a point without leaving the region. An m-connected region is one whose complement has m simply connected components.

multipolar coordinates Multiple polar coordinates. *See* spherical coordinates.

multivalued function *See* multiple-valued function.

multivariable function A function of more than one variable.

multivariate Involving more than one distinct, although not necessarily independent, variables; the term often refers to random variables.

multivariate analysis The study of mathematical methods of constructing optimal designs for the collection, organization and processing of multivariate statistical data with the intention of clarifying the nature and structure of the item in question for the purpose of obtaining practical and scientific inferences.

multivariate analysis of variance Suppose two simple statistical hypotheses, say H_1 and H_2, specify respectively the mean of n k-variate normal populations with common covariance matrix $\Sigma = (\sigma_{ij})$ $i, j = 1, 2, \ldots, k$. For the ith observation, we have the regression model

$$z_i = y_i - Bx_i, \quad i = 1, 2, \ldots, n \qquad (1)$$

where $z_i' = (z_{i1}, z_{i2}, \ldots, z_{ik_2})$, $y_i' = (y_{i1}, y_{i2}, \ldots, y_{ik_2})$, $x_i' = (x_{i1}, x_{i2}, \ldots, x_{ik_1})$, $B = (\beta_{rs})$, $r = 1, 2, \ldots, k_2$, $s = 1, 2, \ldots, k_1$, $k_1 < n$, $k_2 < n$, B of rank $\min(k_1, k_2)$. We may also express (1) as the one over-all regression model

$$Z = Y - XB', \qquad (2)$$

where $Z' = (z_1, z_2, \ldots, z_n)$, $Y' = (y_1, y_2, \ldots, y_n)$, $X' = (x_1, x_2, \ldots, x_n)$ with Z' and Y' $k_2 \times n$ matrices and X' a $k_1 \times n$ matrix.

We assume that:

(a.) the z_i are independent normal random $k_2 \times 1$ matrices (vectors) with zero mean and common covariance matrix Σ,
(b.) the $x_{ij}, i = 1, 2, \ldots, n, j = 1, 2, \ldots, k_1$, are known,
(c.) X is of rank k_1,
(d.) $B = B^1$ and $B = B^2$ are parameter matrices specified respectively by the hypotheses H_1 and H_2,
(e.) the y_i are stochastic $k_2 \times 1$ matrices, and $E_1(Y) = XB^{1'}$, $E_2(Y) = XB^{2'}$.

System (2) together with assumptions (a)–(e) constitute the so-called *multivariate analysis of variance.*

multivariate linear model The system

$$X = A\theta + E,$$

where X is of dimension $n \times k$; A is $n \times p$; θ is $p \times k$ and E is $n \times k$. This model appears in connection with outliers in statistical data.

Murray-von Neumann equivalence Two projections e, f in a $\mathbf{C}^*$ algebra A are *Murray-von Neumann equivalent* if there exists $v \in A$ such that $e = v^*v$, $f = vv^*$.

mutually singular measures *See* singular measure.

N

Napier's number The number e which is the limit of the expression $\left(1+\frac{1}{n}\right)^n$ as n tends to infinity:

$$e = \lim_{n\to\infty}\left(1+\frac{1}{n}\right)^n = 2.71828182845\cdots$$

The number e is the base for the natural logarithm $f(x) = \ln(x)$.

Nash-Moser Implicit Function Theorem If a compact Riemannian manifold V^n is sufficiently close to another compact Riemannian manifold $\overline{V}^n$ which has a free imbedding in E^m ($\mathbf{R}^m$), then V^n also has a free imbedding in E^m ($\mathbf{R}^m$).

natural boundary For a given analytic function $f(z)$, the boundary of the region beyond which $f(z)$ cannot be analytically continued.

natural equation of a surface An equation which specifies a curve independent of any choice of coordinates or parameterization. The study of natural equations began with the following problem: given two functions of one parameter, find the space curve for which the functions are the curvature and torsion.

Euler gave an integral solution for plane curves (which always have torsion $\tau = 0$). Call the angle between the tangent line to the curve and the x-axis ϕ, the tangential angle, then

$$\phi = \int \kappa(s)\,ds,$$

where κ is the curvature. Then the equations

$$\kappa = \kappa(s), \qquad \tau = 0,$$

where τ is the torsion, are solved by the curve with parametric equations

$$x = \int \cos\phi\,ds$$
$$y = \int \sin\phi\,ds.$$

The equations $\kappa = \kappa(s)$ and $\tau = \tau(s)$ are called the natural (or intrinsic) equations of the space curve.

An equation expressing a plane curve in terms of s and radius of curvature R (or κ) is called a Cesàro equation, and an equation expressing a plane curve in terms of s and ϕ is called a Whewell equation.

Among the special planar cases which can be solved in terms of elementary functions are the circle, logarithmic spiral, circle involute, and epicycloid. Enneper showed that each of these is the projection of a helix on a conic surface of revolution along the axis of symmetry. The above cases correspond to the cylinder, cone, paraboloid, and sphere.

natural equations of a curve A system of equations for the given curve which define both the curvature and torsion of the curve as functions of the arc length parameter of the curve.

natural geometry of a surface *See* natural equation of a surface.

natural logarithm The logarithm whose base is the number (Napier's number) e; written variously as $\log_e(z)$, $\ln(z)$ and $\log(z)$, although $\log(z)$ can also denote the logarithm whose base is 10. The inverse to the exponential function with base e; if $\ln(z) = w$, then $z = e^w$. Also given, for real z, by

$$\ln(z) = \int_1^z \frac{1}{t}dt.$$

nearly everywhere Everywhere except for a countable set. Used in the theory of the Denjoy integral.

0-8493-0320-6/00/$0.00+$.50

negative curvature The curvature of a surface at a saddle point. A space with total curvature being negative is a saddle-type space.

negative infinity A formal symbol ($-\infty$), which for certain manipulations can represent the limit of a negative quantity that decreases without bound. *See also* infinity.

negative orientation *See* positive orientation.

negative variation (1.) A function N associated with a given real-valued function f of bounded variation which is defined by

$$N(x) = \frac{1}{2}[T(x) - f(x)],$$

where T is the total variation of f. The *positive variation* of f is defined by

$$P(x) = \frac{1}{2}[T(x) + f(x)].$$

(2.) For a real measure μ, *See* positive variation.

net A directed subset of a topological space, together with a function from the directed set to the space.

Neumann problem A problem in partial differential equations in which a harmonic function is sought that is continuous on a given domain and its boundary, and whose normal derivative reduces to a given continuous function on the boundary of the domain.

Neumann series The series of functions

$$g + Kg + K^2g + \cdots + K^ng + \cdots$$

where

$$Kh(x) = \int_a^b K(x, y)h(y)\,dy$$

is a continuous function of x and $K(x, y)$ is a kernel which is bounded and continuous except perhaps for $y = x$.

Neumann's function The function

$$y_\nu(x) = \frac{\cos(\nu\pi)J_\nu(x) - J_{-\nu}(x)}{\sin(\nu\pi)},$$

where $J_\nu(x)$ is the Bessel function of order ν, is a *Neumann function* of order ν.

Nevanlinna Theory The study of the growth of meromorphic functions by using a growth estimate which Nevanlinna derived from the Poisson-Jensen formula and which is now known as *Nevanlinna's Second Fundamental Theorem.*

Nevanlinna's exceptional values A set of finite linear measure, off of which the given meromorphic function $f(z)$ has a growth of $o(T(r, f))$ as r tends to infinity, where $T(r, f)$ is defined as it is in Nevanlinna's First Fundamental Theorem.

Nevanlinna's First Fundamental Theorem Let f be a meromorphic function, and let $a \in \mathbf{C}$. Further, suppose

$$f(z) - a = \sum_{k=m}^{\infty} c_k z^k$$

for $c_m \neq 0$, $m \in \mathbf{Z}$ is the Laurent series expansion of $f - a$ at the origin. Then,

$$T\left(r, \frac{1}{f-a}\right) = T(r, f) - \log|c_m| + \phi(r, a)$$

where $|\phi(r, a)| \leq \log(2) + \log^+|a|$ and $T(r, f) = m(r, f) + N(r, f)$ is the characteristic function of f with

$$m(r, f) = \frac{1}{2\pi}\int_0^{2\pi} \log^+\left|f\left(re^{i\phi}\right)\right| d\phi$$

and

$$N(r, f) = \int_0^r \frac{n(t, \infty) - n(0, \infty)}{t}\,dt + n(0, \infty)\log(r)$$

where

$$n(t, a) = \sum_{\substack{|z| \leq t \\ f(z) = a}} \begin{pmatrix} \text{multiplicity of an} \\ a\text{-point at } z \end{pmatrix}.$$

Nevanlinna's Second Fundamental Theorem Let f be a non-constant, meromorphic function; let $q \geq 2$, and let $z_1, \ldots, z_q \in \mathbf{C}$ be distinct points. Then

$$m(r, f) + \sum_{n=1}^{q} m\left(r, \frac{1}{f - z_n}\right) \leq 2T(r, f) + S(r, f),$$

where $S(r, d)$ is a function which has growth of $o(T(r, f))$ except on a set of finite measure known as the exceptional set. *See* Nevanlinna's exceptional values. *See* Nevanlinna's First Fundamental Theorem, where $m(r, f)$ and $T(r, f)$ are defined.

Newton's method The iterative formula for approximating a zero of a function $f(x)$ by computing successive approximations

$$x_{n+1} = x_n - \frac{f(x_n)}{f'(x_n)};$$

this method approximates the zeros of the function by using the x-intercept of its tangent line at each successive x_n.

nilmanifold The compact quotient space of a given connected, nilpotent Lie group.

node (1.) A point where two or more branches of a graph meet.
(2.) A vertex of a network, tree or digraph.

non-Euclidean geometry A geometry in which the parallel postulate of Euclidean geometry is replaced with an alternate parallel postulate, significantly changing the properties of the space. If it is assumed that there is more than one parallel line to a given line through a given point, the resulting geometry is known as hyperbolic or Lobachevskian geometry; if there are no parallel lines, the resulting geometry is known as elliptic geometry.

nondecreasing *See* nondecreasing function. An analogous definition can be made for a nondecreasing sequence. *See* monotonically increasing sequence.

nondecreasing function A real-valued function $f(x)$, of a real variable, for which $f(x_1) \geq f(x_2)$ whenever $x_1 \geq x_2$; a function whose values either stay the same or increase but never decrease. *See* monotonically increasing function.

nondegenerate function *See* degenerate mapping.

nondegenerate mapping *See* degenerate mapping.

nonlinear functional analysis The study of nonlinear operators between infinite dimensional vector spaces.

nonlinear integral equation An integral equation where the unknown function appears in a non-linear term; typically the unknown is composed with another function.

nonlinear transformation A transformation, usually between vector spaces, which is not linear.

nonpositive curvature Curvature of a surface at a point which is either zero or negative, indicating that the surface is either flat at that point or that the point is a saddle point.

nontangential approach For a sequence $\{a_j\}$, in the unit disk D in the complex plane, tending to a limit $z_0 \in \partial D$ in such a way that the quantity

$$\frac{|z_0 - a_j|}{1 - |a_j|}$$

remains bounded, as $j \to \infty$.

nontangential maximal function The function $N_\alpha u$ associated with a given complex function $u(z)$, on the open unit disk in the complex plane, defined by

$$(N_\alpha u)\left(e^{it}\right) = \sup\left\{|u(z)| : z \in e^{it}\Omega_\alpha\right\}$$

where Ω_α is defined to be the union of the open disk $D(0, \alpha)$ centered at zero with radius α and the line segments from $z = 1$ to points in $D(0, \alpha)$.

norm A real-valued function $\|\cdot\|$ on a linear space X (over **R** or **C**) such that
(i.) $\|x + y\| \le \|x\| + \|y\|$
(ii.) $\|ax\| = |a| \cdot \|x\|$
(iii.) $\|x\| = 0$ implies $x = 0$
for all $x, y \in X$ and for all $a \in \mathbf{R}$ (or **C**).

normal analytic space An analytic space where the local rings of every point are all normal, or integrally-closed integral domains.

normal coordinates The coordinates of a mechanical system composed of coupled oscillators, each subject to an elastic restoring force which is displaced from its equilibrium configuration. These normal coordinates must be independent of each other: each can be excited while the others remain at rest. *See also* normal point.

normal curvature The curvature of the curve of intersection of a given surface with the plane defined by the normal to the surface at the specified point and the tangent line to that point; the curvature of a normal section of a surface at a specific point. The curvature is positive if the principal normal to the section points in the same direction as the normal to the surface, and is negative otherwise.

normal derivative The directional derivative of a given function, defined on a curve or surface, in the direction of the outward pointing normal to the given curve or surface at the specified point:

$$\frac{\partial h(x)}{\partial n} = \nabla h(\vec{x}) \cdot \vec{n}.$$

normal extension Given an operator T on a Hilbert space H, a normal operator N on a Hilbert space $H' \supset H$ such that the restriction of N to H is T:

$$N|_H = T.$$

See normal operator.

The term is more commonly used in algebra, for a normal field extension.

normal family A collection $\mathcal{F}$ of analytic functions on a common domain D such that every sequence of functions in $\mathcal{F}$ contains a subsequence which converges uniformly on every compact subset of D.

normal form A standard, or canonical, form of a structure or object. For example, the "Jordan normal form" of an $n \times n$ matrix, a block diagonal matrix where each block is either a constant multiple of the identity or a "Jordan block" of the form

$$\begin{pmatrix} a & 1 & 0 & \dots & 0 \\ 0 & a & 1 & \dots & 0 \\ \cdot & \cdot & \cdot & \cdot & \cdot \\ 0 & 0 & \cdots & a & 1 \\ 0 & 0 & \cdots & 0 & a \end{pmatrix}.$$

After a change of basis, every $n \times n$ matrix assumes the Jordan normal form.

normal frame The frame, or trihedral, of a point formed by the tangent, normal and binormal lines to a curve at that point in three-dimensional space.

normal line The line perpendicular to the tangent line at the point of tangency to a curve in $\mathbf{R}^2$.

normal operator An operator A on a Hilbert space, which commutes with its adjoint: $AA^* = A^*A$.

normal plane The plane perpendicular to the tangent of a surface at a given point.

normal point Let T be a tensor field, (M, g) a Riemannian space of dimension n and x a point in M. Choose a neighborhood $W_x \subseteq V_x$ of $O \in TM_x$, for which $\exp_x|_{W_x}$ is injective. The inverse mapping is

$$f: U_x \to TM_x.$$

For a point $y \in U_x$ relative to a basis $e_1, \dots, e_n$ of TM_x, a number z satisfying

$$f(y) = ze_i \quad \text{for some } i,\ i = 1, 2, \dots, n$$

is called a *normal point*. The points z_i, satisfying

$$f(y) = z_i e_i, \quad i = 1, 2, \dots, n$$

are called *normal coordinates*.

normal to surface The line perpendicular to the tangent plane of a given surface at a specified point.

normal vector A vector which is orthogonal to the given plane.

normal vector bundle A morphism $F : Y \to X$ of manifolds induces a natural morphism $\phi : T_Y \to F^*T_X$, from the holomorphic tangent bundle of Y to the pullback of the holomorphic tangent bundle of X. If F is a closed imbedding of complex manifolds, then T_Y is (locally) a direct summand of f^*T_x, and we define the *normal vector bundle* of Y in X (of rank dim X − dim Y) to be F^*T_X/T_Y.

normalization Multiplication of a quantity by a suitable constant so that the resulting product has norm equal to 1, value 1 at a point, or some other desired ("normal") property. For example, a non-zero, real-valued function $f(x)$ on [0, 1] may be normalized so that its integral is 1 or so that its value at 1 is 1.

normed linear space A vector space X, over a field $F(= \mathbf{R}$ or $\mathbf{C})$, together with a norm, defined on that space, making X a topological vector space. The associated metric and the topology on X are defined by the norm. *See* norm, topological vector space.

***n*th derivative** The composition of the derivative function with itself n times:

$$\frac{d^{(n)}f}{dx^{(n)}} = \underbrace{\frac{d}{dx}\left(\frac{d}{dx}\cdots\left(\frac{df}{dx}\right)\cdots\right)}_{n \text{ times}},$$

also denoted by $f^{(n)}(x)$.

***n*th differential** Given an arbitrary small change (increment) dx of the independent variable x (differential of the independent variable x), the corresponding term $dy \equiv df \equiv \frac{dy}{dx}\,dx \equiv f'(x)dx$ in the expansion

$$f(x + \Delta x) - f(x) = f'(x)\,dx + O(dx^2)$$

of a differentiable function $y = f(x)$ is called the first order differential of the dependent variable y at the point x. Similarly, the (first-order) differential of the differentiable function $y = f(x_1, x_2, \dots, x_n)$ of n variables $x_1, x_2, \dots, x_n$ is

$$dy \equiv df \equiv \frac{\partial f}{\partial x_1}dx_1 + \frac{\partial f}{\partial x_2}dx_2 + \cdots + \frac{\partial f}{\partial x_n}dx_n.$$

The differential of each independent variable is regarded as a constant, so that $d^2x \equiv d(dx) \equiv 0$. The differential of a dependent variable is a function of the independent variable or variables. The second-order, third-order, ..., nth-order differentials of suitably differentiable functions by successive differentials of the first-order differential. For example for a function f of two variables x_1, x_2 the nth differential is given by

$$d^n f(x_1, x_2) = \sum_{k=0}^{n} \binom{n}{k} \frac{\partial^n f}{\partial x_1^{n-k}\partial x_2^k} dx_1^{n-k} dx_2^k.$$

Given a problem involving m independent variables $x_1, x_2, \dots, x_m$, any function of the order $dx_1^{n_1} \cdot dx_2^{n_2} \cdots dx_n^{r_n}$ as $dx_1 \to 0$, $dx_2 \to 0, \dots, dx_n \to 0$ is an infinitesimal of order $n_1 + n_2 + \cdots + n_m$. In particular, the nth-order differential $d^n f$ of a suitably differentiable function is an infinitesimal of order n.

nth partial derivative The composition of the partial derivative function with itself n times:

$$\frac{\partial^{(n)} f}{\partial x^{(n)}} = \underbrace{\frac{\partial}{\partial x}\left(\frac{\partial}{\partial x}\cdots\left(\frac{\partial f}{\partial x}\right)\cdots\right)}_{n \text{ times}}.$$

nuclear operator A trace class operator. *See* trace class.

null set *See* set of measure 0.

null space The set of all elements x in the domain of a given function f such that $f(x) = 0$. The term is usually used when f is a linear operator on a vector space. Also *kernel*.

nullity The dimension of the null space, or kernel, of a given matrix or operator.

numerical analysis The study of computation and its accuracy, stability and implementation on a computer. The central concerns of numerical analysis include the determination of appropriate numerical models for applied problems, and the construction and analysis of efficient algorithms.

numerical integration Any one of several different methods of finding a result which approximates a solution of a given differential equation to any required degree of accuracy.

numerical quadrature The evaluation of a definite integral by a formula involving weighted sums of function values at specified points.

numerical range The set of values (Tx, x) for T a given linear operator on a Hilbert space H, and for all $x \in H$ with $\|x\| \leq 1$.

numerical solution of integral equation Any one of several different methods of finding a result which approximates a solution of a given integral equation to any required degree of accuracy.

O

o-convergence A sequence $\{x_l\}$ in a vector lattice X is *O-convergent* to $x \in X$ provided there is a sequence $\{w_j\} \subset X$ such that $w_1 \geq w_2 \geq, \cdots, \wedge_{j \geq 0} w_j = 0$ and $|x - x_j| \leq w_j$.

oblate spheriod The surface of revolution swept out by an ellipse rotated about its minor axis.

oblique coordinates Cartesian coordinates whose axes are not perpendicular to each other.

octant Any one of the eight trihedral sections into which three-dimensional space is divided by the Cartesian coordinate axes.

odd function A function f between two sets on which addition is defined, such that

$$f(-x) = -f(x).$$

If f maps $\mathbf{R}$ to $\mathbf{R}$, it is odd if its graph is symmetric about the origin.

Oka's Principle Holomorphic problems that can be stated in the language of cohomology and solved topologically, can also be solved holomorphically. This principle is not formulated strictly as a theorem, but proves true in many examples, especially on Stein spaces.

one-parameter group of transformations A family $\{\phi_t\}_{t\in\mathbf{R}}$ of transformations of a manifold such that $\phi_{t+s}(x) = \phi_t(\phi_s(x))$ and $\phi_{-t}(x) = \phi_t^{-1}(x)$ for all $t, s \in \mathbf{R}$ and for all $x \in M$.

one-to-one correspondence A one-to-one, onto function; a bijection.

open arc A continuous mapping from an open interval $(a, b) \subset \mathbf{R}$ into $\mathbf{R}^n$.

The range of an open arc is not itself an open set, unless considered relative to some set smaller than $\mathbf{R}^n$. For example, the open arc $\{e^{it} : 0 < t < \pi\} \subset \mathbf{R}^2$ is an open subset of the unit circle.

open cover For a subset A of a topological space X, a collection of open subsets of X, whose union contains A.

open interval The set of all real numbers lying between two endpoints, not including the endpoints; denoted by $]a, b[$ or (a, b). Here $-\infty \leq a < \infty$ and $-\infty < b \leq \infty$.

open map A function $F : A \to B$, between two topological spaces, such that $f(U)$ is open in B, whenever $U \subseteq A$ is open.

Open Mapping Theorem (1.) Every continuous, surjective linear operator $A : X \to Y$ between complete, pseudo-normed linear spaces X, Y is open; that is, the image $A(U)$ of every open set U in X is an open set in Y. (2.) Let $T : \mathbf{R}^n \to \mathbf{R}^n$ be a $\mathbf{C}^1$ mapping with non-zero Jacobian in an open set $D \subset \mathbf{R}^n$. Then $T(D)$ is open.

open topological manifold *See* topological manifold. The term *open* is to distinguish from a manifold *with boundary*.

open tubular neighborhood *See* tubular neighborhood.

operating function Let X be a complex Banach space, λ a complex variable, Λ any subset of the complex λ-plane and denote by $B(X, X)$ the Banach space of bounded linear operators on x. By an *operating function* is meant a mapping

$$S\colon \Lambda \to B(X, X), \quad \lambda \to S_\lambda.$$

operation (1.) Any procedure (such as addition, set union, etc.) that generates a unique value according to some set of rules from one or more given values.

0-8493-0320-6/00/$0.00+$.50

(2.) A function determined by such a procedure.

operational calculus *See* functional calculus.

operator A function that maps a linear space into a linear space.

operator with boundary condition *See* ordinary differential operator.

optimal control The study of certain problems of maximizing or minimizing linear functionals associated with differential equations. For example, given a system of differential equations

$$\frac{dy_j}{dt} = f_j(x_1, \ldots, x_n; u_1, \ldots, u_m)$$
$$x_j(t_0) = x_j, \quad j = 1, \ldots, n,$$

the problem of determining the parameters u_j so as to minimize the functional

$$J[u] = \int_a^b F(x_1, \ldots, x_n; u_1, \ldots, u_m)dt.$$

orbit (1.) The sequence of points generated by repeated composition of the given function f at the associated point x:

$$\{x, f(x), f(f(x)), \ldots\}$$

The orbit is said to be closed if this set is finite.

(2.) The trajectory of an ordinary differential equation.

(3.) [Group Theory] Under an action of a group G on a non-empty set S, the set of products, of all elements of the group with the associated element $x \in S$:

$$x^G = orb_G(x) = \{gx : g \in G\}$$

order of infinitesimal The relative size of an infinitesimal, as compared to another infinitesmial. For infinitesimals u and v viewed as functions of x, u and v have the same order if $\lim_{x\to 0}\left|\frac{u}{v}\right|$ exists and is non-zero; u is of higher order than v if $\lim_{x\to 0}\left|\frac{u}{v}\right|$ equals zero, and u is of lower order than v if $\lim_{x\to 0}\left|\frac{u}{v}\right|$ tends toward infinity.

order of infinity Relative size of one infinite quantity as compared to another. For two quantities u and v viewed as functions of x and tending to ∞ as x tends to 0, u and v have the same order, denoted $u \asymp v$, if $\lim_{x\to 0}\left|\frac{u}{v}\right|$ exists and is non-zero; u is of higher order than v, denoted $v = o(u)$, if $\lim_{x\to 0}\left|\frac{u}{v}\right|$ tends towards infinity; and u is of lower order than v, denoted $u = o(v)$, if $\lim_{x\to 0}\left|\frac{u}{v}\right|$ equals zero.

order of pole The largest value m such that c_{-m} is non-zero in the Laurent series expansion of the function f expanded about the pole a:

$$f(z) = \frac{c_{-m}}{(z-a)^m} + \frac{c_{-m+1}}{(z-a)^{m-1}} + \cdots$$
$$+c_0 + c_1(z-a) + c_2(z-a)^2 + \cdots.$$

order of zero point The multiplicity of a complex number a, as a zero of an analytic function $f(z)$; the lowest order of the derivative of f which is nonvanishing at a: $\min\{n : f^{(n)}(a) \neq 0\}$.

ordered n-tuple An element of the n-fold Cartesian product $S \times S \times \cdots \times S$, of a given set S. Hence a set of n elements $(s_1, \ldots, s_n)$ of S, with $(s_1, \ldots, s_i, \ldots, s_j, \ldots, s_n)$ and $(s_1, \ldots, s_j, \ldots, s_i, \ldots, s_n)$ considered different n-tuples, unless $s_i = s_j$.

ordered pair An ordered n-tuple, with $n = 2$. *See* ordered n-tuple.

ordered triple An ordered n-tuple, with $n = 3$. *See* ordered n-tuple.

ordinary curve The locus of a point with one degree of freedom.

ordinary differential equation An equation in an unknown function of one variable, and its derivatives. The abbreviation ODE is sometimes used.

ordinary differential operator An operator defined on a vector space of functions of one variable, as a polynomial in the differentiation operator $D = d/dt$. Typically, differential operators are unbounded operators, defined on dense domains. Initial conditions of a problem may figure in the definition of the domain. For example, the Sturm-Liouville problem

$$y'' + \lambda y = 0, \quad y(0) = y(\pi) = 0,$$

on $[0, \pi]$ corresponds to the differential operator $D^2 + \lambda D$ on $L^2[0, \pi]$, defined on the domain $D = \{y : y \text{ continuous}, y, y' \in BV \text{ and } y(0) = y(\pi) = 0\}$, (BV= functions of bounded variation on $[0, \pi]$).

ordinary Dirichlet series *See* Dirichlet series.

ordinary helix A curve lying on a right circular cylinder which cuts the elements of the cylinder under a constant angle; also known as a circular helix; in parameter form it is the curve given by the equations

$$x = a \sin\theta, \quad y = a\cos\theta, \quad \text{and} \quad z = b\theta$$

where a and b are constants and θ is the parameter.

ordinary point (1.) For a second order differential equation $y''+P(x)y'+Q(x)y = 0$, a point a, such that both P and Q are analytic in a neighborhood of a.
(2.) A non-isolated point of a curve where the tangent is smooth and the curve is non-intersecting; a point which is not a singular point; a point at which a given function is analytic.

ordinary singularity A point at which a given function is not analytic.

ordinate The vertical or y-coordinate of a point in a two-dimensional Cartesian coordinate system, equal to the distance from the point to the x-axis measured along a line parallel to the y-axis. *See also* abscissa.

ordinate set The set of all points in two-dimensional space such that the ordinate, or y-coordinate, is less than the value of the associated function evaluated at the x-coordinate.

orientable manifold A manifold M on which it is possible to define a $\mathbf{C}^\infty$ n-form, which is not zero at any point. Equivalently, M can be covered by charts (U_p, ϕ_p) in such a way that whenever $U_p \cap U_q \neq \emptyset$, then $\phi_p \circ \phi_q^{-1}$ is orientation preserving. *See also* atlas.

orientation *See* oriented manifold.

oriented atlas An atlas on a manifold M, consisting of charts (U_p, ϕ_p) such that whenever $U_p \cap U_q \neq \emptyset$, then $\phi_p \circ \phi_q^{-1}$ is orientation preserving.

oriented element An element of a graph with an orientation which is assigned to it by the ordering of its vertices.

oriented manifold *See* orientable manifold. M is given an *orientation*, whenever a $\mathbf{C}^\infty$ n-form, which is not zero at any point of M, is chosen.

origin The point in $\mathbf{R}^n$ or $\mathbf{C}^n$ having the coordinates $(0, \ldots, 0)$.

Orlicz class The set L_M of all functions $x(t)$ such that

$$\int_G M(x(t))\, dt < \infty$$

where G is a bounded closed set in $\mathbf{R}^n$, dt represents the Lebesgue measure, and $M(u)$ is a given N-function, an even complex function which is increasing for positive u and such that

$$\lim_{u\to 0}\frac{u}{M(u)} = \lim_{u\to 0}\frac{M(u)}{u} = 0.$$

Orlicz space The Banach space L^*_M of Lebesgue measurable functions x on a bounded closed set G in $\mathbf{R}^n$ such that

$$\|x\|_M = \sup\left\{\int_G x(t)y(t)\,dt : \int_G N(y(t))\,dt \le 1\right\} < \infty$$

where $M(u)$ is the given N-function, an even complex function which is increasing for positive u and such that

$$\lim_{u\to 0}\frac{u}{M(u)} = \lim_{u\to 0}\frac{M(u)}{u} = 0,$$

and $N(u)$ is the N-function complementary to M.

orthogonal complement The subspace consisting of all vectors which are orthogonal (perpendicular) to each vector in a given set. Orthogonal complements are defined in terms of an inner product on the vector space V. In symbols, $U^\perp = \{v \in V : (v, w) = 0 \text{ for all } w \in U\}$. When U is a closed subspace, V is the direct sum of U and $U^\perp$.

orthogonal curvilinear coordinates A coordinate system in three-dimensional space such that, given any point, the three coordinate surfaces passing through the point are mutually orthogonal and the three coordinate curves passing through the point are mutually orthogonal. (A coordinate surface is obtained by holding the value of one coordinate fixed and allowing the other two to vary; a coordinate curve is obtained by holding the values of two coordinates fixed and allowing the remaining one to vary freely.) The term also applies to coordinate systems for the plane, in which case the requirement is that, given any point, the two coordinate curves passing through the point are orthogonal to each other.

orthogonal frame A set of three mutually orthogonal unit vectors at a point in three-dimensional Euclidean space, or such a set at each point of a curve in Euclidean space. A Frenet frame for a smooth curve consists of the unit tangent vector, the principal unit normal vector, and the unit binormal vector.

orthogonal polynomials Given real numbers a and $b > a$, and a nonnegative function $w(x)$, continuous on $[a, b]$ and such that $\int_a^b w(x)|x|^n dx < \infty$, for $n = 0, 1, 2, \ldots$, a sequence of polynomials $\{\phi_j(x)\}$, satisfying

$$\int_a^b \phi_i(x)\phi_j(x)w(x)dx = \delta_{i,j}.$$

Equivalently, the functions $\{\phi_j\}$ form an orthonormal set in the inner product space $L^2(\mu)$, where μ is the measure $d\mu = w(x)dx$ on $[a, b]$. *See* orthonormal set.

Examples are:

Legendre polynomials. $a = -1, b = 1$, $w(x) \equiv 1$;
Jacobi polynomials. $a = -1, b = 1$, $w(x) = (1-x)^\alpha(1+x)^\beta$;
Laguerre polynomials. $a = 0, b = \infty$, $w(x) = e^{-x}x^\alpha$;
Hermite polynomials. $a = -\infty, b = \infty$, $w(x) = e^{-x^2}$.

orthogonal set A subset of an inner product space with the property that any two distinct vectors in the subset are orthogonal to each other (have 0 inner product). *See also* orthonormal set.

orthogonal trajectory A planar curve which intersects every member of a given family of curves at right angles. For example, a straight line through a point p is an orthogonal trajectory to the family of concentric circles with center p.

orthonormal basis A complete orthonormal set. *See* complete orthonormal set.

orthonormal set A set, S, of vectors in an inner product space satisfying two properties: 1.) $\|v\| = 1$, for all $v \in S$, 2.) $(v, w) = 0$, for all $v, w \in S$ with $v \neq w$. Each vector in S is a unit vector and any two distinct vectors in S are orthogonal. *See also* orthogonal set, complete orthonormal set.

oscillation A measure of the local variation of a real-valued function. If f is a real-valued function defined on a metric space, the oscillation of f at a point p of the space is the difference between

$$\lim_{\epsilon \to 0} \{\sup\{f(x) : d(x, p) < \epsilon\}\}$$

and

$$\lim_{\epsilon \to 0} \{\inf\{f(x) : d(x, p) < \epsilon\}\},$$

provided these two limits exist.

osculating circle A circle which is tangent to a smooth curve at a point, lies on the same side of the tangent line to the curve as the curve does, and has radius equal to the radius of curvature of the curve at the point. For a space curve, the osculating circle lies in the osculating plane. *See also* radius of curvature, osculating plane.

osculating plane The plane spanned by the unit tangent vector and the principal unit normal vector to a smooth curve at a given point of the curve. This is the plane which "best fits" the curve at the point.

outer area An upper estimate for the area of a bounded subset of the plane. Defined as a limit of a decreasing sequence of overestimates; each overestimate is obtained by partitioning a rectangle containing the subset into a grid of (small) squares and adding the areas of all those squares which intersect the subset. For "nice" subsets, the outer area will equal the inner area, yielding a fairly general definition of area. *See also* inner area.

outer function An analytic function on the unit disk of the form:

$$F(z) = e^{it} \exp\left[\frac{1}{2\pi} \int_{-\pi}^{\pi} \frac{e^{i\theta} + z}{e^{i\theta} - z} k(\theta)\, d\theta\right],$$

where k is a real-valued integrable function on $[-\pi, \pi]$ and t is a real constant. Non-zero functions in the Hardy space H^1 have an essentially unique factorization as the product of an inner function and an outer function. *See also* inner function.

outer measure A set function μ^*, defined on a hereditary σ-ring of subsets of a set X and taking values in the extended nonnegative real numbers $\mathbf{R}_+ \cup \{\infty\}$ which is

(i.) monotone: $E \subseteq F \Rightarrow \mu^*(E) \leq \mu^*(F)$; and

(ii.) countably subadditive: $\mu^*(\cup_1^\infty E_j) \leq \sum_1^\infty \mu^*(E_j)$.

Also called *Carathéodory outer measure.*

Given a non-negative, (finitely) additive, finite set function μ, defined on the elementary sets in $\mathbf{R}^n$, one can define an outer measure μ^* corresponding to μ, for any subset $E \subseteq \mathbf{R}^n$, by

$$\mu^*(E) = \text{glb} \sum_{j=1}^{\infty} \mu(A_j),$$

where the glb is over all coverings $\{A_j\}$ of E by elementary sets. When μ is the volume of an elementary set, μ^* is called *Lebesgue outer measure.*

outer product *See* vector product of vectors.

overdetermined system of differential equations A system of differential equations in which the number of equations exceeds the number of unknown functions. Such a system cannot generally be solved unless some compatibility condition is satisfied.

P

P ($\wp$) function *See* Weierstrass $\wp$ function.

P-function of Riemann Suppose the differential equation

$$p_n(z)\frac{d^n y}{dz^n} + p_{n-1}(z)\frac{d^{(n-1)}y}{dz^{(n-1)}} + \cdots$$
$$+p_0(z)y = 0,$$

having polynomial coefficients, has ν singular points, $a_1, \ldots, a_\nu$, all regular, and has exponents (roots of the indicial equation) $a_{j1}, \ldots, a_{jn}$ at the regular singular point a_j, $j = 1, \ldots, \nu$ and $b_1, \ldots, b_n$ at ∞. The n solutions corresponding to each of the regular singular points are grouped together under the term *Riemann P-function*, with the notation

$$P\left\{\begin{array}{cccc} a_1 & \cdots & a_\nu & \infty \\ a_{11} & \cdots & a_{\nu 1} & b_1 \; z \\ \cdot & \cdot & \cdot & \\ a_{1n} & \cdots & a_{\nu n} & b_n \end{array}\right\}.$$

p.p. French for *almost everywhere. See* almost everywhere.

Padé approximation An approximation to a transcendental function by a rational function. The Padé approximant, $f_{p,q}$, of f will be the best approximation to f in a neighborhood of 0 by a rational function whose denominator has degree at most p and whose numerator has degree at most q. The power series for $f_{p,q}$ will agree with the power series for f for more terms than does the power series of any other rational function with the same degree constraints.

Padé table An infinite matrix (or table) whose (p, q)-entry ($p = 0, 1, 2, \ldots$, $q = 0, 1, 2, \ldots$) is the Padé approximant $f_{p,q}$ of a given function f. *See also* Padé approximation.

Painlevé's Theorem Let E be a compact subset of the complex plane, $\mathbf{C}$, and let $\Omega = (\mathbf{C} \setminus E) \cup \{\infty\}$, the complement of E in the Riemann sphere (the one point compactification of $\mathbf{C}$). Suppose that for each $\epsilon > 0$, E has a cover consisting of disks whose radii sum to at most ϵ. Then any bounded, analytic function on Ω is constant.

Paley-Weiner Theorem **(1.)** Let f be an analytic function in the upper half plane, $y = \Re z > 0$. If

$$\sup_{0<y<\infty} \frac{1}{2\pi}\int_{-\infty}^{\infty} |f(x+iy)|^2\,dx = C < \infty,$$

then there is a square integrable function g defined on $(0, \infty)$ such that

$$f(z) = \int_0^\infty g(t)e^{itz}\,dt, \quad \Re z > 0$$

and $\int_0^\infty |g(t)|^2\,dt = C.$

(2.) Let f be an entire function such that $|f(z)| \leq Ae^{B|z|}$, for all z. (A and B are positive constants.) If

$$\int_{-\infty}^{\infty} |f(x)|^2\,dx < \infty,$$

then there is a square integrable function g defined on $(-B, B)$ such that

$$f(z) = \int_{-B}^{B} g(t)e^{itz}\,dt.$$

Pappus's Theorem If a region $R = \{(x, y) : a \leq x \leq b, f(x) \leq y \leq g(x)\}$ has an area A and centroid $(\overline{x}, \overline{y})$, then the volume of the solid obtained by rotating R about the x-axis is $2\pi A\overline{y}$. In words: the volume obtained by rotating a planar region about an axis which does not intersect the region is equal to the product of the area of the region and the distance moved by the centroid of the region.

0-8493-0320-6/00/$0.00+$.50

parabola One of the conic sections. A parabola is obtained by intersecting a right circular cone with a plane which is parallel to a straight line in the surface of the cone. Alternately, a parabola is obtained by taking all points equidistant to a fixed point (the focus) and a fixed straight line (the directrix). In a Cartesian coordinate plane, after a suitable rotation, a parabola is the graph of a quadratic function, $y = ax^2 + bx + c$. *See also* conic section.

parabolic coordinates Coordinates (u, v) for the plane which are related to Cartesian coordinates by the formulas:

$$x = \frac{u^2 - v^2}{2}, \qquad y = uv.$$

The coordinate curves for parabolic coordinates consist of two families of mutually orthogonal parabolas.

parabolic cylinder A generalized cylinder set whose cross section is a parabola. Example: $\{(x, y, z) : y = ax^2 + bx + c\}$, where $a \neq 0$, b, and c are constants.

parabolic cylinder function Solutions of the differential equation

$$\frac{d^2 f}{dx^2} = \left(\frac{x^2}{4} - \nu - \frac{1}{2}\right) f$$

(order ν). Also known as Weber-Hermite functions. Parabolic cylinder functions may be expressed in terms of confluent hypergeometric functions and they have a variety of integral representations.

parabolic cylindrical coordinates Coordinates (u, v, z) for three-dimensional Euclidean space which are related to Cartesian coordinates by the formulas:

$$x = \frac{1}{2}(u^2 - v^2), \qquad y = uv, \qquad z = z.$$

The coordinate surfaces for parabolic cylindrical coordinates intersect the xy-plane in two families of mutually orthogonal parabolas.

parabolic differential operator An operator of the form

$$a\frac{\partial^2}{\partial x^2} + b\frac{\partial^2}{\partial x \partial y} + c\frac{\partial^2}{\partial y^2} + d\frac{\partial}{\partial x} + e\frac{\partial}{\partial y} + f,$$

where a, b, c are not all 0, $b^2 - 4ac = 0$, and either $2cd \neq be$ or $2ae \neq bd$. *See also* parabolic equation, elliptic differential operator, hyperbolic differential operator.

parabolic equation A second order partial differential equation of the form

$$a\frac{\partial^2 u}{\partial x^2} + b\frac{\partial^2 u}{\partial x \partial y} + c\frac{\partial^2 u}{\partial y^2} + d\frac{\partial u}{\partial x} + e\frac{\partial u}{\partial y} + fu = F(x, y),$$

where a, b, c are not all 0, $b^2 - 4ac = 0$, and either $2cd \neq be$ or $2ae \neq bd$. The heat equation is a primary example of a parabolic equation. *See also* elliptic equation, hyperbolic equation.

parabolic point A point $\mathbf{p}$ on a regular surface $M \subset \mathbf{R}^3$ is said to be parabolic if the Gaussian curvature $K(\mathbf{p}) = 0$ but $S(\mathbf{p}) \neq 0$ (where S is the shape operator), or, equivalently, exactly one of the principal curvatures κ_1 and κ_2 is 0.

parabolic segment The arc length of the parabolic segment is given by

$$s = \sqrt{4x^2 + y^2} + \frac{y^2}{2x} \ln\left(\frac{2x + \sqrt{4x^2 + y^2}}{y}\right).$$

The area contained between the curves

$$y = x^2, \qquad y = ax + b$$

can be found by eliminating y,

$$x^2 - ax - b = 0,$$

so the points of intersection are

$$x_\pm = \tfrac{1}{2}\left(a \pm \sqrt{a^2 + 4b}\right).$$

Therefore, for the area to be nonnegative, $a^2+4b>0$, and

$$\begin{aligned} x_\pm &= \tfrac{1}{4}\left(a^2 \pm 2a\sqrt{a^2+b^2}+a^2+4b\right) \\ &= \tfrac{1}{4}\left(2a^2+4b \pm 2a\sqrt{a^2+4b}\right) \\ &= \tfrac{1}{2}\left(a^2+2b \pm a\sqrt{a^2+4b}\right), \end{aligned}$$

so the area is

$$\begin{aligned} A &= \int_{x_-}^{x_+} [(ax+b)-x^2]\,dx \\ &= \left[\tfrac{1}{2}ax^2+bx-\tfrac{1}{3}x^3\right]_{(a-\sqrt{a^2+4b})/2}^{(a+\sqrt{a^2+4b})/2}. \end{aligned}$$

Now,

$$\begin{aligned} & x_+^2 - x_-^2 \\ &\quad = \tfrac{1}{4}\left[(a^2+2a\sqrt{a^2+4b}+a^2+4b)\right. \\ &\quad = \left.-\ (a^2-2a\sqrt{a^2+4b}+a^2+4b)\right] \\ &\quad = \tfrac{1}{4}\left[4a\sqrt{a^2+4b}\right] = a\sqrt{a^2+4b}. \\ & x_+^3 - x_-^3 = (x_+ - x_-)(x_+^2 + x_-x_+ + x_-^2) \\ &= \sqrt{a^2+4b}\left\{\tfrac{1}{4}(a^2+2a\sqrt{a^2+4b}\right. \\ &\qquad +a^2+4b) \\ &\qquad +\tfrac{1}{4}[a^2-(a^2+4b)] \\ &\qquad \left.+\tfrac{1}{4}(a^2-2a\sqrt{a^2+4b}+a^2+4b)\right\} \\ &= \tfrac{1}{4}\sqrt{a^2+4b}(4a^2+4b) \\ &= \sqrt{a^2+4b}(a^2+b). \end{aligned}$$

So,

$$\begin{aligned} A &= \tfrac{1}{2}a^2\sqrt{a^2+4b}+b\sqrt{a^2+4b} \\ &\qquad - \tfrac{1}{3}(a^2+b)\sqrt{a^2+4b} \\ &= \sqrt{a^2+4b}\left[\left(\tfrac{1}{2}-\tfrac{1}{3}\right)a^2+b\left(1-\tfrac{1}{3}\right)\right] \\ &\qquad = \left(\tfrac{1}{6}a^2+\tfrac{2}{3}b\right)\sqrt{a^2+4b} \\ &= \tfrac{1}{6}(a^2+4b)\sqrt{a^2+4b} \\ &= \tfrac{1}{6}(a^2+4b)^{3/2}. \end{aligned}$$

paraboloid A surface which, when in standard position, is the solution set of an equation of one of two forms:

$$\frac{z}{c} = \frac{x^2}{a^2} + \frac{y^2}{b^2} \quad \text{(elliptic paraboloid)}$$

or

$$\frac{z}{c} = \frac{x^2}{a^2} - \frac{y^2}{b^2} \quad \text{(hyperbolic paraboloid)}.$$

paraboloid of revolution An elliptic paraboloid (*See* paraboloid) with a circular cross section. A paraboloid of revolution is obtained from a parabola in a plane by rotating the parabola around its axis of symmetry. A mirror in the shape of a paraboloid of revolution will reflect light emanating from a source at the focus as a beam parallel to the axis of symmetry. In reverse, an antenna in the shape of a paraboloid of revolution can be used to concentrate at the focus: electromagnetic radiation from a distant source.

paracompact space A Hausdorff topological space with the property that every open covering has an open, locally finite refinement.

paracompact topological space A Hausdorff space X such that every open cover $\{U_\alpha\}_{\alpha\in A}$ of X has a locally finite open refinement $\{V_\beta\}_{\beta\in B}$. That is, there is a mapping $\tau : B \to A$ with $V_\beta \subset U_{\tau(\beta)}$ and every point has a neighborhood W meeting only finitely many V_β.

parallel translation along a curve Given a curve $c : [a,b] \to M$ and an initial vector $u \in T_{c(a)}M$, there is a unique parallel vector field $U(t)$ along c with $U(a) = u$. (*See* parallel vector field.) By *parallel translation along c*, we mean the linear map $P(c)_a^b : T_{c(a)}M \to T_{c(b)}M$, defined by assigning $U(b)$ to u.

parallel vector field For a manifold M, a vector field $Y(t)$ along a C^∞ curve $c : [a,b] \to M$ such that

$$\nabla_{\partial/\partial t}Y = \frac{dY^i}{dt} + \Gamma_j^{ik}\dot{x}^j Y^k = 0$$

for $i = 1, \ldots, m$, where $\Gamma_i^{kj}\partial_k$ are the Christoffel symbols.

parallelogram law The identity,

$$\|v+w\|^2+\|v-w\|^2=2\|v\|^2+2\|w\|^2,$$

valid in any Hilbert space. An analogue of the statement in geometry that the sum of the squares of the lengths of the two diagonals of a parallelogram is equal to the sum of the squares of the lengths of the four sides.

parameter (1.) A variable which is used to express location along a curve; one of two variables used to express location on a surface; one of several variables used to express location on a hypersurface. *See* parametric equations.
(2.) A variable, or one of several variables, used to characterize the elements of a family (e.g., a family of differential equations, or a family of curves).

parametric equations A set of equations which expresses the coordinates of a point on a curve in terms of one parameter, or the coordinates of a point on a surface in terms of two parameters. Example: The unit circle is given by $x = \cos t$, $y = \sin t$, $0 \le t \le 2\pi$. More generally, a k-dimensional hypersurface in n-dimensional space can be described by n equations in k parameters.

parametric representation A description of a curve or surface in terms of parametric equations.

Parseval's equality (1.) The equality $\int_a^b |f(x)|^2\,dx = \sum_{k=1}^{\infty} |c_k|^2$, where $f \in L^2[a,b]$, $\phi_1, \phi_2, \dots$ is a complete orthonormal set of functions in $L^2[a,b]$, and $c_k = \int_a^b f(x)\overline{\phi(x)}\,dx$. Analogous equalities in other L^2-spaces are also referred to as Parseval's equality.

(2.) The equality $\|v\|^2 = \sum_{k=1}^{\infty} |(v, e_k)|^2$, where $e_1, e_2, \dots$ is an orthonormal basis in an abstract Hilbert space.

Parseval's identity The identity

$$\frac{1}{2\pi}\int_0^{2\pi} |f(re^{it})|^2\,dt = \sum_{n=0}^{\infty} |a_n|^2 r^{2n},$$

where $f(z) = \sum_{n=0}^{\infty} a_n z^n$ is convergent on an open disk of radius R and $r < R$. *See also* Parseval's equality.

partial derivative The derivative of a function of two or more variables taken with respect to a single variable (the other variables being held fixed). For example, if $f(x, y)$ is a function of two variables,

$$\frac{\partial f}{\partial x} = \lim_{h \to 0} \frac{f(x+h, y) - f(x, y)}{h}.$$

$D_x f$ and f_x are common notational alternatives to $\frac{\partial f}{\partial x}$.

partial differential equation An equation in which the unknown is a function of two or more variables and in which partial derivatives of the unknown function appear. *See also* boundary value problem. Important examples of partial differential equations include:

$$\frac{\partial^2 u}{\partial x^2}+\frac{\partial^2 u}{\partial y^2}+\frac{\partial^2 u}{\partial z^2}=0$$

Laplace's equation,

$$\frac{\partial^2 u}{\partial t^2}=c^2\frac{\partial^2 u}{\partial x^2}+\frac{\partial^2 u}{\partial y^2}+\frac{\partial^2 u}{\partial z^2}$$

the wave equation,

$$a^2\frac{\partial u}{\partial t}=\frac{\partial^2 u}{\partial x^2}+\frac{\partial^2 u}{\partial y^2}+\frac{\partial^2 u}{\partial z^2}$$

the heat equation.

A partial differential equation (PDE) has order n if n is the greatest of the orders of the partial derivatives which appear in the equation.

partial differential operator An operator whose definition involves one or more partial derivatives. *See* operator. One of the most significant partial differential operators is the Laplace operator, which, in Cartesian coordinates in 3 dimensions, has the form:

$$\nabla = \frac{\partial^2}{\partial x^2} + \frac{\partial^2}{\partial y^2} + \frac{\partial^2}{\partial z^2}.$$

partial differentiation The process of finding a partial derivative. *See* partial derivative.

partial sum The sum of the first N terms of an infinite series; thus, $a_1 + \ldots + a_N$ is a partial sum for $\sum_{k=1}^{\infty} a_k$.

partially isometric operator An operator, V, acting on a Hilbert space, with the property that V^*V and VV^* are orthogonal projections. V maps the range of V^*V isometrically onto the range of VV^*; V vanishes on the orthogonal complement of the range of V^*V.

partition (1.) A finite subset $\{t_0, t_1, \ldots, t_n\}$ of an interval $[a, b]$ such that $a = t_0 < t_1 < \ldots < t_n = b$.
(2.) A family of nonempty, disjoint subsets of a set S whose union is S.

partition of unity (1.) A family of nonnegative, continuous, real valued functions on a topological space, X, such that, for each $x \in X$, there is a neighborhood of x on which all but finitely many functions in the family vanish and such that $\sum f(x) = 1$ (the sum taken over all f in the family).
(2.) Often restricted to a compact Hausdorff space X with a finite open covering $\{U_1, \ldots U_n\}$: a family $\{f_1, \ldots, f_n\}$ of nonnegative, continuous, real valued functions such that the support of f_j is a subset of U_j, for each j, and $\sum_{j=1}^{n} f_j(x) = 1$.

Pascal line The line through the three points of intersection of the opposite sides of a hexagon inscribed in a conic. *See* Pascal's Theorem.

Pascal's configuration Each subsequent row of Pascal's triangle is obtained by adding the two entries diagonally above. This follows immediately from the binomial coefficient identity

$$\begin{aligned} \binom{n}{r} &\equiv \frac{n!}{(n-r)!r!} = \frac{(n-1)!n}{(n-r)!r!} \\ &= \frac{(n-1)!(n-r)}{(n-r)!r!} + \frac{(n-1)!r}{(n-r)!r!} \\ &= \frac{(n-1)!}{(n-r-1)!r!} + \frac{(n-1)!}{(n-r)!(r-1)!} \\ &= \binom{n-1}{r} + \binom{n-1}{r-1}. \end{aligned}$$

Pascal's Theorem If the six vertices of a hexagon lie on a conic, then the three points of intersection of pairs of opposite sides are collinear.

path A continuous function which maps a closed interval $[a, b]$ into an open subset of n-dimensional real or complex Euclidean space.

path of integration The path along which a line integral is defined. *See* line integral, path.

Peano continuum A continuous image of the unit interval, $[0, 1]$.

Peano curve A continuous function on the interval $[0, 1]$ whose range is the square $[0, 1] \times [0, 1]$.

Peano's Existence Theorem A theorem asserting the existence of at least one solution to the problem

$$y' = f(t, y), \; y(t_0) = y_0,$$

on the interval $[t_0, t_0 + \alpha]$, where $f(t, y)$ is assumed to be continuous and $|f(t, y)| \le M$ on $\{(t, y) : t \in [t_0, t_0 + a], y \in [y_0 - b, y_0 + b]\}$ and $\alpha = \min(a, b/M)$.

pedal curve A planar curve determined by a fixed smooth curve, C, and a fixed point, P, in the following fashion: On each tangent line to C select the point closest to P; the locus of all such points is the pedal curve of C with respect to P.

pentagamma function The fourth derivative of the natural logarithm of the gamma function:

$$\psi^{(3)}(x) = \frac{d^3}{dx^3}\psi(x) = \frac{d^4}{dx^4}\ln(\Gamma(x)).$$

An alternate definition is:

$$\psi^{(3)}(x) = -\int_0^\infty \frac{t^3 e^{-xt}}{t-1}\,dt.$$

See also digamma function, polygamma function.

period The minimal positive number, p, for which a periodic function, f, satisfies the equation, $f(x+p) = f(x)$, for all x.

period matrix A $2n \times n$ matrix of complex numbers whose rows, viewed as vectors in $\mathbf{C}^n$, are periods for a periodic meromorphic function on $\mathbf{C}^n$. The $2n$ rows are linearly independent over $\mathbf{R}$.

period parallelogram A parallelogram in the plane, or in complex space, which, for a doubly periodic function, f, has the property that the values of f everywhere are determined by the values of f in the parallelogram and the periodicity of f. If $P_1 = (x_1, y_1)$ and $P_2 = (x_2, y_2)$ (or $P_1 = x_1 + iy_1$ and $P_2 = x_2 + iy_2$) are primitive periods of f (so P_1 and P_2 are not colinear, or P_1/P_2 is not real), then the fundamental period parallelogram of f is the parallelogram whose vertices are 0, P_1, P_2 and $P_1 + P_2$.

period relation The Weierstrass zeta function $\zeta(u)$, corresponding to a lattice 2Ω, with generators ω, ω', is not elliptic, but satisfies

$$\zeta(u+2\omega) = \zeta(\omega) + 2\eta$$
$$\zeta(u+2\omega') = \zeta(u) + 2\eta'.$$

The (Legendre-Weierstrass) period relation is

$$\eta\omega - \eta'\omega' = \frac{1}{2}\pi i.$$

periodic function (**1.**) A function, f, of a real variable for which there exists a nonzero number, p, such that $f(x+p) = f(x)$, for all real numbers x.
(**2.**) A function, f, defined on an Abelian group for which there exists a group element, g, not the identity, such that $f(t+g) = f(t)$, for all elements t in the group.

periodicity modulus An elliptic function of order 2 has four double values, ∞, e_1, e_2, and e_3. The cross ratios $(e_i - e_j)/(e_i - e_k)$ are all equal and the square root of their common value is called the *(periodicity) modulus* of the period lattice.

Perron integrable function Let $f(x)$ be an extended real valued function on $[a, b]$, and denote upper and lower derivates of a function g by $\overline{D}g$ and $\underline{D}g$. *See* derivate. Then f is *Perron integrable* on $[a, b]$ provided
(i.) there is a real-valued function U on $[a, b]$ (called a *major function* of f), such that $\underline{D}U(x) > -\infty$ and $\underline{D}U(x) \geq f(x)$ on $[a, b]$;
(ii.) there is a real-valued function V on $[a, b]$ (called a *minor function* of f), such that $\overline{D}V(x) < \infty$ and $\overline{D}V(x) \leq f(x)$ on $[a, b]$; and
(iii.) The numbers

$$\inf\{U(b) - U(a) : U \text{ is a major function of } f \text{ on } [a, b]\}$$
$$\sup\{V(b) - V(a) : V \text{ is a minor function of } f \text{ on } [a, b]\}$$

are equal. In this case, the common value in (iii) is called the *Perron integral* of f.

Perron integral *See* Perron integrable function.

Perron's method A very general method for solving the Dirichlet problem on a nearly arbitrary bounded region Ω in the complex plane. Given boundary values f, a harmonic function u, on Ω which agrees with f on the boundary of Ω may generally be obtained by taking the least upper bound of all subharmonic functions, v, on Ω for which $\limsup_{z\to w} v(z) \leq f(w)$, for all w on the boundary of Ω.

perturbation theory A group of theorems, concerning relations between two Hilbert space operators T and $T + V$, where V is "small" in some sense. For example, the norm of V may be small, V may be compact, etc.

Petrovski's Theorem All distribution solutions u of the partial differential equation $P(D)u = 0$ on $\mathbf{R}^n$ are analytic functions on $\mathbf{R}^n$ if and only if the homogeneous part P_m of highest order of P is nonvanishing on $\mathbf{R}^n$.

Pfaffian form A differential form of order 1. *See* differential form.

phase space The collection of all vectors, v, representing possible states of a physical system of n particles. The components of a vector in phase space give the position and momentum of each of the n particles at one instant in time.

Phragmén-Lindelöf Theorem Let $D = \{re^{i\theta} : r \geq 0, |\theta| < \alpha < \pi\}$, for some α. Let A, B, C, δ be positive constants with $0 < \delta < \pi/(2\alpha)$. Assume f is analytic on D and continuous on $\overline{D}$. If $|f(z)| \leq A$ on the boundary of D and $|f(z)| \leq B\exp(C|z|^\delta)$ on D, then $|f(z)| \leq A$ on D. An extension of the maximum modulus principle.

Picard group The group of locally principal divisors on a variety $\mathcal{V}$ modulo the subgroup of principal divisors; denoted $Pic(\mathcal{V})$.

Picard variety Let V be a variety, and write $G(V)$ for the set of divisors, $G_l(V)$ for the set of divisors linearly equivalent to 0, and $G_a(V)$ for the group of divisors algebraically equal to 0. Then $G_a(V)/G_l(V)$ is called the *Picard variety*. The Albanese variety is dual to the Picard variety.

Picard's Theorem Let f be an analytic function with an essential singularity at a and let D be a neighborhood of a. Then the image, $f(D)$, of D contains every complex number except possibly one number; furthermore, on D, f assumes each complex number, except possibly one, an infinite number of times.

Picard-Lindelöf Theorem A theorem asserting uniqueness of the solution to the problem

$$y' = f(t, y), \; y(t_0) = y_0,$$

under a uniform Lipschitz condition, in y, on $f(t, y)$, in addition to the assumptions assuring existence (of, say, Peano's Existence Theorem). *See* Peano's Existence Theorem.

piecewise continuous function A function defined on an interval which is continuous except for jump discontinuities at finitely many points.

Poincaré metric The hermitian metric $d\mu = (1 - |\zeta|^2)^{-2} d\zeta \otimes d\bar{\zeta}$, on the unit disk in the complex plane.

Plancherel Theorem (1.) The Fourier transform is an isometry of $L^1(\mathbf{R}) \cap L^2(\mathbf{R})$ onto a dense subspace of $L^2(\mathbf{R})$ and therefore has a unique extension to an isometry of $L^2(\mathbf{R})$ onto $L^2(\mathbf{R})$.
(2.) If G is a locally compact Abelian group with Haar measure and $\hat{G}$ is its dual group, the Fourier transform is an isometry of $L^1(G) \cap L^2(G)$ onto a dense subspace of $L^2(\hat{G})$ and therefore has a unique extension to an isometry of $L^2(G)$ onto $L^2(\hat{G})$.

plane The solution set in $\mathbf{R}^n$ of an equation of the form:

$$A_1x_1 + \ldots + A_nx_n = C,$$

where at least one $A_k \neq 0$.

plane coordinates A coordinate system for a plane. The Cartesian coordinates of analytic geometry are the simplest such system. *See* coordinate system.

plane curve A curve (or path) whose range is a subset of a plane. *See also* space curve.

plot *See* graph.

plotting *See* curve tracing.

Plücker coordinates Consider a (straight) line L on the affine plane over a field K. L is the set of all solutions of an algebraic equation of degree 1, such as

$$u_0 + u_1x_1 + u_2x_2 = 0.$$

The coefficients $u_0, u_1, u_2 \in K$ are the *Plücker coordinates* of L. All triplets (u_0, u_1, u_2) are permitted, except those with $u_1 = u_2 = 0$; these would not describe a nontrivial condition. On the other hand, the coordinates of a straight line are not unique, since multiplication of the u_i by a fixed nonzero scalar $\lambda \in K$ merely replaces the above polynomial equation of the line by an equivalent one. The transition from (u_0, u_1, u_2) to $(\lambda u_0, \lambda u_1, \lambda u_2)$ with $\lambda \in K^*$ is called rescaling. Plücker coordinates are only determined up to an arbitrary rescaling.

Plücker's relations Relationships between the number of singularities of plane algebraic curves. For a given plane curve,

$$m = n(n-1) - 2\delta - 3\kappa$$
$$n = m(m-1) - 2\tau - 3\iota$$
$$\iota = 3n(n-2) - 6\delta - 8\kappa$$
$$\kappa = 3m(m-2) - 6\tau - 8\iota,$$

where m is the class, n the order, δ the number of nodes, κ the number of cusps, ι the number of stationary tangents (inflection points), and τ the number of bitangents. Only three of these equations are linearly independent.

pluriharmonic function A function, u, defined on a domain $D \subseteq \mathbf{C}^n$ which has continuous second derivatives and which satisfies the equations:

$$\frac{\partial}{\partial z_j}\frac{\partial}{\partial \bar{z}_k}u = 0, \quad j, k = 1, \ldots, n.$$

This is equivalent to the requirement that on every complex line, $\{a + bz : z \in \mathbf{C}\}$, determined by $a, b \in \mathbf{C}^n$, the function $z \to u(a + bz)$ is harmonic on $\{z : a + bz \in D\}$.

plurisubharmonic function A real valued function, u, on a domain $D \subseteq \mathbf{C}^n$ such that:

1) $-\infty \leq u(z) < \infty$, all $z \leq 0$;

2) $u(z)$ is upper semi-continuous and not constantly equal to $-\infty$; and

3) if $z, w \in \mathbf{C}^n$ and if $r > 0$ is such that $\{z + sw : s \in \mathbf{C} \text{ and } |s| \leq r, \} \subseteq D$, then $u(z) \leq \frac{1}{2\pi}\int_0^{2\pi} u(z + re^{i\theta}w)\, d\theta$.

By condition 3), on every complex line, $\{z + sw : s \in \mathbf{C}\}$, determined by $z, w \in \mathbf{C}^n$, the function $s \to u(z + sw)$ is subharmonic on $\{s \in \mathbf{C} : z + sw \in D\}$.

Plurisubharmonic functions play a role in several complex variables analogous to the role of subharmonic functions in the theory of one complex variable. Examples include $\log|f|$ and $|f|^c$, $c > 0$, where f is holomorphic on D. For functions on D with continuous second derivatives, this definition is equivalent to the condition:

$$\sum_{j,k=1}^{n} \frac{\partial^2 u}{\partial z_j \partial \bar{z}_k}(z) w_j \overline{w}_k \geq 0,$$

for all $z \in D, w \in \mathbf{C}^n$.

Poincaré distance For $\alpha, \beta \in D$, the open unit disk, the quantity

$$\Pi(\alpha, \beta) = \tan^{-1}\left|\frac{\beta - \alpha}{1 - \bar{\alpha}\beta}\right|.$$

Poincaré series The function $\theta_q\alpha(z)$ defined by

$$\theta_q\alpha(z) = \sum_{g \in G} \alpha(g(z))g'(z)^q$$

where G is a Kleinian group, D a G-invariant open set in $\mathbf{C}$, $(z \in D)$, α an analytic function on D and q is an integer > 2.

Poincaré condition That a space X have a trivial fundamental group. Equivalently, X is simply-connected, or, every closed path in X can be shrunk to a point within X.

Poincaré manifold A non-simply connected 3-manifold; also called a dodecahedral space.

point (1.) In geometry, an undefined term understood to be the constituent element of a line, plane, etc. and characterized as having zero dimensionality.
(2.) An element of a set, especially if the set is subject to some geometric or spatial interpretation.

point at infinity The additional point adjoined to a locally compact topological space, X, in forming the one point compactification of X. In particular, when X is the complex plane, the addition of a point at infinity yields a space homeomorphic to a sphere.

point of density For a subset, E, of $\mathbf{R}$, a point for which

$$\lim_{\delta \to 0} \frac{\mu(E \cap (x-\delta, x+\delta))}{2\delta} = 1,$$

where μ is Lebesgue measure on $\mathbf{R}$.

See also Lebesgue's density function.

point of tangency A point at which two differentiable curves cross and at which their tangent lines also coincide. *See* tangent line.

point spectrum The set of eigenvalues of a linear operator T from a complex linear space X to itself. That is, the set of $x \in X$ such that $Tx = \lambda x$, for some complex number λ.

point-slope equation of a line An equation of the form,

$$y - y_1 = m(x - x_1),$$

for a straight line which passes through the point (x_1, y_1) and whose slope is m.

Poisson bracket A skew symmetric, bilinear form which assigns a smooth, real-valued function, $\{F, G\}$, to each pair of smooth, real-valued functions on a smooth manifold and which satisfies the Jacobi identity:

$$\{\{F, G\}, H\} + \{\{H, F\}, G\} + \{\{G, H\}, F\} = 0$$

and Leibnitz' rule:

$$\{F, GH\} = \{F, G\}H + G\{F, H\}.$$

When F and G are functions of configuration and momentum variables, $q = (q_1, \ldots, q_n)$ and $p = (p_1, \ldots, p_n)$, the Poisson bracket is given by:

$$\{F, G\} = \sum_{i=1}^{n} \left(\frac{\partial F}{\partial p_i}\frac{\partial G}{\partial q_i} - \frac{\partial F}{\partial q_i}\frac{\partial G}{\partial p_i} \right).$$

Poisson integral The integral appearing in Poisson's integral formula. *See* Poisson's integral formula.

Poisson kernel The family of functions, P_r, given by the formula:

$$P_r(\theta) = \frac{1 - r^2}{1 - 2r\cos\theta + r^2} = \sum_{n=-\infty}^{\infty} r^{|n|}e^{in\theta}.$$

The values of a real harmonic function on the unit disk can be computed using a convolution formula involving values of the harmonic function on the boundary of the unit disk and the Poisson kernel. Poisson kernels can be written for other domains: For example, the

Poisson kernel on the upper half plane (of the complex plane) is:

$$\frac{1}{1+t^2}\Re\left[\frac{itw-1}{it-w}\right].$$

Poisson's integral formula The formula,

$$u(re^{i\theta}) = \frac{1}{2\pi}\int_{-\pi}^{\pi} u(e^{it})P_r(\theta - t)\,dt,$$

where u is a real harmonic function on the closed disk and P_r is the Poisson kernel for the closed disk. *See* Poisson kernel. A somewhat more general version, for u harmonic on an open disk D with radius R and center a and continuous on $\bar{D}$ is:

$$u(a+re^{i\theta}) = \frac{1}{2\pi}\int_{-\pi}^{\pi}\left[\frac{R^2-r^2}{R^2-2rR\cos(\theta-t)+r^2}\right] u(a+Re^{it})\,dt.$$

Poisson's summation formula The formula

$$\sum_{n=-\infty}^{\infty}\phi(2\pi n) = \frac{1}{2\pi}\sum_{k=-\infty}^{\infty}\int_{-\infty}^{\infty}\phi(t)e^{-ikt}\,dt,$$

where ϕ is a continuously differentiable function on $\mathbf{R}$, all the integrals in the formula exist, and $\sum_{n=-\infty}^{\infty}\phi(2\pi n + t)$ converges uniformly for t in $[0, 2\pi)$ to a function which can be expanded in a Fourier series.

polar (1.) Related to or expressed in terms of polar coordinates.
(2.) A convex set, S^o, associated to an arbitrary subset, S, of $\mathbf{R}^n$ or, more generally, to a subset of a locally convex linear topological space. If $S \subseteq \mathbf{R}^n$, then $S^o = \{x \in \mathbf{R}^n : (x, s) \leq 1, \text{ for all } s \in S\}$. If $S \subseteq V$, a locally convex linear space, $S^o = \{x \in V^* : |x(s)| \leq 1, \text{ for all } s \in S\}$, where V^* is the dual space of V.

polar coordinates A coordinate system, (r, θ), for the plane, related to Cartesian coordinates by the formulas: $x = r\cos\theta$, $y = r\sin\theta$. If (r, θ) are the coordinates of a point P, then r is the distance from P to the origin and θ is the angle (measured counterclockwise, in radians) between the positive x-axis and the ray from the origin through P.

polar element A point in the polar. *See* polar of coordinate system, pole of coordinate system.

polar equation An equation expressed in polar coordinates.

polar of coordinate system Points (coordinate systems) $P{:}p$ and $Q{:}q$ are conjugate with respect to the conic $xAx = 0$ (plane curve of order two) if $pAq = 0$. The self conjugate points are the points of the conic. The polar of P is the set of all points conjugate to P. *See also* pole of coordinate system, polar element.

polar triangle Let T be a spherical triangle, with vertices P_1, P_2, P_3, on a sphere S. Let A_1, A_2, A_3 be arcs of the great circles on S having P_1, P_2, P_3, as poles and forming a spherical triangle T'. Then T' is called the *polar triangle* of T. *See* pole of a circle, spherical triangle.

polarization identity For a Hilbert space with inner product $(\cdot, \cdot)$ and norm $\|\cdot\|$, the identity

$$(x, y) = \frac{1}{2}[\|x+y\|^2 - \|x+iy\|^2],$$

for vectors x, y. As a result, the inner product is determined by the norm and an isometry (which preserves norms) also preserves inner products.

pole A singularity, a, of an analytic function, f, such that f is analytic at every point of some neighborhood of a except at a itself and $\lim_{z\to a} |f(z)| = \infty$. Example: a is a pole for $1/(z-a)^m$, for m a positive integer.

pole of a circle Given a circle C on the surface of a sphere and a diameter d of the sphere perpendicular to the plane of C, then the endpoints of d are called the *poles* of C.

pole of coordinate system The dual notion of the polar of a point. *See* polar of coordinate system, polar element.

Polish space A topological space which is homeomorphic to a separable complete metric space.

polydisk The set, $\{(z_1, \ldots, z_n) : |z_1| \leq 1, \ldots, |z_n| \leq 1\}$, in $\mathbf{C}^n$. More generally, if $B_{r_i}(a_i) = \{z \in \mathbf{C} : |z - a_i| \leq r_i\}$, then $B_{r_i}(a_i) \times \ldots \times B_{r_n}(a_n)$ is a polydisk in $\mathbf{C}^n$.

polygamma function The derivative of order n, $n = 1, 2, 3, \ldots$, of the natural logarithm of the gamma function; a generalization of the digamma function:

$$\psi^{(n)}(x) = \frac{d^n}{dx^n}\psi(x) = \frac{d^{n+1}}{dx^{n+1}} \ln(\Gamma(x)).$$

An alternate definition is:

$$\psi^{(n)}(x) = (-1)^n \int_0^\infty \frac{t^n e^{-xt}}{t-1}\, dt.$$

See also digamma function.

polyharmonic function A function, u, defined on a region, D, in $\mathbf{R}^n$, which satisfies the equation $\Delta^m u = 0$, for some positive integer, m (the order of the function). Here, Δ is the Laplace operator and u has continuous partial derivatives of order $2m$.

Polynomial Approximation Theorem Every continuous function on a compact subset, X, of R^n can be uniformly approximated on X by a polynomial function. *See* Weierstrass's Theorem. For a far reaching generalization, *See* Stone-Weierstrass Theorem.

polynomial function A function of the form $f(x) = a_n x^n + \ldots + a_1 x + a_0$, where $a_n, \ldots, a_0$ are either real or complex constants and x is either a real or a complex variable. There are extensions of this concept to multi-variable functions and to functions whose domains are more complicated systems, such as matrices or operators on Hilbert space. For example, a monomial in n variables is a function of the form $f(x_1, \ldots, x_n) = c x_1^{k_1} x_2^{k_2} \cdot \ldots \cdot x_n^{k_n}$, where the k_i are non-negative integers; a polynomial of n variables is a finite sum of monomials.

polynomially convex region A region R in $\mathbf{C}^n$ which is equal to its polynomially convex hull

$$\hat{R} = \{z \in R : |f(z)| \leq \sup_{w \in R} |f(w)|, \text{ for all polynomials } f\}.$$

Pontryagin Duality Theorem Let G be a locally compact group and $\hat{G}$, the dual group (the group of continuous homomorphism of G into the unit circle group, provided with the topology of uniform convergence on compacta). Then the dual group of $\hat{G}$ is (isomorphic to) G.

porter The constant appearing in formulas for the efficiency of the Euclidean algorithm,

$$C = \frac{6 \ln 2}{\pi^2}\left[3 \ln 2 + 4\gamma - \frac{24}{\pi^2}\zeta'(2) - 2\right] - \frac{1}{2} = 1.4670780794\ldots,$$

where γ is Euler's constant and $\zeta(z)$ is the Riemann zeta function.

positive definite function (1.) A complex-valued function, f, defined on $\mathbf{R}$ such that, for all real numbers $x_1, \ldots, x_N$ and complex numbers $z_1, \ldots, z_N$ (with N any positive integer),

$$\sum_{j,k=1}^{N} f(x_j - x_k) z_j \bar{z}_k \geq 0.$$

See Bochner's Theorem.

(2.) A complex-valued function, f, defined on the dual group, $\hat{G}$, of a locally compact Abelian group such that, for all $g_1, \ldots, g_N \in \hat{G}$ and all complex numbers $z_1, \ldots, z_N$,

$$\sum_{j,k=1}^{N} f(g_j - g_k) z_j \bar{z}_k \geq 0.$$

The continuous positive definite functions are precisely the Fourier transforms of positive measures on G.

positive definite kernel A complex valued function, k, of two variables such that, for any positive integer n, $x_1, \ldots, x_n$, and $\lambda_1, \ldots, \lambda_n \in \mathbf{C}$,

$$\sum_{i,j=1}^{n} k(x_i, x_j) \lambda_i \bar{\lambda}_j \geq 0.$$

If k is measurable and positive definite on a measure space, (X, μ), then the corresponding integral operator on $L^2(X, \mu)$ is a positive operator.

positive distribution A distribution T such that $T(\phi) \geq 0$ for every nonnegative $\phi \in \mathbf{C}_0^\infty(\mathbf{R}^n)$. *See* generalized function.

positive harmonic function *See* harmonic function.

positive infinity A concept useful for describing the limit of a quantity that grows without bound. For example, a function $f(x)$ is said to approach *positive infinity* ($+\infty$) as x approaches a, written

$$\lim_{x \to a} f(x) = +\infty$$

provided, given any number M, there exists a $\delta > 0$, for which $f(x) > M$ holds for all x satisfying $|x - a| < \delta$.

Other uses of $+\infty$ are to describe the cardinality or size of a non-finite set or to indicate the sum of an infinite series. For example, the notation $\sum_{n=1}^{+\infty} \frac{1}{2^n} = 1$ means that the sum of:

$$\frac{1}{2}, \frac{1}{4}, \frac{1}{8}, \ldots, \frac{1}{2^N}$$

converges to the number 1, as N approaches $+\infty$.

positive orientation Given a circle C in the plane, or, more generally, any simple closed curve C, we can think of a particle as traveling around the curve (without stopping) in one of two possible directions. If the particle travels in the counter-clockwise direction we say that C has *positive orientation,* otherwise it has *negative orientation.*

More precisely, the movement of the particle should be described by a parametric representation $(x(t), y(t))$ of C and then the orientation is determined by these representing functions. *See* Green's Theorem and Cauchy's integral theorem for important examples using this concept.

positive semidefinite kernel Let (X, Ω, μ) be a measure space and suppose that $k : X \times X \to \mathbf{C}$ is an $\Omega \times \Omega$-measurable function. Define a linear operator

$$(Kf)(x) = \int_X k(x, y) f(y) \, d\mu(y)$$

on $L^2(\mu)$. If the operator K is bounded on the Hilbert space $L^2(\mu)$ and satisfies:

$$(Kf, f) = \int_X \int_X k(x, y) f(y) \overline{f(x)} \, dy dx \geq 0$$

for all $f \in L^2(\mu)$, then we say that the operator K is a *positive semidefinite operator* and the function k is a *positive semidefinite kernel.*

positive variation (1.) A function P associated with a given function f of bounded variation which is defined by $P(x) = \frac{1}{2}[T(x) + f(x)]$, where T is the total variation of f.

(2.) Let μ be a real measure on a σ-algebra $\mathcal{M}$. Associated with μ are three *positive* measures: the positive variation μ^+, the negative variation μ^- and the total variation $|\mu|$. They have the following properties:

$$\mu = \mu^+ - \mu^-, \quad |\mu| = \mu^+ + \mu^-$$

and for every measurable set E

$$-|\mu|(E) \leq \mu(E) \leq |\mu|(E).$$

This representation of μ as a difference of positive measures is known as the *Jordan decomposition of* μ.

power series A formal infinite series of the form:

$$a_0 + a_1(x-a) + a_2(x-a)^2 + \dots.$$

If the series converges for any $x \neq a$, then it typically converges absolutely on an interval $a-\rho < x < a+\rho$, if x is a real variable, and in an open disk $|x-a| < \rho$, if x is a complex variable. The optimum value ρ is called the *radius of convergence* of the power series.

Some important examples are:

$$e^x = 1 + x + \frac{x^2}{2!} + \frac{x^3}{3!} + \dots$$
$$\sin x = x - \frac{x^3}{3!} + \frac{x^5}{5!} + \dots$$

where $\rho = +\infty$ in both cases, i.e., the formula holds for all real x. The formulas also hold if x is replaced by z and z is allowed to be any complex number.

pre-Hilbert space A complex vector space with an *inner product*. That is, a complex vector space H with a complex function (x, y), called the *inner product* of x and y, defined on $H \times H$. The inner product should be linear in the first variable and satisfy: $(x, y) = \overline{(y, x)}$. Also, $(x, x) \geq 0$ with the value 0 only for the zero vector $x = 0$.

The *norm* of a vector x is defined by $\|x\| = (x, x)^{1/2}$. Thanks to the triangle inequality

$$\|x + y\| \leq \|x\| + \|y\|$$

an inner product space becomes a metric space with the metric:

$$d(x, y) = \|x - y\|.$$

A *Hilbert space* is a normed linear space which is complete in this metric.

Preparation Theorem Let $f(z)$ be an analytic function in a neighborhood of the origin. If $f(0) = 0$ but f is not identically zero, then

$$f(z) = z^k g(z)$$

where $g(z)$ is analytic in a neighborhood of the origin and $g(0) \neq 0$. This decomposition is unique and the integer k is called the *order* of the zero of f at the origin.

See Weierstrass's Preparation Theorem for the analogous situation for analytic functions in several complex variables.

prime divisor Let n be a natural number. A natural number m is said to be a *divisor* of n, written $m|n$, if $n = km$ for some natural number k. A *prime divisor* of n is a divisor of n which is also prime.

Prime Number Theorem The famous Prime Number Theorem was proved in the late 1800s and concerns the distribution of prime numbers (natural numbers p, with no factors other than p and 1). Let $N(x)$ be the number of primes between 1 and x. The theorem says that:

$$\lim_{x\to\infty} \frac{N(x)\log x}{x} = 1.$$

Thus, the number of primes less than x is approximately $\frac{x}{\log x}$ for large x. So the number of primes is infinite but represents a very small proportion of the integers less than x for large values of x.

primitive function The primitive of a function $f(x)$ is any function $F(x)$ satisfying $F'(x) = f(x)$. In other words, a primitive function is an antiderivative or indefinite integral of $f(x)$. In symbols,

$$F(x) = \int f(x)\,dx.$$

See Fundamental Theorem of Calculus for the existence and uniqueness of the antiderivative.

principal analytic set An analytic set in $\mathbf{C}^n$ is a set which is locally the set where finitely many holomorphic functions vanish. In the case that the analytic set is generated by a *single* holomorphic function, then the set is called a principal analytic set. *See* first Cousin problem and second Cousin problem for more information on the structure of analytic sets.

principal axis Let A be a real symmetric $n \times n$ matrix with n distinct eigenvalues. Consider the quadratic form:

$$q(\mathbf{x}) = \mathbf{x} \cdot A\mathbf{x}.$$

Thus, for

$$A = \begin{pmatrix} a & b \\ b & c \end{pmatrix} \quad \mathbf{x} = \begin{pmatrix} x_1 \\ x_2 \end{pmatrix}$$

we have $q(\mathbf{x}) = ax_1^2 + 2bx_1x_2 + cx_2^2$.

The curve

$$ax_1^2 + 2bx_1x_2 + cx_2^2 = 1$$

will be an ellipse with major and minor axes obtained by rotating the x_1 and x_2 axes through some angle. The rotated axes are called the principal axes for the quadratic form.

The vector directions of these axes turn out to be eigenvectors for A. In general, the eigenspaces of A are called the principal axes of q.

principal bundle Let G be a Lie group. A (smooth) principal bundle with structure group G is a pair (P, T), where

(i.) $P = (Q, \pi, B, G)$ is a smooth fiber bundle.

(ii.) $T: Q \times G \to Q$ is a right action of G on P.

(iii.) P admits a coordinate representation $\{(U_a, \psi_a)\}$ such that

$$\psi_a(x, ab) = \psi_a(x, a) \cdot b,$$

$x \in U_a$, $a, b \in G$.

(Note that we write $T(z, a) = z \cdot a$.)

principal component analysis Let A be a Banach algebra. The set of invertible elements forms a group G under multiplication. In fact, this is a topological group since multiplication and inversion are continuous in G.

As a topological space, G has a connected component G_1 containing the identity element of A. This set is called the principal component of G and consists of all connected subsets of G that contain the identity element e in A.

An elementary example of principal component analysis is the following simplified version of the Arens-Royden Theorem for commutative Banach algebras. Suppose that A is the Banach algebra of all continuous complex valued functions on the unit circle in the complex plane. Then, the quotient group G/G_1, in this case, is isomorphic to the additive group of integers. Moreover, two invertible members of A are in the same coset of G_1 if and only if they are homotopic mappings.

principal curvature A particle traveling in space with position vector $\mathbf{R}(t)$, at time t, will have a unit tangent vector $\mathbf{T}(t)$, principal normal vector $\mathbf{N}(t)$ and curvature κ. *See* principal normal. For a unit speed curve, κ is the length of the component of $\mathbf{R}''(t)$ in the direction of $\mathbf{N}(t)$.

Now suppose we are given a surface S and point $p_0 \in S$. Let $\mathbf{n}_0$ be the surface normal at p_0. Consider particles traveling on S at unit speed and passing through the point p_0 at $t = 0$. The normal curvature of the surface S at p_0, in the direction of the tangent vector $\mathbf{T}(0)$, is the component of $\mathbf{R}''(0)$ in the direction $\mathbf{n}_0$. It turns out that this quantity, denoted by κ_n, depends only on the direction of the path at p_0 and hence is a property of the surface.

The maximum and minimum values κ_n, corresponding to all tangent vectors, are called the *principal curvatures* of the surface at the point p_0. The corresponding directions on which these extreme values are attained are called the *principal directions*.

principal directions *See* principal curvature.

principal divisor Let X be a complex manifold and $\mathcal{O}_X$ its structural sheaf. Let $\mathcal{M}$ be the sheaf of meromorphic functions on X, $Pic(X)$ the Picard group of X and $Div(X)$ the group of Weil divisors on X. For each meromorphic function on X, we can associate a Weil divisor by taking its zeros and poles and so we get a map

$$H^0(X, \mathcal{M}^*/\mathcal{O}_X^*) \to Div(X).$$

The map is an isomorphism and gives rise to the exact sequence

$$H^0(X, \mathcal{M}^*) \to Div(X) \to H^1(X, \mathcal{O}_X^*) \to H^1(X, \mathcal{M}^*).$$

The image of $H^0(X, \mathcal{M}^*)$ in this sequence is the *group of principal divisors.*

principal normal A particle traveling in space with position vector $\mathbf{R}(t)$, at time t, will have a unit tangent vector $\mathbf{T}(t)$. The *principal normal vector,* written $\mathbf{N}(t)$ is defined by:

$$\mathbf{N}(t) = \frac{\mathbf{T}'(t)}{\|\mathbf{T}'(t)\|}.$$

It turns out that these two vectors are perpendicular to each other and the plane they determine is called the *osculating plane.*

This plane has the greatest contact with the curve at the point $\mathbf{R}(t)$ and best describes the motion at time t. The bending of the curved path in this plane is called the *curvature.* It is defined by

$$\kappa = \left\| \frac{d\mathbf{T}}{ds} \right\|$$

and can be used to determine the centipetal force on the particle by means of the formula:

$$\mathbf{F}_N = m\kappa \left(\frac{ds}{dt}\right)^2 \mathbf{N}.$$

principal part Suppose that we are given an nth order linear partial differential operator on $\mathbf{R}^m$

$$D(f) = \sum_{0 \le |\alpha| \le n} c_\alpha \partial^\alpha f$$

where $\alpha = (\alpha_1, \dots, \alpha_m)$ and $\partial^\alpha f$ denote the derivative of f

$$\frac{\partial_1^\alpha}{\partial x_{1_1}^\alpha} \frac{\partial_2^\alpha}{\partial x_{2_2}^\alpha} \cdots \frac{\partial_n^\alpha f}{\partial x_{n_n}^\alpha}.$$

Then, the principal part of the operator is the sum of the nth order terms, i.e.,

$$D_1(f) = \sum_{|\alpha| = n} c_\alpha \partial^\alpha f$$

where $|\alpha| = \sum_1^n \alpha_j$. The idea is that the main properties of the operator can be derived from the highest order terms.

principal value For a multiple-valued function, a designated choice of values in a particular subset of the domain where the restriction is single-valued.

For example, for a fixed value y, $-1 \le y \le 1$, there are infinitely many solutions to the equation: $\sin x = y$. But, there is a unique solution with $-\frac{\pi}{2} \le x \le \frac{\pi}{2}$. This solution is called the *principal value* for the inverse sine function (or arcsine function). Thus, the principal value for $\arcsin\frac{1}{2}$ is $\frac{\pi}{6}$ and yet $\sin(\frac{5\pi}{6}) = \frac{1}{2}$.

For the complex logarithm, the domain for the principal value is usually chosen to be the complex plane with the negative real axis removed, and the principal value of $\log(z)$ in this domain, sometimes denoted $\text{Log}(z)$ is $\log|z| + i \arg(z)$, with $-\pi < arg(z) \le \pi$. *See also* Cauchy principal value.

principle of linearized stability A principle giving conditions under which a nonlinear dynamical system is asymptotically stable near a fixed point. In other words, when a slightly perturbed initial state will eventually return to the equilibrium state. As an example, the linear system

$$\frac{d\vec{x}}{dt} = A\vec{x}(t)$$

where A is an $n \times n$ real matrix, will be asymptotically stable at the origin provided that each eigenvalue λ satisfies $\Re\lambda < 0$.

principle of localization If two functions (real valued, on $[0, 2\pi]$) are equal in an interval $I \subset [0, 2\pi]$ then the difference of their Fourier series converges to 0 on any interval I' contained in the interior of I. The difference of their conjugate series also converges in I', but not necessarily to 0. Thus the convergence of Fourier series is a local phenomenon (Riemann's Principle of Localization).

principle of nested intervals A sequence of closed bounded intervals $[a_n, b_n]$ of real numbers, where $n = 1, 2, \ldots$, such that

$$-\infty < a_{n-1} \le a_n \le b_n \le b_{n-1} < \infty.$$

A fundamental property of the real numbers is that there exists a real number x_0 which belongs to *all* the intervals, i.e., $a_n \le x_0 \le b_n$ for all $n = 1, 2, \ldots$.

principle of superposition Consider a second order linear differential equation of the form

$$y'' + p(x)y' + q(x)y = 0.$$

A solution to this equation is a twice differentiable function $y = y(x)$ which satisfies the above equation for all x. Given two solutions $y_1(x)$ and $y_2(x)$ one sees that a third solution is obtained from the linear combination

$$y = c_1 y_(x) + c_2 y_2(x),$$

where c_1, c_2 are arbitrary real constants.

The *principle of superposition* states that every solution to the equation is of the above form provided that neither y_1 nor y_2 is a constant multiple of the other. For example, every solution to $y'' + y = 0$ is a linear combination of $\sin x$ and $\cos x$.

More generally, an nth order, linear, homogeneous differential equation can be solved by finding a system of n linearly independent solutions $y_1, \ldots, y_n$ and forming the linear combination

$$y = c_1 y_1 + \cdots + c_n y_n.$$

Pringsheim's Theorem A continued fraction is an expression of the form

$$\frac{a_1}{b_1 + \frac{a_2}{b_2 + \ldots}},$$

which is understood to be the limit of the sequence:

$$a_1, \frac{a_1}{b_1 + a_2}, \frac{a_1}{b_1 + \frac{a_2}{b_2 + a_3}}, \ldots.$$

Pringsheim's Theorem states that the above sequence converges provided $|b_n| \ge |a_n| + 1$ for all $n = 1, 2, \ldots$. A well-known example is to take $a_n = 1$, $b_n = 2$ for all n; then the resulting continued fractions converge to $\sqrt{2} - 1$.

progression Two common propagation patterns for numerical sequences are
(i.) *arithmetic progressions*, which are of the form

$$a, a + x, a + 2x, a + 3x, \ldots$$

and
(ii.) *geometric progressions*, which are of the form

$$a, ax, ax^2, ax^3, \ldots,$$

where in both cases a and x are arbitrary numbers.

For example, the sequence $1, 3, 5, 7, \ldots$ is an arithmetic progression whereas the sequence $1, 2, 4, 8, \ldots$ is a geometric progression. Thus, the infinite series

$$\sum_{n=0}^{\infty} \frac{1}{2^n} = 1 + \frac{1}{2} + \frac{1}{4} + \ldots = 2$$

is called a *geometric series* because the terms in the series are a geometric progression.

projection on a line The projection of a point p_0 on a line l is the unique point $p \in l$ which minimizes the distance between p_0 and l. In geometry, this point is determined by constructing a line perpendicular to l and passing through p_0. The point of intersection between these lines is the point p.

projection on plane The projection of a point p_0 on a plane Π is the unique point $p \in \Pi$ which minimizes the distance between p_0 and Π. In geometry, this point is determined by constructing a line perpendicular to Π and passing through p_0. The point of intersection between this line and the plane is the point p.

projection operator An operator from a vector space X onto a linear subspace E, which satisfies $Px = x$, for all $x \in E$, and $P^2x = Px$, for all $x \in X$. If X is a Hilbert space and P is a self-adjoint operator on X, then P is called an *orthogonal projection* operator onto E. A Hilbert space projection which is not an orthogonal projection is sometimes called a *projector.*

projective coordinates The projective plane $\mathbf{P}$ is an object of study in the theory of projective geometry. It can defined as the set of lines through the origin or directions in three-dimensional space $\mathbf{R}^3$. In other words, the *point* $S \in \mathbf{P}$ of the projective plane is determined by a triple (x_1, x_2, x_3) of real numbers, where this triple is assumed not to be the origin. We call these triples, with the identification

$$(x_1, x_2, x_3) = (cx_1, cx_2, cx_3), \quad c \neq 0$$

homogeneous coordinates of the point S. This mapping of $\mathbf{P}$ into the set of homogeneous coordinates is called the *projective coordinate system.*

The projective plane can be thought of as the projective space associated with the vector space $\mathbf{R}^3$. In a similar manner, there are projective spaces associated with any vector space.

projective geometry Given two distinct planes Π, Π' and a central point O not belonging to either, one can project a point $p \in \Pi$ to a unique point $p' \in \Pi'$. The point p' is the intersection of the line determined by the points O, p, and the plane Π.

Projections of this kind are called *central projections.* Notice that lines project to lines and triangles project to triangles. But a circle typically projects to an ellipse.

Projective geometry is the study of geometric properties of figures that are preserved under central projections. As an example, it can be proved that every conic projects to another conic (although not necessarily of the same type).

projective line element Let K be a ground field. If we denote by P^1K $(= P^1(K))$ $(= K \cup \{\infty\})$ the one-dimensional projective line (space) over K, then its points are the one-dimensional subspaces of K^2 called *projective line elements.*

projective set A member of the σ-algebra generated by the projections of coanalytic sets in a Polish space. *See* analytic set.

projective transformation Given a vector space E, we can define the projective space $\mathbf{P}(E)$ associated with E as the set of lines through the origin in E. *See* projective coordinates. Given a one-to-one linear transformation $u : E \to F$ to a vector space F we get a mapping

$$\mathbf{P}(u) : \mathbf{P}(E) \to \mathbf{P}(F)$$

determined by the homogeneous coordinates.

This mapping is called a *projective transformation* if it is bijective, that is, if dim $\mathbf{P}(E)$ equals dim $\mathbf{P}(F)$.

prolate spheroid An ellipsoid produced by rotating an ellipse through a complete revolution about its longer axis is called a *prolate spheroid* or a *prolate ellipsoid.*

proper convex function A convex function on $\mathbf{R}^n$ is any continuous function real-valued function $f(\mathbf{x})$ satisfying:

$$f\left(\frac{\mathbf{x}+\mathbf{y}}{2}\right) \leq \frac{f(\mathbf{x}) + f(\mathbf{y})}{2}$$

for all $\mathbf{x}, \mathbf{y} \in \mathbf{R}^n$.

For example, any linear (or even affine linear) function is convex because the above relation holds with an equality.

A *proper* convex function is one that satisfies the above relation with a strict inequality. For $n = 1$, this condition will hold for any twice differentiable function with $f''(x) > 0$ for all $x \in \mathbf{R}$.

proper equivalence relation *See* equivalence relation. A *proper* equivalence relation is one which makes an actual distinction between objects.

Thus, the relation among integers that a is related to b if and only if their difference is either an odd or an even integer makes no distinction at all. On the other hand, if we require the difference to be an *even* integer, then we partition the set of all integers into two distinct classes; i.e., the even integers and the odd integers.

proportionality Two quantities x and y are said to be *proportional* provided each one is a constant multiple of the other, i.e., the relation $y = kx$ holds, with $k \neq 0$.

As an example, one says that the circumference C of a circle is proportional to its radius r. The familiar formula is $C = 2\pi r$ so the *constant of proportionality* in this case is $k = 2\pi$.

pseudo-differential operator A pseudo-differential operator is a mapping $f \rightarrow T(f)$ given by

$$(Tf)(x) = \int_{\mathbf{r}^n} a(x, \xi)\hat{f}(\xi)e^{2\pi i x\cdot\xi}\, d\xi$$

where

$$\hat{f}(\xi) = \int_{\mathbf{r}^n} f(x)e^{-2\pi i x\cdot\xi}\, dx$$

is the Fourier transform of f, and where $a(x, \xi)$ is the *symbol* of T.

In the special case

$$a(x\xi) = \sum_{|\alpha|\leq m} a_\alpha(x)(2\pi i\xi)^\alpha$$

the properties of the Fourier transform show that

$$(Tf)(x) = \sum_{|\alpha|\leq m} a_\alpha(x)\left(\frac{\partial}{\partial x}\right)^\alpha f(x).$$

Note that in the 2 dimensional setting, $n = 2$, the multi-index $\alpha = (\alpha_1, \alpha_2)$ and

$$\left(\frac{\partial}{\partial x}\right)^\alpha f(x) = \frac{\partial^{\alpha_1+\alpha_2}}{\partial x_1^{\alpha_1}\partial x_2^{\alpha_2}} f(x).$$

pseudo-function The Fourier transform of a function in $\mathbf{C}_0(\mathbf{R})$.

pseudo-Riemannian metric A *Riemannian metric* is an inner product on the tangent vectors to a smooth manifold or surface which gives a way of measuring "lengths" and "angles" of tangent vectors. *See* Riemannian metric.

A *pseudo-Riemannian metric* is similar except that the restriction that the inner product be positive is dropped. Instead, it assumed that the bilinear form is nondegenerate, meaning that the only vector orthogonal to everything is the zero vector. Lorentz metrics form an important example of a pseudo-Riemannian metric.

pseudoconvex domain *See* Levi pseudoconvex domain.

pseudodistance A real-valued function d on $X \times X$, for a topological space X, which satisfies the axioms of a metric, except that $d(p, q)$ may $= 0$ for certain $p, q \in X$ with $p \neq q$. *See* Carathéodory pseudodistance, Kobayashi pseudodistance.

pseudolength A nonnegative real-valued function F on a vector space X satisfying $F(av) = |a|F(v)$, for $a \in \mathbf{C}$ and $v \in X$.

pseudosphere The pseudosphere is the surface of revolution obtained by rotating the parametric curve

$$x(t) = \sin t \quad z(t) = \log\tan\frac{t}{2} + \cos t$$

for $0 < t < \frac{\pi}{2}$ about the z-axis. The resulting surface has constant Gaussian curvature -1 and is used as a model for hyperbolic or non-Euclidean geometry.

pseudotensorial form *See* tensorial form.

pullback (1.) Pullback diagram. Given sets A_1, A_2, K and maps $\alpha_1 : A_1 \rightarrow K, \alpha_2 : A_2 \rightarrow K$, the *pullback* is a set B, together with maps $\beta_1 : B \rightarrow A_1, \beta_2 : B \rightarrow A_2$ such that
(i.) $\alpha_1\beta_1 = \alpha_2\beta_2$, and
(ii.) if C is another set, with maps $\gamma_j : C \rightarrow A_j (j = 1, 2)$ satisfying $\alpha_1\gamma_1 = \alpha_2\gamma_2$, then there is a unique map $\delta : C \rightarrow B$ with $\beta_j\delta = \gamma_j (j = 1, 2)$.

Normally, the sets are objects and the maps are morphisms in some category. Pullbacks exist in Abelian categories and in some others such as the category of Hilbert modules.

(2.) Pullback bundle. Given a map $F : Y \rightarrow X$ between two complex manifolds and E a vector bundle over X, defined by a covering $\{U_\alpha\}$ of X and corresponding cocycle $\{g_{\alpha\beta}\}$, the *pullback bundle* on Y, denoted F^*E, is defined by the cover $\{F^{-1}(U_\alpha)\}$ and the cocycle $\{g_{\alpha\beta} \circ F\}$.

purely contractive analytic operator function A function $C(z)$, whose values are contraction operators from one Hilbert space H to another Hilbert space H_*, analytic in an open set D and satisfying $\|C(z)\| < 1$ in D.

purely contractive part Suppose that we are given an analytic contraction-valued function $C(z)$ mapping one Hilbert space H into another H_*. (*See* purely contractive analytic operator function.) If we have the direct sum $C(z) = D(z) \oplus V$, where $D(z)$ is a purely contractive analytic function from L to L_* and V is an isometric isomorphism from $H \ominus L$ onto $H_* \ominus L_*$, then $D(z)$ is called the purely contractive part of $C(z)$.

purely singular measure Given a positive measure μ and an arbitrary measure λ on a σ-algebra Ω, λ is *purely singular with respect to* μ (written $\lambda \ll \mu$) provided $E \in \Omega$ and $\mu(E) = 0$ imply $\lambda(E) = 0$. When μ is not specified, it is understood to be Lebesgue measure.

pushout Given sets A_1, A_2, K and maps $\alpha_1 : K \rightarrow A_1, \alpha_2 : K \rightarrow A_2$, the *pushout* is a set B, together with maps $\beta_1 : A_1 \rightarrow B, \beta_2 : A_2 \rightarrow B$ such that
(i.) $\beta_1\alpha_1 = \beta_2\alpha_2$, and
(ii.) if C is another set, with maps $\gamma_j : A_j \rightarrow C (j = 1, 2)$ satisfying $\gamma_1\alpha_1 = \gamma_2\alpha_2$, then there is a unique map $\delta : B \rightarrow C$ with $\delta\beta_j = \gamma_j (j = 1, 2)$.

Normally, the sets are objects and the maps are morphisms in some category. Pushouts exist in Abelian categories and in some others such as the category of Hilbert modules.

Putnam's Theorem Suppose that M, N, and T are bounded linear operators on a Hilbert space H. Assume also that M and N are both normal operators, i.e., they commute with their adjoints operators:

$$MM^* = M^*M \quad \text{and} \quad NN^* = N^*N.$$

The Fuglede-Putnam Theorem states that if $MT = TN$, then $M^*T = TN^*$.

Fuglede proved the case $M = N$ and later Putnam extended the theorem to the general case.

Q

quadrant The Cartesian plane is divided in half by the x-axis and again by the y-axis. This forms the four *quadrants* of the xy-plane. The quadrant where $x > 0$ and $y > 0$ is called the first quadrant. The other three are numbered in a counterclockwise direction.

quadrant of angle The Cartesian plane is divided into four quadrants. *See* quadrant. If an angle is oriented so that its initial side is the positive x-axis and it opens in the counterclockwise direction, then the *quadrant of the angle* is the quadrant in which the terminal side belongs. In cases where the terminal side is an axis, then the angle is a *quadrantal angle.*

quadratic differential Given a rectifiable curve C in space, described by

$$x = x(t), y = y(t), z = z(t)$$

for $a \leq t \leq b$, the length of C is given by the formula

$$L(C) = \int_a^b \sqrt{\frac{dx}{dt}^2 + \frac{dy}{dt}^2 + \frac{dz}{dt}^2}\, dt.$$

Thus, the element of arclength ds can be thought of as satisfying the *quadratic differential* relationship

$$ds^2 = dx^2 + dy^2 + dz^2.$$

More generally, quadratic differentials are used to describe the length element in a Riemannian manifold.

quadratic form A *homogeneous quadratic form* in the variables $x_1, \ldots, x_n$ is a polynomial

$$q(x_1, \ldots, x_n) = \sum_{i=1}^{n} \sum_{j=1}^{n} b_{i,j} x_i x_j$$

in which each term is degree two. This form may be written as a matrix product XBX^T. *See also* principal axis.

quadratic function A quadratic function in the variable x is a polynomial of degree 2 in x. Thus, $f(x) = ax^2 + bx + c$ represents the general quadratic function (if $a \neq 0$). The roots $f(x) = 0$ can be found using the quadratic formula,

$$x = \frac{-b \pm \sqrt{b^2 - 4ac}}{2a}.$$

quadratic inequality An inequality of the form $ax^2 + bx + c \geq 0$, where $a \neq 0$. A quadratic inequality can be easily solved using the quadratic formula and the graph of $y = ax^2 + bx + c$. For example, if $a < 0$, then the solution is

$$\frac{-b - \sqrt{b^2 - 4ac}}{2a} \leq x \leq \frac{-b + \sqrt{b^2 - 4ac}}{2a}.$$

quadrature (**1.**) The *quadrature of a circle* refers to the ancient problem of constructing a square with the same area.

(**2.**) In a more modern context, the name given to the problem of finding areas by solving a definite integral:

$$\text{area} = \int_a^b f(x)\, dx.$$

(**3.**) The method of solving a differential equation of the form

$$y' = f(x)g(y),$$

where $y' = \frac{dy}{dx}$, by separation of variables. *See* separation of variables.

quadric curve A quadric curve is the graph of a second degree polynomial equation in x and y

$$ax^2 + bxy + cy^2 + dx + ey + f = 0.$$

Such curves include the equations of circles, ellipses, parabolas, and hyperbolas. The

0-8493-0320-6/00/$0.00+$.50

above curve is also called a *conic* because it represents all possible intersections of a plane with a right circular cone.

quadric surface The graph of a second degree polynomial in x, y, and z. For example, the graph

$$\frac{x^2}{a^2}+\frac{y^2}{b^2}+\frac{z^2}{c^2}=1$$

yields a "football" shaped surface called an *ellipsoid*. Other interesting surfaces are hyperboloids and paraboloids. These names are derived from the conic curves that are obtained by intersecting the surface with a plane parallel to the coordinate planes: $x = 0$, $y = 0$, and $z = 0$.

quasi-analytic class A class of infinitely differentiable functions on the circle **T** such that the only function in the class which vanishes, together with all its derivatives, at any point $t \in \mathbf{T}$ is the identically 0 function. *See* Denjoy-Carleman Theorem.

quasi-analytic function *See* quasi-analytic class.

quasi-conformal mapping A map between two differentiable manifolds which, in local coordinates $dz = dx + idy$ and $d\bar{z} = dx - idy$, has the spherical line element $ds^2 = |g_1dz + q_2d\bar{z}|^2$ such that

$$D(z)=\frac{|g_1|+|g_2|}{|g_1|-|g_2|}$$

is bounded for $|z| < R$, for some $R > 0$. Note that $D(z) = 1$ in the conformal (analytic) case.

quasi-nilpotent operator A bounded linear operator from a Hilbert space H to itself, satisfying

$$\lim_{n\to\infty}\|T^n\|^{1/n}=0.$$

The notion generalizes that of a nilpotent operator, satisfying $T^n = 0$ for some positive integer n.

The spectral radius is 0 for a quasi-nilpotent operator. *See* spectral radius.

quasi-norm A real-valued function $\|\cdot\|$, defined on a vector space **V** over a scalar field **F** (usually the real or complex numbers) satisfying
(i.) $\|x\| \geq 0$, for $x \in \mathbf{V}$, with equality if and only if $x = 0$;
(ii.) $\|x+y\| \leq \|x\|+\|y\|$, for $x, y \in \mathbf{V}$; and
(iii.) $\|-x\| = \|x\|$ and

$$\lim_{\alpha_j\to 0}\|\alpha_j x\|=0 \qquad \lim_{\|x_j\|\to 0}\|\alpha x_j\|=0,$$

where $x, x_j \in \mathbf{V}$ and $\alpha, \alpha_j \in \mathbf{F}$. *See* Frechet space.

quotient norm *See* quotient space.

quotient space A common situation in linear algebra is to have a vector space X and a vector subspace Y. One then obtains the vector quotient space X/Y whose elements are the cosets $x + Y$, for $x \in X$. Here, the coset $x_1 + Y$ is equal to the coset $x_2 + Y$ if and only if $x_1 - x_2 \in Y$. One can then define the usual vector operations and show that the quotient space is again a vector space.

If X is a normed vector space, the quotient space can be given the *quotient norm*, defined by

$$\|x+Y\|=\inf_{y\in Y}\|x+y\|.$$

If X is a Hilbert space, the quotient norm of $x+Y$ is equal to the norm of the projection of x onto the orthogonal complement $Y^{\perp}$ of Y and the quotient space X/Y can be naturally identified with $Y^{\perp}$.

R

radial maximal function The function defined on the unit circle by

$$\tilde{u}(e^{i\theta}) = \sup_{0 \le r < 1} |u(re^{i\theta})|$$

where $u(z)$ is a function defined on the unit disk. The radial maximal function has important applications to the theory of Hardy Spaces. For example, if the function u is harmonic for $|z| < 1$ and $0 < p < \infty$ then $u = \Re f$ for some analytic function $f(z)$ in the Hardy Space H^p if and only if the radial maximal function is in the space L^p.

radius of convergence *See* power series.

radius of curvature A particle traveling in space has three associated vectors at time t, namely: the unit tangent **T**, the principle normal **N**, and binormal **B**. *See* principal normal.

The osculating plane determined by **T** and **N** best describes the motion at time t and the bending of the curved path in this plane is called the curvature κ. *See* curvature. The reciprocal of k is called the *radius of curvature R*. The basic idea is that the motion of the particle is approximately uniform motion along a circle of radius R in the osculating plane.

radius of principal curvature The principal curvatures for a surface are the maximum and minimum values for the normal curvatures. *See* principal curvature. Analogous to radius of curvature for curves, the *radii of principal curvature* for a point on a surface are just the reciprocals of the principal curvatures.

radius of torsion A particle traveling in space has three associated vectors at time t, namely: the unit tangent **T**, the principle normal **N**, and binormal **B**. *See* principal normal. These form the Frenet frame $\{\mathbf{T}, \mathbf{N}, \mathbf{B}\}$ and satisfy

$\mathbf{T}' = \kappa \mathbf{N}$
$\mathbf{N}' = -\kappa \mathbf{T} + \tau \mathbf{B}$
$\mathbf{B}' = -\tau \mathbf{N},$

where the scalar quantities are the curvature κ and torsion τ. *See* curvature, torsion.

The *radius of torsion* is defined in similar manner to the radius of curvature; namely, it is the reciprocal of τ.

radius vector The motion of a particle traveling in space is described by a *position* or *radius* vector $\mathbf{R}(t) = x(t)\mathbf{i} + y(t)\mathbf{j} + z(t)\mathbf{k}$. The velocity and acceleration vectors are then obtained by differentiating $\mathbf{R}(t)$.

Radon transform The operator

$$R_\theta f(t) = \int_{(x,\theta)=t} f(x) d\mu_{n-1},$$

on $f \in L_0^2(\mathbf{R}^n)$, where μ_{n-1} is Lebesgue measure on the plane $(x, \theta) = t$.

Radon-Nikodym derivative *See* Radon-Nikodym Theorem.

Radon-Nikodym Theorem Given a positive σ-finite measure μ and an arbitrary complex measure λ, on the same σ algebra Ω, with λ absolutely continuous with respect to μ, there exists a unique function $h \in L^1(\mu)$ such that

$$\lambda(E) = \int_E h d\mu,$$

for $E \in \Omega$. The function h is called the *Radon-Nikodym derivative* of λ with respect to μ, denoted $h = d\lambda/d\mu$.

Ramanujan conjecture Also known as Ramanujan's hypothesis. Ramanujan proposed that

$$\tau(n) \sim O(n^{11/2+\epsilon}),$$

where $\tau(n)$ is the tau function. This was proved by Deligne (1974), in the course of

0-8493-0320-6/00/$0.00+$.50

proving the more general Petersson conjecture.

Ramanujan-Petersson conjecture A conjecture about the eigenvalues of modular forms under Hecke operators.

ramification number Let $f : M \to N$ be a nonconstant, holomorphic mapping between Riemann surfaces. Suppose that we can choose local coordinates z, vanishing at a point $p \in M$, and ζ, vanishing at $f(p)$ in N, such that $\zeta = f(z) = \sum_{j \geq n} a_j z^j$ $(a_n \neq 0)$. The number n is called the *ramification number* of f at p.

ramification point (1.) *See* ramification number, for a map f between manifolds.
(2.) For a finite mapping $f : X \to Y$, of irreducible varieties, a point $p \in Y$ at which the number of inverse images is not equal to the degree of f.

Ramification Theorem Let k be an algebraically closed field and x an indeterminate. Given y, separably algebraic over $k(x)$, suppose that the ramification index e at a place of $k(x, y)$ over the place $x = 0$ of $k(x)$ is not divisible by char k. Then y can be written as a power series in $x^{1/e}$.

ramified covering space A manifold M^* is a *ramified covering* of a manifold M provided there is a continuous, surjective map (the *ramified covering map*) $f : M^* \to M$ such that, for every $p^* \in M^*$ there are local coordinates z^* on M^*, vanishing at p^*, and z on M, vanishing at $f(p^*)$, and an integer $n > 1$ such that f is $z = z^{*n}$, in terms of these coordinates.

ramified element Let A be a ring of integers which is a UFD (unique factorization domain). Then a *ramified element* is a rational prime p such that there are repeated factors in the complete factorization of p in A.

range of function Let $f : X \to Y$ be a function mapping the set X into the set Y. The range of f is the set

$$R_f = \{y \in Y : y = f(x) \text{ for some } x \in X\}.$$

Also called *image* of $f(x)$.

rank (1.) The row rank of an $n \times m$ matrix A with entries in a field F is the maximum number of linearly independent rows. Here, linear independence of the row vectors $A_1, \ldots, A_r$ over F means that the equation:

$$c_1 A_1 + \cdots + c_r A_r = 0$$

where each $c_i \in k$, has only the trivial solution $c_1 = c_2 = \ldots = 0$.

It is a theorem that the row rank is equal to the column rank and hence this common value is called the rank of a matrix.
(2.) For a linear operator $T : X \to Y$, between two (possibly infinite dimensional) vector spaces, the dimension of the range $T(X)$ of T.

rapidly decreasing distribution *See* tempered distribution.

rapidly decreasing sequence A sequence $\{x_n\}$ satisfying

$$\sup_n n^m |x_n| < \infty$$

for all $m = 1, 2, \ldots$

ratio of geometric progression A geometric progression is a sequence of the form:

$$a, ar, ar^2, ar^3, \ldots$$

In general, a sequence $\{a_n\}$ is a geometric progression provided the ratios satisfy: $a_n/a_{n-1} = r$ for all $n = 1, 2, \ldots$. The common value for the ratios is called the ratio of a geometric progression.

ratio test A criterion for the convergence of an infinite series $\sum_{n=1}^{\infty} a_n$ of positive terms

a_n; namely, the series converges if

$$\limsup_{n\to\infty} \frac{a_n}{a_{n-1}} < 1.$$

The simplest case is when the series of terms is a geometric progression with constant ratio $0 \le r < 1$. *See* geometric series.

rational function The ratio of two polynomials

$$P(x) = a_0 + a_1 x + a_2 x^2 + \ldots + a_n x^n$$

and

$$Q(x) = b_0 + b_1 x + b_2 x^2 + \ldots + b_m z^m$$

where n, m are nonnegative integers and the a_j and b_j are real or complex numbers.

Rayleigh-Ritz method Calculus of variations problems involve minimizing an integral expression over a class of admissible functions. For example, according to Fermat's principle the path of a light ray in an inhomogeneous 2 dimensional medium in which the velocity of light is $c(x, y)$ solves the variational problem

$$\int_{x_0}^{x_1} \frac{\sqrt{1+y'^2}}{c(x,y)} dx = \min$$

where the admissible functions are all continuous curves, which have piecewise continuous derivatives and join the end points of the path.

The Rayleigh-Ritz method is a numerical method for constructing a sequence of functions $y_1, y_2, \ldots$ with the property that $y_n \to y$, where y is the minimizing function.

real analytic function A real-valued function $f(x)$, defined on an interval (a, b) of real numbers such that, for each x_0, $a < x_0 < b$, there is a $\delta > 0$ and a sequence of real numbers $\{a_n\}$ such that the power series:

$$\sum_{n=0}^{\infty} a_n (x - x_0)^n$$

converges for $x_0 - \delta < x < x_0 + \delta$, and is equal to $f(x)$ there.

Similar definitions can be given for real analytic functions in two or more variables.

real analytic manifold An n-dimensional real manifold on which there is defined an atlas with charts having transition functions which are real analytic. *See* atlas, real analytic function.

real analytic structure An atlas on a topological space consisting of local homeomorphisms and transition mappings that make it into a real analytic manifold. *See* real analytic manifold.

real axis The complex numbers **C** are usually depicted as the points in the Cartesian plane by means of the identification:

$$a + bi \longleftrightarrow (a, b).$$

With this identification, the x-axis in the Cartesian plane becomes the *real axis* in **C**.

real function A real-valued function. *See* real-valued function.

real Hilbert space A vector space over **R** with inner product for which the metric:

$$d(x, y) = \|x - y\|$$

is complete. *See* inner product, pre-Hilbert space.

real linear space A vector space in which the scalar field is the real numbers. *See* vector space.

The most basic example is n-dimensional Euclidean space $\mathbf{R}^n$ but other interesting examples are function spaces such as the real linear space of continuously differentiable real-valued functions defined on an interval, $\mathbf{C}^1(I)$.

real variable Suppose that a function $f(x_1, \ldots, x_n)$ is defined on n-dimensional Euclidean space $\mathbf{R}^n$. f is then said to be a function of the n real variables $x_1, \ldots, x_n$. Thus, the term *real variable* refers to the variables used to define a function on **R** or $\mathbf{R}^n$.

real-valued function A function defined on a set A and taking values in the set of real numbers. In calculus, the typical examples include one-variable functions denoted $y = f(x)$ or several variable functions, say $T(x, y, z)$, which measure the temperature or other physical quantity of a solid 3-dimensional object.

rearrangement (1.) A series $\sum_{j=1}^{\infty} b_j$ is a *rearrangement* of $\sum_{j=1}^{\infty} a_j$, provided there is a one-to-one map σ of the positive integers onto itself such that $a_j = b_{\sigma(j)}$, for $j = 1, 2, \ldots$. The sum of the series $\sum a_j$ is the same as the sum of each of its rearrangements if and only if $\sum |a_j|$ converges. (2.) Two real-valued functions $f(x)$ and $g(x)$ defined on the real line **R** are said to be *rearrangements* of each other provided:

$$m(\{x \mid f(x) > t\}) = m(\{x : g(x) > t\})$$

holds for all real numbers t. Here $m(E)$ denotes the Lebesgue measure or length of the set E.

Intuitively, the graph $y = f(x)$ has been rearranged by interchanging different parts of it to form the graph $y = g(x)$. One simple example would be $g(x) = f(x - a)$, where a is any real number. The graph of $f(x)$ is then translated to the left or right to form the graph of $g(x)$.

rearrangement invariant A *rearrangement invariant* is a quantity that is defined on functions, which yields the same value for any two functions which are rearrangements of each other. *See* rearrangement.

As an example, if two nonnegative functions $f(x)$ and $g(x)$ are rearrangements of each other, then they have the same integrals:

$$\int_{-\infty}^{\infty} f(x)\,dx = \int_{-\infty}^{\infty} g(x)\,dx.$$

rectangular coordinates The system of Cartesian coordinates which assigns each point in the plane an x and y coordinate is also called a rectangular coordinate system because of the geometry involved in locating the point (x, y). By comparison, one can also use a system of *polar coordinates* when a problem has a certain amount of circular symmetry.

In 3-dimensional space there are also cylindrical and spherical coordinates that can be used, depending on the needs of the problem.

rectangular hyperbola A hyperbola in rectangular coordinates whose axes of symmetry are parallel to the coordinate axes. Hence the locus of the equation

$$\frac{(x-h)^2}{a^2} - \frac{(y-k)^2}{b^2} = 1$$

where the point (h, k) is the center of the hyperbola.

Some authors add the additional requirement that $a = b$ in the above.

rectifiable curve A space curve, $\mathbf{r}(t) = x(t)\mathbf{i} + y(t)\mathbf{j} + z(t)\mathbf{k}$, where $a \le t \le b$, of finite length. If all the functions are continuously differentiable, then the length is finite and is given by the formula

$$L(C) = \int_a^b \sqrt{(x')^2 + (y')^2 + (z')^2}\,dt = \int_a^b \|\mathbf{r}'\|\,dt.$$

In case $\mathbf{r}$ is not continuously differentiable, then the curve is said to be *rectifiable* provided the least upper bound of the set of all lengths of polygonal paths inscribed in C is finite. The length $L(C)$ is defined to be this least upper bound in this case.

rectifying plane A particle traveling in space has three associated vectors at time t, namely: the unit tangent **T**, the principle normal **N**, and binormal **B**. *See* principal normal. These form the Frenet frame $\{\mathbf{T}, \mathbf{N}, \mathbf{B}\}$.

The Frenet frame at a point on a curve defines, locally, three planes given by the pairs of these vectors. The *osculating plane* is determined by the vectors **T** and **N**. The *normal plane* is determined by the vectors **N** and **B** and finally, the *rectifying plane* is determined by the vectors **T** and **B**.

recurrence formula A *recurrence formula* is a way of defining a sequence $\{x_j\}$, in which, for each j, x_j is defined as a fixed function of x_{j-1}. For example, the Newton method for finding roots is to substitute x_{j-1} in the function $x - f(x)/f'(x)$ over and over to produce a sequence

$$x_j = x_{j-1} - \frac{f(x_{j-1})}{f'(x_{j-1})}$$

which converges to a root, given an appropriate starting value x_0.

Closely related is a *recursion formula*, which may change with repeated computations based on previous values of the variables. For example, $j! = j\cdot(j-1)!$ expresses j factorial as a formula involving $j-1$ factorial.

recursion formula *See* recurrence formula.

reduced extremal distance Let Ω be a domain in the complex plane. Given a subset $E \subset \partial\Omega$ and a point $z_0 \in \Omega$, one would like a quantity similar to the extremal distance between E and z_0. Unfortunately, this is ∞. *See* extremal distance.

To get around this problem, one computes the extremal distance between E and a small circle of radius r centered at z_0 and then subtracts the quantity

$$\frac{1}{2\pi}\log r.$$

It turns out that the limit exists as r tends to zero and gives a quantity which we denote by $d(z_0, E)$.

This number is not conformally invariant so that instead, the reduced extremal distance between E and z_0 is defined to be

$$\delta(z_0, E) = d(z_0, E) - d(z_0, \partial\Omega).$$

This quantity turns out to be nonnegative and conformally invariant.

reducible germ of analytic set This topic refers to the local behavior of the zero set for an analytic function of several complex variables. For example, the local properties of the zero set of an irreducible polynomial in several complex variables would be an *irreducible* germ of an analytic set.

reduction formula In integral calculus, a reduction formula for an integral is one in which the integral is expressed by a similar integral but with smaller values of a parameter. Thus, the complete answer can be obtained by repeated use of the formula. A typical example is

$$\int x^n e^x\,dx = x^n e^x - n\int x^{n-1}e^x\,dx.$$

This formula shows how to compute an integral above for any positive integer n because after repeatedly using the formula we eventually arrive at the elementary integral $\int e^x\,dx = e^x + C$.

reference angle *See* related angle.

reflection in a line Given a line through the origin in Euclidean space $\mathbf{R}^n$, it is possible to express an arbitrary position vector $\mathbf{r}$ as a sum

$$\mathbf{r} = \mathbf{r}_1 + \mathbf{r}_2$$

where $\mathbf{r}_1$ is parallel to the line and $\mathbf{r}_2$ is perpendicular to it.

The reflected point is then obtained by reversing the direction of $\mathbf{r}_2$, i.e.,

$$\mathbf{r}^* = \mathbf{r}_1 - \mathbf{r}_2.$$

In case the line does not contain the origin, then translate the origin so that it does.

reflection in a plane Given a plane through the origin in Euclidean space $\mathbf{R}^n$, it is possible to express an arbitrary position vector **r** as a sum

$$\mathbf{r} = \mathbf{r}_1 + \mathbf{r}_2$$

where $\mathbf{r}_1$ is parallel to the plane and $\mathbf{r}_2$ is perpendicular to it.

The reflected point is then obtained by reversing the direction of $\mathbf{r}_2$, i.e.,

$$\mathbf{r}^* = \mathbf{r}_1 - \mathbf{r}_2.$$

In case the plane does not contain the origin, then translate the origin so that it does.

reflection in a point In Euclidean space $\mathbf{R}^n$, the reflection of a point with position vector **r**, about the origin, is simply its negative $\mathbf{r}^* = -\mathbf{r}$. The reflection in a point is obtained by translating the origin to the given point and proceeding as above.

reflection property of conic section Each of the conics has an interesting focal property. For example, a parabolic mirror is based on the reflection property that rays of light parallel to the axis of the parabola are reflected to its focus. *See* parabola. *See also* ellipse, hyperbola for similar properties.

reflexive relation A relation R satisfying $(x, x) \in R$, for all x. Thus $x \sim x$ for all x, when the relation $\sim$ is reflexive. *See* relation.

regular boundary point A point x_0 in the boundary $\partial\Omega$ of an open set in $\mathbf{R}^n$ such that, near x_0, $\partial\Omega$ can be described as the level set of a smooth function $g(x) = 0$ where the gradient satisfies $|\nabla g(x_0)| > 0$.

regular closed set A closed subset of $\mathbf{R}^n$ with the property that, locally, it is the intersection of a finite number of regular level sets, $\{x : g_i(x) = 0\}$, where the functions g_i are differentiable and have nonzero gradients.

regular embedding A differentiable function $f : \mathbf{R}^m \to \mathbf{R}^n$ which is a one-to-one mapping and such that its derivative matrix has constant rank m (here $m \leq n$).

For example, a smooth curve $x = x(t)$, $y = y(t)$ is a regular embedding of the line into the plane provided the curve does not intersect itself and

$$x'^2(t) + y'^2(t) > 0$$

for all $-\infty < t < \infty$.

The definition can be generalized to smooth manifolds.

regular factorization Suppose $\Theta(z) : H_1 \to H_2$ is a contractive, analytic, Hilbert space operator-valued function for $|z| < 1$. Let $\Theta(z) = \Theta_2(z)\Theta_1(z)$, where $\Theta_1(z) : H_1 \to H'$ and $\Theta_2 : H' \to H_2$ be a factorization of Θ into the product of contractive, analytic, operator-valued functions. Let $\Delta(t) : H_1 \to H_1$ be defined by $\Delta(t) = [I - \Theta(e^{it})^*\Theta(e^{it})]^{\frac{1}{2}}$, and let $\Delta_1(t)$ and $\Delta_2(t)$ be analogously defined. The operator $Z : \Delta v \to \Delta_2\Theta_1 v \oplus \Delta_1 v$ is an isometry from $\Delta L^2(H_1)$ into $\Delta_2 L^2(H') \oplus \Delta_1 L^2(H_1)$, and can be extended by continuity to the closures of those spaces. The factorization $\Theta = \Theta_2\Theta_1$ is called a *regular factorization* if the above (extended) Z is unitary. B. Sz.-Nagy and C. Foias have proved that if Θ is the characteristic operator function of a contraction operator, then the regular factorizations of Θ are in one-to-one correspondence with the invariant subspaces of T.

regular linear map A densely defined linear map between locally convex, topological vector spaces with zero singularity. *See* singularity of linear map.

regular measure A Borel measure μ on a space X such that, for every Borel set $E \subseteq X$,
(i.) $|\mu|(E) = \sup\{|\mu|(K) : K$ compact and $K \subseteq E\}$, and
(ii.) $|\mu|(E) = \inf\{|\mu|(V) : V$ open and $V \supseteq E\}$.

regular point (1.) A point, in a neighborhood of which a given function of a complex variable is analytic.
(2.) A point of a complex space that has a neighborhood which is a complex manifold.

See complex space.

(3.) A point at which a quasi-conformal map w is differentiable and has positive Jacobian $\text{Jac}(w) = (|w_z|^2 - |w_{\bar{w}}|^2) > 0$.

regular singular point (1.) For the system $y' = A(t)y$, where A is an $n \times n$ matrix of functions, a point at which there is a fundamental matrix (i.e., a matrix $Y(t)$ satisfying $Y'(t) = A(t)Y(t)$) which does not have an essential singularity.

(2.) A linear nth order differential equation

$$p_n(x)\frac{d^n y}{dx^n} + p_{n-1}(x)\frac{d^{n-1}y}{dx^{n-1}} + \dots + p_0(x) = 0,$$

with polynomial coefficients, is given. Let x_0 be a point which is a zero of the leading coefficient $p_n(x)$. If $y = (x - x_0)^r(a_0 + a_1(x-x_0)+\dots)$ is substituted in the equation, then the coefficient of the lowest power of $x - x_0$ is a polynomial in r of degree n. Then x_0 is called a *regular singular point* of the differential equation.

regular submanifold A manifold M is called a submanifold of the manifold M' if:

(1) The set M is a subset of M'.

(2) The identity map i from M into M' is an imbedding of M into M'.

Moreover, if the topology of M coincides with the topology of M as a subspace of M', then M is called a regular submanifold of M'.

regular value A point at which a given function of a complex variable is analytic.

regularity Analyticity. *See* analytic function.

regularization Given a distribution $u \in D'(\mathbf{R}^n)$, the family $u * \phi_\epsilon$ of $\mathbf{C}^\infty$ functions, where $\phi(x) = \exp\{1/(\sum x_i^2 - 1)\}$ and $\phi_\epsilon(x) = \epsilon^{-n}\phi(x/\epsilon)$. As $\epsilon \to 0$, $u * \phi_\epsilon \to u$, in $D'(\mathbf{R}^n)$, as $\epsilon \to 0$.

Reinhardt domain A set $D \subseteq \mathbf{C}^n$ such that $z = (z_1, \dots, z_n) \in D$ implies $(e^{it_1}z_1, \dots, e^{it_n}z_n) \in D$, for arbitrary $t_1, \dots, t_n \in [0, 2\pi)$. Sometimes the implication is only required for the case $t_1 = \dots = t_n$.

related angle For any angle θ, the angle ϕ with $0 \le \phi \le \pi/2$ such that $|\sin\theta| = \sin\phi$. It follows that, as long as it is defined, the absolute value of any trigonometric function of θ equals the same trigonometric function of ϕ. The *related angle* is also known as the *reference angle*.

relation A relation $\sim$, *between two sets A* and B, is any subset of $A \times B$. We write $a \sim b$, when the pair (a, b) belongs to $\sim$. A *relation on a set A* is a relation between A and A.

relative cohomology group Let $\mathcal{F}$ be a sheaf of vector spaces over a $\mathbf{C}^\infty$-manifold M and A a closed subset of M. Let $\mathcal{F}^{(0)}$ denote the sheaf of germs of continuous sections of $\mathcal{F}$ with supports contained in A and define $\mathcal{F}^{(n)}$ inductively by: $\mathcal{F}^{(1)}$ is the sheaf of germs of continuous sections of $\mathcal{Z}^{(1)} = \mathcal{F}^{(0)}/\mathcal{F}$ and $\mathcal{F}^{(n+1)}$ the sheaf of sections of $\mathcal{Z}^{(n+1)} = \mathcal{F}^{(n)}/\mathcal{Z}^{(n)}$. We have the natural exact (quotient) sequences

$$0 \to \mathcal{F} \to \mathcal{F}^{(0)} \to \mathcal{Z}^{(1)} \to 0$$

$$0 \to \mathcal{Z}^{(n)} \to \mathcal{F}^{(n)} \to \mathcal{Z}^{(n+1)} \to 0, n \ge 1.$$

We obtain the long exact sequence

$$0 \to \mathcal{F} \to \mathcal{F}^{(0)} \to \mathcal{F}^{(1)} \to \dots$$

where $i : \mathcal{F} \to \mathcal{F}^{(0)}$ is inclusion and $d_n : \mathcal{F}^{(n)} \to \mathcal{F}^{(n+1)}$ is the composition of the maps $\mathcal{F}^{(n)} \to \mathcal{Z}^{(n+1)} \to \mathcal{F}^{(n+1)}$. The *cohomology groups of M, relative to A with values in* $\mathcal{F}$ are $H^0(M, \mathcal{F})$ = the continuous sections in $\mathcal{F}$, with supports contained in A and, for $q \ge 1$,

$$H^q(M, \mathcal{F}) = \text{ker} d_q/\text{im} d_{q-1}.$$

relative extremum A local maximum or a local minimum.

relative maximum *See* local maximum.

relative minimum *See* local minimum.

Rellich's Uniqueness Theorem Consider a physical system in which incoming waves are scattered by an obstacle and generate outgoing waves. Suppose $U(t) = e^{At}$ is the operator that maps the incoming waves to the outgoing waves, and let $U_0(t)$ be the corresponding operator for the system in which no scattering obstacle is present. A function f is called *eventually outgoing* if there is a constant t_0 such that $[U_0(t)f](x) = 0$ for $|x| < t - t_0$. *Rellich's Uniqueness Theorem* asserts that if f is a solution of the eigenvalue problem $Af = i\sigma f$ for a real value σ in some exterior domain G, and if f is eventually outgoing, then $f = 0$ on G.

remainder (1.) In a ring with Euclidian algorithm, given two elements a, b, there exist q and r such that $a = qb + r$, where, according to a given partial ordering, $r < b$. Then r is called the *remainder.*
(2.) In approximation problems, the difference of the approximating quantity and the quantity being approximated. Thus, it is advantageous to be able to show how small the remainder is, in such problems.

Remmert's Theorem (1.) Remmert's Mapping Theorem: Let $f : X \to Y$ be a proper, holomorphic mapping of complex spaces, and let $A \subset X$ be an analytic set. Then $f(A)$ is analytic in Y.
(2.) Remmert's Embedding Theorem: A connected, n-dimensional Stein space X can be mapped homeomorphically by a holomorphic f, onto a closed subspace of some $\mathbf{C}^m$. If X is a manifold, then f can be chosen to be an embedding with $m = 2n$ (except $m = 3$, if $n = 1$).

Remmert-Stein Theorem Let Ω be an open set in $\mathbf{C}^n$ and suppose that there is an analytic disk $\Delta \subset \partial\Omega$ centered at P. Denote by $T_P(\Delta)$ the tangent space to Δ at P and by N_Δ its real orthogonal complement in $T_P(\partial\Omega)$. Assume there is a small neighborhood U of $0 \in N_\Delta$ such that $\Delta + U \subseteq \partial U$. Let $\mathcal{D} \subseteq \mathbf{C}^m$ be strongly pseudoconvex with C^4 boundary. Then there is no proper holomorphic map $\rho : \Omega \to \mathcal{D}$.

A special case is the fact that there is no proper map of the ball to the polydisk or of the polydisk to the ball, in $\mathbf{C}^n$.

removable singularity A limit point x_0 of the domain of a function $f(x)$, such that, for a suitable value y_0, the function f_1, defined by

$$f_1(x) = f(x), x \neq x_0; \; f_1(x_0) = y_0$$

has no singularity at x_0.

representation A continuous homomorphism of a topological group into the multiplicative group of operators from a topological vector space to itself.

representation in terms of arc length A parameterization of a curve γ in $\mathbf{R}^2$ or $\mathbf{R}^3$ of the form $x = x(s), y = y(s), z = z(s), s \in [a, b]$, where the length of γ from $(x(a), x(a), x(a))$ to $(x(s), y(s), z(s))$ is s. For example, $x = \cos\theta, y = \sin\theta, 0 \leq \theta \leq 2\pi$ is a *representation of the unit circle in* $\mathbf{R}^2$ *in terms of arc length.*

representation problem To determine all the representations, up to unitary equivalence, of a given topological group. *See* representation.

reproducing kernel In a Hilbert space H, whose elements are functions with domain containing a set D, a family of elements of H, $\{k_z, z \in D\}$, satisfying $(f, k_z) = f(z)$, for $f \in H$. The term *kernel* derives from the fact that the inner product in H is often given by an integral.

residue For a meromorphic function $f(z)$ with a pole at $a \in \mathbf{C}$, one can write

$$f(z) = \sum_{k=1}^{m} c_k(z - a)^{-k} + g(z),$$

where $g(z)$ is analytic at a. The coefficient c_1 is called the residue of f at a, sometimes written $c_1 = Res(f; a)$.

Residue Theorem If Ω is a simply connected open set in **C**, f is meromorphic in Ω and γ is a closed curve lying in Ω and free of poles of f, then

$$\frac{1}{2\pi i}\int_\gamma f(z)dz = \sum Res(f; a) Ind_\gamma(a),$$

where the sum is over the poles of f in Ω, $Ind_\gamma(a)$ is the index (winding number) of γ with respect to a, and $Res(f; a)$ is the residue of f at a. *See* residue, winding number.

resolution of the identity A projection-valued function $E(\lambda)$, $\lambda \in \mathbf{R}$, on a Hilbert space H, satisfying
(i.) $E(\lambda_1)H \subseteq E(\lambda_2)H$, for $\lambda_1 \leq \lambda_2$;
(ii.) $E(\lambda)$ is continuous from the right:

$$\lim_{\lambda \downarrow \lambda_0} E(\lambda) = E(\lambda_0)$$

(strong limit); and
(iii.) $0 = \lim_{\lambda \downarrow -\infty} E(\lambda)$ and $I = \lim_{\lambda \uparrow \infty} E(\lambda)$.

A version of the Spectral Theorem states that with every self-adjoint operator $T : H \to H$, there is associated a unique resolution of the identity $E(\lambda)$, called the *spectral resolution of* T, such that

$$(Tx, y) = \int_{-\infty}^{\infty} \lambda d(E(\lambda)x, y).$$

Similarly, for a unitary operator $U : H \to H$, there is a unique spectral resolution, supported on $[0, 2\pi)$, such that

$$(Ux, y) = \int_0^{2\pi} e^{i\lambda} d(E(\lambda)x, y).$$

There is also a complex-variable version for a normal operator on H.

resolvent The operator $R_\lambda = (\lambda I - T)^{-1}$, where T is a linear operator from a complex vector space to itself and λ is a complex number.

resolvent equation If T is a linear operator on a Banach space and λ and μ lie in the resolvent set of T, then

$$\begin{aligned}(\lambda I - T)^{-1} &- (\mu I - T)^{-1} \\ &= (\mu - \lambda)(\lambda I - T)^{-1}(\mu I - T)^{-1}.\end{aligned}$$

This is often called *Hilbert's* resolvent equation.

resolvent set For a linear operator T from a complex vector space to itself, the set of complex λ such that $\lambda I - T$ has a bounded inverse. It is the set-theoretical complement of the *spectrum* of T.

restriction Given a function $f : A \to B$ and a subset $C \subset A$, the *restriction of* f *to* C is the function $f_1 : C \to B$, defined by $f_1(x) = f(x)$, for $x \in C$. The usual notation for the restriction of f to C is $f|_C$.

retarded differential equation The differential equation describing rectilinear motion of the form

$$a = \frac{d^2y}{dt^2},$$

where the acceleration on a is negative.

Ricatti equation The nonlinear ordinary differential equation

$$x' = p(t)x^2 + q(t)x + r(t),$$

on an interval $[a, b]$, where p, q and r are continuous functions and $' = \frac{d}{dt}$.

Ricci curvature On a Riemannian manifold (M, g),

$$R_{i\bar{j}} = -\frac{\partial^2}{\partial z^i \partial \overline{z^j}} \log\det(g_{k\bar{l}}).$$

Ricci form The form ρ, defined by

$$\rho = \frac{\sqrt{-1}}{2\pi} R_{i\bar{j}} dz^i \wedge \overline{dz^j},$$

where $R_{i\bar{j}}$ is the Ricci curvature. *See* Ricci curvature.

Riemann integral A Riemann sum is a sum of the form

$$S = \sum_{j=1}^{n} f(c_j)[x_j - x_{j-1}],$$

where $f(x)$ is a real-valued function on $[a, b]$, $\{x_j\}$, $j = 0, 1, \ldots, n$, is a partition of $[a, b]$, that is, $a = x_0 < x_1 < \ldots < x_n = b$, and $x_{j-1} \leq c_j \leq x_j$, for $j = 1, \ldots, n$. If, for every $\epsilon > 0$ there is a $\delta > 0$, and a real number L such that $\max_{1 \leq j \leq n}(x_j - x_{j-1}) < \delta$ implies $|S - L| < \epsilon$, then we say that f is Riemann integrable, with integral L and write

$$L = \int_a^b f(x)dx.$$

Riemann lower integral The least upper bound of the lower Darboux sums. *See* Darboux sum.

Riemann Mapping Theorem Every simply connected, open, proper subset of the plane is conformally equivalent to the open unit disk. Also called *Riemann's Existence Theorem, Riemann's Theorem.*

Riemann Roch Theorem For a smooth projective curve C, the equation

$$\dim D - \dim(K \backslash D) = \deg D - g - 1,$$

where D is an arbitrary divisor on C, K its canonical class, and g its genus.

Riemann sphere The unit sphere in $\mathbf{R}^3$, regarded as the complex plane, with the point at ∞ added. To map the sphere to $\mathbf{C} \cup \{\infty\}$, place the sphere tangent to $\mathbf{C}$ with the south pole at the origin in $\mathbf{C}$, and draw the straight line from the north pole to the point $z \in \mathbf{C}$. The point of intersection of this line with the sphere is the point identified with z. The north pole itself is identified with ∞.

Riemann sum *See* Riemann integral.

Riemann surface (1.) A complex analytic manifold is called an *abstract Riemann surface.*

(2.) The *Riemann surface of a function* is a specific complex analytic manifold, forming a branched covering of the complex plane, on which a given analytic function (possibly multiple-valued, with algebraic singularities) lifts to a single-valued analytic function. That is, given an analytic function $f(z)$, which can be analytically continued to a possibly multiple-valued function in a subset D of the complex plane, the *Riemann surface of* f is a complex analytic manifold R, together with a (projection) map $P : R \to \mathbf{D}$, and a single-valued analytic function $F : R \to \mathbf{C}$, such that, for every $z \in R$, $f(Pz) = F(z)$, for some branch of f.

Riemann upper integral The greatest lower bound of the upper Darboux sums. *See* Darboux sum.

Riemann's Continuation Theorem If $\mathcal{S}$ is a coherent sheaf on a manifold M of which free resolutions of length $\leq p$ exist locally, then, for every analytic subset $A \subset M$, the homomorphisms

$$\rho_j : H^j(M, \mathcal{S}) \to H^j(M \backslash A, \mathcal{S})$$

are bijective, for $j \leq \text{codim}_M A - (p + 2)$, and injective for $j = \text{codim}_M A - (p + 1)$.

Riemann's Existence Theorem *See* Riemann Mapping Theorem.

Riemann's problem To find a Riemann P-function

$$P \left\{ \begin{matrix} a_1 & a_2 & a_3 & \\ a_{11} & a_{21} & a_{31} & z \\ a_{12} & a_{22} & a_{32} & \end{matrix} \right\},$$

that is, a solution of an equation $p_2(x)y'' + p_1(x)y' + p_0(x) = 0$, with regular singular points at a_1, a_2, a_3, and exponents a_{j1}, a_{j2} corresponding to a_j, $j = 1, 2, 3$, such that

1. it is single-valued and continuous in the whole plane, except at a_1, a_2, a_3;

2. any two determinations f_1, f_2, f_3, of the function are linearly dependent:

$$c_1 f_1 + c_2 f_2 + c_3 f_3 = 0,$$

where c_1, c_2, c_3 are constants; and
3. for $j = 1, 2, 3$, in a neighborhood of a_j, there are two distinct determinations

$$(z - a_j)^{b_j} g_{j1}(z), (z - a_j)^{b_j'} g_{j2}(z),$$

where g_{j1} and g_{j2} are analytic and nonzero in the neighborhood of a_j. *See* P-function of Riemann.

Riemann's Theorem *See* Riemann Mapping Theorem.

Riemann-Lebesgue Lemma The Fourier coefficients of an L^1 function tend to 0. That is, if $f \in L^1[-\pi, \pi]$, then $a_n = \hat{f}(n) = (2\pi)^{-1} \int_{-\pi}^{\pi} f(t)e^{-int}dt \to 0$, as $|n| \to \infty$.

Riemann-Roch Theorem Let M be a compact Riemann surface of genus g. Let $\mathcal{A}$ be an integral divisor on M with dimension $r(\mathcal{A})$. Let $\Omega(\mathcal{A})$ be the set of Abelian differentials ω with divisor (ω), satisfying $(\omega) \geq \mathcal{A}$ and let $i(\mathcal{A}) = \dim \Omega(\mathcal{A})$. Then

$$r(\mathcal{A}^{-1}) = \deg\mathcal{A} - g + 1 + i(\mathcal{A}).$$

See Abelian differential, divisor, integral divisor.

Riemann-Stieltjes integral Let f and g be real-valued functions on a real interval $[a, b]$. For a partition $\{x_j, j = 0, 1, \ldots, n\}$ of $[a, b]$, i.e., for $a = x_0 < x_1 < \ldots < x_n = b$, and for $c_j \in [a_{j-1}, a_j]$, $j = 1, \ldots, n$, write the sum

$$S(\{x_j\}) = \sum_{j=1}^{n} f(c_j)[g(x_j) - g(x_{j-1})].$$

If, for every $\epsilon > 0$, there is a $\delta > 0$ and a number L such that, for every partition $\{x_j\}$ with $\max_{1 \leq j \leq n}[x_j - x_{j-1}] < \delta$, we have $|S(\{x_j\}) - L| < \epsilon$, then we say that f is *Riemann-Stieltjes integrable*, with respect to g and write

$$L = \int_a^b f(x)dg(x).$$

Riemannian connection A Riemannian manifold M with scalar product g, possesses a unique connection ∇, which is torsion-free and for which g is parallel ($\nabla g = 0$). This ∇ is called the *Riemannian connection* for g.

Riemannian manifold A manifold on which there is defined a Riemannian metric. *See* Riemannian metric.

Riemannian metric A positive-definite inner product, $(\cdot, \cdot)_x$, on $T_x(M)$, the tangent space to a manifold M at x, for each $x \in M$, which varies continuously with x.

Riemannian submersion Let M, N be Riemannian manifolds and $\pi : M \to N$ be a submersion. For $p \in M$, let V_p and $H_p = V_p^{\perp}$ be the vertical and horizontal space at p. (*See* vertical space.) Then π is a *Riemannian submersion* if $D\pi(p) : H_p \to T_{\pi(p)}N$ is a linear isometry, for every $p \in M$.

Riesz decomposition Let Ω be a bounded open set in $\mathbf{R}^n$ with $\mathbf{C}^2$ boundary and $u \in \mathbf{C}^2(\bar{\Omega})$ be subharmonic on Ω. Then

$$u(x) = \int_{\partial\Omega} P(x, y)u(y)d\sigma(y) - \int_{\Omega} (\Delta u(y))G(x, y)dV(y),$$

where P and G are the Poisson kernel and Green's function for Ω.

Riesz potentials The singular integral operators

$$I_\alpha(f) = \frac{1}{\gamma(\alpha)} \int_{\mathbf{R}^n} |x - y|^{-n+\alpha} f(y)dy,$$

where $\gamma(\alpha) = \pi^{n/2}2^\alpha\Gamma(\alpha/2)/\Gamma([n - \alpha]/2)$.

Riesz transform For $f \in L^p(\mathbf{R}^n)$, $1 \leq p < \infty$, the singular integral operators

$$R_j(f)(x) = \lim_{\epsilon \to 0} c_n \int_{|y| \geq \epsilon} \frac{y_j}{|y|^{n+1}} f(x-y)dy$$

where

$$c_n = \frac{\Gamma\left(\frac{n+1}{2}\right)}{\pi^{(n+1)/2}}$$

for each $j = 1, \ldots, n$.

Riesz's Theorem (1.) The projection from $L^p[-\pi, \pi]$ to H^p is bounded if $1 < p < \infty$. Thus if $f \sim \sum_{-\infty}^{\infty} a_n e^{int} \in L^p$, then $f_+ \sim \sum_0^{\infty} a_n e^{int} \in H^p$. This is a theorem of Marcel Riesz. The result is false if $p = 1$ or $p = \infty$.

(2.) If X is a locally compact Hausdorff space, then every bounded linear functional Φ on $\mathbf{C}_0(X)$ has the form $\Phi f = \int_X f d\mu$, $(f \in \mathbf{C}_0(X))$, where μ is a regular, complex, Borel measure on X. This theorem of Frigyes Riesz is called the Riesz Representation Theorem.

Riesz-Fischer Theorem Let $\{u_\alpha, \alpha \in A\}$ be an orthonormal set in a Hilbert space H. Then for every element $(c_\alpha) \in L^2(A)$, there is an $f \in H$ with $(f, u_\alpha) = c_\alpha$, for $\alpha \in H$. Together with Bessel's inequality, this implies that the map sending f to (f, u_α) maps H onto $L^2(A)$.

Riesz-Thorin Theorem Let (X, Ω, μ), (Y, Λ, ν) be measure spaces, with positive measures μ, ν and D a subspace of $L^p(\mu)$ for all $p > 0$. We say an operator T, defined on D, taking values in the set of ν-measurable functions on Y is *of (strong) type* (p, q) if

$$\|Tf\|_{q,\nu} \leq K\|f\|_{p,\mu}$$

for $f \in D$; that is, if T is bounded from $L^p(\mu)$ to $L^q(\nu)$ (with domain D).

The *Riesz-Thorin Theorem* states that if D contains all the μ-measurable simple functions and if T is of type (p_1, q_1) and (p_2, q_2), $1 \leq p_i, q_i \leq \infty$, then T is of type (p, q), whenever

$$\frac{1}{p} = \frac{1-t}{p_1} + \frac{t}{p_2}, \frac{1}{q} = \frac{1-t}{q_1} + \frac{t}{q_2}.$$

The theorem is an example of *interpolation of operators*, since the action of T is *interpolated* between the two pairs of spaces. *See also* M. Riesz's Convexity Theorem.

right adjoint linear mapping *See* left adjoint linear mapping.

right conoid A surface consisting of all lines parallel to a given plane that intersect a given line perpendicular to the plane and pass through a given curve.

right derivative *See* left derivative.

right differentiable Having a right derivative. *See* left derivative.

right helicoid A surface given by the parametric equations $x = u \cos v$, $y = u \sin v$, and $z = mv$.

right linear space A vector space over a noncommutative field, in which scalar multiplication is permitted on the right only. *See* vector space.

rigid motion A mapping of $\mathbf{R}^2$ (or $\mathbf{R}^3$) onto itself, consisting of a parallel translation, a rotation about a point (or, in $\mathbf{R}^3$, about an axis), or a composite of the two.

rigidity of a sphere The sphere is rigid in the following sense. Let $\varphi: \Sigma \to S$ be an isometry of a sphere $\Sigma \subset \mathbf{R}^3$ onto a regular surface $S = \varphi(\varepsilon) \subseteq \mathbf{R}^3$. Then S is a sphere. Intuitively, this means that it is not possible to deform a sphere made of a flexible but inelastic material.

Rigidity Theorem If the faces of a convex polyhedron were made of metal plates and the edges were replaced by hinges, the polyhedron would be rigid. The theorem was stated by Cauchy (1813), although a mistake in this paper went unnoticed for more than 50 years.

ring of sets A collection $\mathcal{R}$ of subsets of a set S such that if $E, F \in \mathcal{R}$, then (i.) $E \cup F \in \mathcal{R}$, and (ii.) $E \backslash F \in \mathcal{R}$. If $\mathcal{R}$ is also closed under the taking of countable unions, then $\mathcal{R}$ is called a *σ-ring.*

Ritz's method *See* Rayleigh-Ritz method. The method was referred to by Ritz's name alone for a period of time following a 1909 paper in which he reapplied Rayleigh's earlier work.

Rolle's Theorem Let $f(x)$ be a real valued function, continuous on $[a, b]$, differentiable on (a, b), and satisfying $f(a) = f(b)$. Then there is a point $c \in (a, b)$ with $f'(c) = 0$.

rolling curve *See* roulette.

rotation number Let Γ be a Jordan curve and $S : \Gamma \to \Gamma$ an orientation preserving homeomorphism. Parameterize Γ by $\gamma = \gamma(y), 0 \leq y \leq 1$ and extended to $-\infty < y < \infty$, with y and y' identified whenever $y - y'$ is an integer. Then $\gamma_1 = S\gamma$ can be represented by a real valued function $y_1 = f(y)$, where $f(y)$ is continuous and strictly increasing and $f(y+1) = f(y) + 1$ (so that $f(y) - y$ is periodic with period 1). The *rotation number*, α of the map S is given by

$$\alpha = \lim_{|n| \to \infty} \frac{f^n(y)}{n},$$

for all y. α is also said to be the *rotation number* of the *flow* S^n (i.e., of the discrete group of homeomorphisms $S^n, 0, \pm 1, \ldots$).

Rouché's Theorem Let $\Omega \subset \mathbf{C}$ be an open set and γ a simple closed curve in Ω. Suppose $f(z)$ and $g(z)$ are analytic in Ω and $|f(z) - g(z)| < |f(z)|$ in γ. Then f and g have the same number of zeros inside γ.

roulette The locus of a point P, keeping a fixed position relative to a curve C' that rolls along a fixed curve C. Also *trochoid.* For example, if C and C' are equal circles, the corresponding roulette is the *limacon of Pascal* or *epitrochoid.*

ruled surface A surface generated by the motion of a straight line.

S

saddle point A critical point of an autonomous system

$$\frac{dx}{dt} = F(x,t), \quad \frac{dy}{dt} = G(x,t)$$

(i.e., a point where $F(x,t) = G(x,t) = 0$), that is approached and entered by two half-line paths L_1 and L_2, as $t \to \infty$, which lie on a line. It is also approached and entered by two half-line paths L_3 and L_4, as $t \to -\infty$, lying on a different line. Each of the four regions between these lines contains a family of paths resembling hyperbolas. These paths are asymptotic to one another, and do not approach the critical point.

saddle surface The *hyperbolic paraboloid* in $\mathbf{R}^3$ given by $z = ax^2 + by^2$, $ab < 0$. The cross section in the xz-plane is a downward parabola, while the cross section in the yz-plane is an upward parabola.

Sard's Theorem Let $f : M \to N$ be a smooth map between smooth manifolds. Then the set of critical values of f has measure 0 in N.

Measure 0 in N means that any coordinate chart applied to the set of critical points has measure 0 in $\mathbf{R}^n$.

Sard-Smale Theorem Let f be a C^q Fredholm mapping of a separable Banach space X into a separable Banach space Y. Then, if $q >$ max (index $f, 0$), the critical values of f are nowhere dense in Y.

scalar field *See* vector space.

scalar multiple A vector multiplied by a scalar by scalar multiplication. *See* vector space.

scalar product of vectors *See* inner product.

scalar quantity An element of the field of scalars. *See* vector space.

scale (**1.**) For A, a subset of a Polish space X, a sequence of norms ϕ_j on A such that if $x_n \in A$, $x_n \to x$ and $\lim_{n\to\infty} \phi_j(x_n) = \mu_j$, for all j, then $x \in A$ and $\phi_j(x) \le \mu_j$, for all j.

(**2.**) For an operator algebra A, the pair $(K_0(A), D(A))$, where $K_0(A)$ is the Grothendieck group of A and $D(A)$ is the set of equivalence classes of $P(A)$, the projections in A, under the Murray-von Neumann equivalence: $e \sim f \iff e = v^*v, f = vv^*$, for some $v \in A$. *See* Grothendieck group.

The scale is a complete invariant for AF-algebras. For UHF algebras, the pair $(K_0(A), [1])$ is a complete invariant. *See also* asymptotic sequence.

Schauder's Fixed-Point Theorem Let X be a Banach space, $S \subset X$ a closed, convex set, and T a continuous map of S into S, such that $T(S)$ has compact closure. Then T has a fixed point in S.

Schottky's Theorem Let $\mathcal{F}$ be the set of all $f(z)$ analytic in $|z| \le 1$ and with $f(z) \ne 0, 1$ for $|z| \le 1$. Then there is a constant C, depending only on $f(0)$ and $|z|$, such that

$$|f(z)| \le C$$

for $|z| < 1$.
The constant C is called Schottky's constant.

Schroedinger equation The partial differential equation

$$\frac{\partial f}{\partial t} = i\Delta f,$$

where $i = \sqrt{-1}$ and $\Delta = \frac{\partial^2}{\partial x^2} + \frac{\partial^2}{\partial y^2}$, is the Laplacian.

Also, the above equation on a Riemannian manifold, when Δ is the Laplace-Beltrami operator.

0-8493-0320-6/00/$0.00+$.50

Schwarz inequality (1.) The inequality

$$|\sum_{j=1}^{n} a_j \bar{b}_j|^2 \leq \sum_{j=1}^{n} |a_j|^2 \sum_{j=1}^{n} |b_j|^2,$$

for complex numbers $\{a_j\}$, $\{b_j\}$, $j = 1, \ldots, n$.
(2.) The above inequality with n replaced by ∞.
(3.) The inequality

$$|\int f\bar{g}d\mu|^2 \leq \int |f|^2 d\mu^2 \int |g|^2 d\mu,$$

for a positive measure μ and f and g integrable with respect to μ. Schwarz inequalities (1.) and (2.) are special cases of this.
(4.) The inequality

$$|(f, g)| \leq \|f\| \|g\|,$$

where f and g are elements of a Hilbert space, with inner product $(\cdot, \cdot)$ and norm $\|\cdot\|$.

Schwarz's Lemma Let $f(z)$ be analytic in the disk $D = \{|z| < 1\}$, $f(0) = 0$ and $|f(z)| \leq 1$ in D, then $|f(z)| \leq |z|$ in D and $|f'(0)| \leq 1$ in D. Furthermore, if $|F(z_0)| = |z_0|$, for some nonzero $z_0 \in D$, or if $|f'(0)| = 1$, then $f(z) = cz$, with c a constant of modulus 1.

Schwarz's principle of reflection Let Ω_+ be a region in the upper half plane of the complex plane and suppose L is an interval on the real axis such that, for every point $x \in L$, there is a disk $D_x = \{z : |z - x| < \epsilon_x\}$ such that $D_x \cap \Omega_+ = D_x^+$, the open upper half-disk of D_x. If $\Omega_- = \{\bar{z} : z \in \Omega_+\}$ is the reflection of Ω_+, and if $F(z) = u(z) + iv(z)$ is analytic in Ω_+ and $\lim_{n\to\infty} v(z_n) = 0$, whenever $\{z_n\} \subset \Omega_+$ and $z_n \to x \in L$, then there is a holomorphic function $F(z)$ on $\Omega_+ \cup L \cup \Omega_-$, with $F(z) = f(z)$ in Ω_+, $F(z) = \overline{f(\bar{z})}$ in Ω_-, and $F(z)$ is real on L.

There is a corresponding formulation for reflection over an arc of a circle.

secant line A straight line, intersecting a curve at a point, but not coinciding with the tangent line to the curve, at the point.

second Cousin problem Let Ω be a domain in $\mathbf{C}^n$ and $\{G_\alpha : \alpha \in \Lambda\}$ be an open cover of Ω. For each $\alpha, \beta \in \Lambda$ such that $G_\alpha \cap G_\beta \neq \emptyset$, let there be given a nonvanishing holomorphic $f_{\alpha\beta} : G_\alpha \cap G_\beta \to \mathbf{C}$ such that (1.) $f_{\alpha\beta} \cdot f_{\beta\alpha} = 1$ and (2.) $f_{\alpha\beta} \cdot f_{\beta\gamma} \cdot f_{\gamma\alpha} = 1$ on $G_\alpha \cap G_\beta \cap G_\gamma$. Find nonvanishing holomorphic $f_\alpha : G_\alpha \to \mathbf{C}$ so that $f_{\alpha\beta} = f_\alpha / f_\beta$ on $G_\alpha \cap G_\beta$, for $\alpha, \beta \in \Lambda$. *See also* first Cousin problem.

Second derivative The derivative of the first derivative, when it exists:

$$\frac{d^2 f(x)}{dx^2} = \frac{d}{dx}\frac{df(x)}{dx}.$$

Also denoted $f''(x)$, $D_{xx}f(x)$, and $D^2 f(x)$.

second fundamental form If X and Y are vector fields on a manifold M, let $\alpha_x(X, Y)$ denote the normal component of the covariant derivative $(\nabla'_X Y)_x$, for $x \in M$. Then $\alpha(X, Y)$ is a differentiable field of normal vectors to M, called the *second fundamental form*.

Second Mean Value Theorem Let $f(x)$ and $g(x)$ be integrable over (a, b), with $g(x)$ positive and monotonically increasing. Then there is $c \in [a, b]$ such that

$$\int_a^b f(x)g(x)dx = g(a) \int_a^c f(x)dx.$$

The theorem is also called *Bonnet's form of the Second Mean Value Theorem*.

second quadrant of Cartesian plane The subset of $\mathbf{R}^2$ of points (x, y), where $x < 0$ and $y > 0$.

segment In a metric space M, with metric d, a continuous image $x([a, b])$ of a closed interval satisfying

$$d(x(t_1), x(t_2)) + d(x(t_2), x(t_3))$$

$$= d(x(t_1), x(t_3)),$$

whenever $a \leq t_1 \leq t_2 \leq t_3 \leq b$.

self-adjoint differential equation An equation of the form $Lf = 0$, where L is a differential operator which satisfies

$$\int_X (Lf)\bar{g}dx = \int_X f\overline{Lg}dx,$$

for f, g belonging to an appropriate (dense) domain in $L^2(X)$, for some $X \subseteq \mathbf{R}$.

self-adjoint operator An operator T on a Hilbert space such that $T = T^*$. In case T is unbounded, this means that the domains of T and T^* are the same and $(Tx, y) = (x, Ty)$, for all x and y in the common domain. *See* adjoint operator.

self-commutator The operator $[T^*, T] = T^*T - TT^*$, where T is an operator on a Hilbert space.

self-polar triangle *See* polar triangle.

self-reciprocal function A function $f(x)$ that is invariant under an integral operator. For example, if

$$f(x) = \int_0^\infty \sqrt{xy}J_\nu(xy)f(y)dy,$$

where J_ν is the Bessel function, then $f(x)$ is called a *self-reciprocal function* with respect to the Hankel transform.

semi-axis The line segments joining the center of an ellipse or ellipsoid to the vertices. For the ellipsoid with equation

$$\frac{x^2}{a^2} + \frac{y^2}{b^2} + \frac{z^2}{c^2} = 1,$$

the line segments from $(0, 0, 0)$ to $(\pm a, 0, 0)$, $(0, \pm b, 0)$ and $(0, 0, \pm c)$.

semi-Fredholm operator An operator $A : X \to Y$, where X and Y are normed linear spaces, such that $R(A)$ is closed and $\dim(\mathrm{Ker}(A)) < \infty$, where $R(A)$ is the range of A and $\mathrm{Ker}(A)$ is its kernel.

semi-norm A real-valued function $p(x)$, defined on a linear space X over a field F, satisfying

$$p(x + y) \leq p(x) + p(y)$$

and

$$p(\alpha x) = |\alpha| p(x),$$

for $x, y \in X, \alpha \in F$.

semicontinuous function at point x_0 A function that is either upper semicontinuous or lower semicontinuous at x_0. *See* upper semicontinuous function at point x_0, lower semicontinuous function at point x_0.

semilinear mapping *See* semilinear transformation.

semilinear transformation A function $F : X \to Y$ between two vector spaces over $\mathbf{C}$ satisfying $F(\lambda x + \mu y) = \bar{\lambda}F(x) + \bar{\mu}F(y)$.

semilog coordinates In semilog coordinates, one coordinate of a point in the plane is the distance of the point from one of the axes, whereas the logarithm (in some base) of the other coordinate is the distance to the other axis. *Semilog coordinates* are usually used in conjunction with semilog graph paper, which has a uniform scale on one axis and a logarithmic scale on the other axis.

separable space A metric space with a countable dense subset.

separating family A family Φ, of functions, with a set S as domain such that, for any two points $p_1, p_2 \in S$, there is a function $f \in \Phi$ such that $f(p_1) \neq f(p_2)$.

separation of variables A technique for solving a first order differential equation

$$y' = f(t, y),$$

($y' = \frac{dy}{dt}$), where $f(t, y)$ is separable, in the sense that $f(t, y) = g(t)h(y)$. The technique is then to write $\int dy/h(y) = \int g(t)dt$, antidifferentiate, and solve for y.

sequence A mapping a from a subset of the integers (usually the positive, or nonnegative, integers) into a set S. The image $a(n)$, of the integer n, is usually written a_n.

sequence of functions *See* sequence. For a sequence of functions, the set S is a set of functions.

sequence of numbers *See* sequence. For a sequence of numbers, the set S is $\mathbf{C}$, the complex numbers, or a subset of $\mathbf{C}$.

sequence of points *See* sequence. For a sequence of points, the set S is some set of points.

sequence of sets *See* sequence. For a sequence of sets, S is a set of sets.

sequence space A vector space whose elements are sequences of real or complex numbers. Also called *coordinate space.* Of special significance are the Banach spaces $l^p(1 \le p \le \infty)$, or c_0. Here

$$l^p = \{(a_0, a_1, \ldots) : \sum_{n=0}^{\infty} |a_n|^p < \infty\},$$
$$1 \le p < \infty,$$
$$l^\infty = \{(a_0, a_1, \ldots) : \sup_n |a_n| < \infty\},$$
$$c_0 = \{(a_0, a_1, \ldots) : a_n \to 0\}.$$

series A formal infinite sum $f_1 + f_2 + \ldots$, of elements, normally in a normed space, where additional definitions may designate some element of the space as the sum. The notation $\sum_{n=1}^{\infty} f_n$ is usual for the series $f_1 + f_2 + \ldots$.

sesquilinear form A function $f : \mathbf{C} \times \mathbf{C} \to \mathbf{C}$ satisfying
(1.) $f(z_1 + z_2, w) = f(z_1, w) + f(z_2, w)$;
(2.) $f(cz, w) = cf(z, w)$, for $c \in \mathbf{C}$; and
(3.) $f(z, w) = \overline{f(w, z)}$.

set Usually, the term is used for any collection of elements. However, some care is needed to avoid such paradoxes as: *if Ψ is the set of all sets X such that $X \notin X$ then is $\Psi \in \Psi$?* Because of this, some versions of symbolic logic begin with the definition: *a set is any object that can appear on either side of the symbol* $\in$. Objects which appear only on the right of $\in$ are called *classes.*

set of degeneracy *See* degeneracy.

set of Jordan content 0 A subset $A \subset \mathbf{R}^n$ such that, for every $\epsilon > 0$, there is a finite collection of cubes $\{C_1, \ldots, C_n\}$ such that $A \subset \cup_j C_j$ and the sum of the volumes of the C_j is $< \epsilon$.

set of measure 0 For a measure space (X, Ω, μ), a set $E \in \Omega$ with $\mu(E) = 0$. Also called *null set, zero set.*

set of ordered pairs Any subset of a Cartesian product of two sets. Hence, any set of elements of the form (x, y), where $(x, y) = (z, w)$ if and only if $x = z$ and $y = w$.

set of quasi-analytic functions *See* quasi-analytic class.

set of uniqueness For a class H of functions on a set S, a set $E \subset S$ such that, $f_1, f_2 \in H$ and $f_1(x) = f_2(x)$, for $x \in E$, implies $f_1(x) = f_2(x)$, everywhere on S.

sheaf A triple (Ω, X, π), where X is a Hausdorff space, $\pi : \Omega \to X$ is a surjective map, and every point $\omega \in \Omega$ has a neighborhood $W \subset \Omega$ such that $\pi|_W$ is a homeomorphism.

sheaf of divisors Let $\mathcal{M}$ be the sheaf of (germs of) meromorphic functions on an open set $\Omega \subseteq \mathbf{C}^n$, let $\mathcal{M}^*$ be the sheaf obtained by removing the zero section from $\mathcal{M}$ and let $\mathcal{O}^* \subseteq \mathcal{M}^*$ be the invertible holomorphic elements. The *sheaf of* (germs of) *divisors* on Ω is the sheaf $\mathcal{M}^*/\mathcal{O}^*$. A section of $\mathcal{M}^*/\mathcal{O}^*$ is called a *divisor.*

A divisor is *integral* if the germ at each point is the germ of a holomorphic function.

sheaf of germs of C^∞ functions Let M be an open C^∞ manifold. For $m \in M$, let $C^\infty(m)$ denote the set of all functions f, defined and of class C^∞ in some neighborhood (depending on f) of m. For $f, g \in C^\infty(m)$, write $f \sim g$ if $f = g$ in some neighborhood of m and let $C^\infty_m = C^\infty(m)/\sim$ (called the *ring of germs* of C^∞ functions at m). Let $\Omega = \cup_{m\in M} C^\infty_m$. Topologize Ω by taking $\omega \in \Omega$, w a representative of ω, and $N \subset M$ an open neighborhood of m on which w is defined; then an open neighborhood of ω is $\{[\omega]_x : x \in N\}$, where $[\omega]_x$ is the residue class of ω in C^∞_x. The projection π is given by $\pi\omega = m$, for $\omega \in \Omega$. *See* sheaf.

sheaf of germs of holomorphic functions Let M be an open analytic manifold. For $m \in M$, let $\mathcal{H}(m)$ denote the set of all functions f, defined and analytic in some neighborhood (depending on f) of m. For $f, g \in \mathcal{H}(m)$, write $f \sim g$ if $f = g$ in some neighborhood of m and let $\mathcal{H}_m = \mathcal{H}(m)/\sim$ (called the *ring of germs* of holomorphic functions at m). Let $\Omega = \cup_{m\in M}\mathcal{H}_m$. Topologize Ω by taking $\omega \in \Omega$, w a representative of ω, and $N \subset M$ an open neighborhood of m on which w is defined; then an open neighborhood of ω is $\{[\omega]_x : x \in N\}$, where $[\omega]_x$ is the residue class of ω in $\mathcal{H}_x$. The projection π is given by $\pi\omega = m$, for $\omega \in \Omega$. *See* sheaf.

shift operator The operator S on the Hilbert space l^2, of all square summable, complex sequences $(\dots, a_{-1}, (a_0), a_1, \dots)$, sending $(\dots, a_{-1}, (a_0), a_1, \dots)$ to $(\dots, (a_{-1}), a_0, a_1, \dots)$ is called the *bilateral* shift operator. The restriction of S to the subspace of sequences with $a_j = 0$, for $j < 0$, is called the *unilateral* shift. Often the shift operators are studied (via their unitarily equivalent copies) on $L^2[-\pi, \pi]$ and H^2, where they act as multiplication by e^{it} and by z, respectively.

Shilov boundary Let A be an algebra of continuous complex functions on a compact Hausdorff space X. A boundary for A is a set $S \subset X$ such that $\sup_X |f| = \max_S |f|$, for $f \in A$. The smallest such boundary (i.e., the intersection of all boundaries of A) is the *Shilov boundary* of A.

Sidon set Let G be a compact Abelian group with dual Γ and let $E \subset \Gamma$. An *E-function* is an $f \in L^1(G)$ such that its Fourier transform $\hat{f}(\gamma) = 0$ for $\gamma \in E$. An *E-polynomial* is a trigonometric polynomial on G which is also an E-function. E is a *Sidon set* if $\sum_{\gamma\in\Gamma} |\hat{f}(\gamma)| \le B\|f\|_\infty$, for every E-polynomial f on G.

Siegel modular function A Γ_n-invariant meromorphic function on the space of all $n \times n$ complex symmetric matrices with positive imaginary parts. In 1984, H. Umemura expressed the roots of an arbitrary polynomial in terms of elliptic Siegel functions.

Siegel upper half space The group of $n \times n$ matrices $Z \in GL(n, \mathbb{C})$ such that ${}^tZ = Z$ and $\Im Z$ is positive definite.

sigma finite A measure space (S, Ω, μ) is σ-finite if S can be written as a countable union $S = \cup_{n=1}^\infty S_n$, where $\mu(S_n) < \infty$, for every n.

sigma-algebra of sets *See* sigma-field of sets.

sigma-field of sets *See* field of sets.

sigma-ring of sets *See* ring of sets.

similar linear operators A pair of linear operators S, T on a Banach space X such that $S = LTL^{-1}$ for some bounded and invertible operator $L : X \to X$. *See* unitarily equivalent operators.

simple function A function taking only finitely many values.

simple singular point Of a system $y' = A(t)y$, where $A(t)$ is an $n \times n$ matrix of functions, a point where every entry of A has at most a simple pole.

simply periodic function A function $f(z)$ of a complex variable such that, for some complex number $\lambda \neq 0$, we have $f(z+\lambda) = f(z)$, and, whenever $f(z+\mu) = f(z)$, there is a real number c such that $\mu = c\lambda$. *See* elliptic function.

Simpson's Rule An approximation to the Riemann integral of a function over a real interval $[a, b]$. Starting with a partition $a = x_0 < x_1 < \cdots < a_{2n} = b$, with an even number of subintervals, one takes, in the interval $[x_{2i}, x_{2i+2}]$, not the area $f(c_i)[x_{2i+2} - x_{2i}]$, as in a Riemann sum, but a quadratic approximation to the area under $f(x)$. If each subinterval $[x_{j-1}, x_j]$ has length h and if $y_j = f(x_j)$, for $j = 0, 1, \cdots, n$, then the *Simpson's Rule* approximation to $\int_a^b f(x)dx$ is

$$\frac{1}{3}h[y_0 + 4y_1 + 2y_2 + 4y_3 + \cdots + 2y_{2n-2} + 4y_{2n-1} + y_{2n}].$$

simultaneous equations Any set of equations, which must all hold at the same time. Also *system of equations.*

simultaneous inequalities Any set of inequalities, all of which must hold at the same time.

sine integral The function on $[0, \infty)$ given by

$$Si(x) = \int_0^x \frac{\sin t}{t} dt.$$

single-valued function A function $f(z)$, usually on some open set in the complex plane $\mathbf{C}$, which takes only one value for each value of the variable, z. Strictly speaking, a function *must be* single-valued, but in some situations, where multiple-valued "functions" must be considered, the adjective *single-valued* is added for clarity.

singular inner function A bounded analytic function in $|z| < 1$ with no zeros in the unit disk and with nontangential limits of modulus 1 almost everywhere on $|z| = 1$. Such a function has a representation

$$f(z) = \exp\{-\int_0^{2\pi} \frac{e^{i\theta} + z}{e^{i\theta} - z} d\sigma(\theta)\},$$

where σ is a finite positive measure on $[0, 2\pi]$, singular with respect to Lebesgue measure.

singular integral An integral which can be assigned a value, as a limit of standard integrals, even though the integrand is not absolutely integrable.

singular integral equation An equation of the form

$$F(f(t), \int k(s,t)f(s)ds) = 0,$$

where $f(t)$ is an unknown function from a certain class (continuous functions, for example) and the integral is a singular integral for some functions $f(t)$ in the class. *See* singular integral.

singular integral operator A linear operator $L : H \to K$, where H and K are spaces of functions, having the form

$$Lf(t) = \int k(s,t)f(s)ds,$$

where, for some functions $f \in H$, the integral is a singular integral. *See* singular integral.

singular kernel A function $k(s, t)$ appearing in integrals of the form

$$\int_E k(s,t)f(s)ds,$$

for a certain class of functions $f(t)$, such that, for some functions in the class, the integral is a singular integral. *See* singular integral.

singular mapping (1.) A mapping which does not have an inverse.
(2.) A densely defined, linear map between two locally convex vector spaces having a non-zero singularity. *See* singularity of linear map.

singular measure Two measures μ and ν are *mutually singular*, written $\mu \perp \nu$, provided there is a Borel set E with $\mu(F) = 0$, whenever F is a μ-measurable subset of E and $\nu(G) = 0$, whenever G is a ν-measurable set, disjoint from E.

The statement μ *is a singular measure* refers to the case where ν is Lebesgue measure.

singular part (1.) Suppose the measure λ has the Lebesgue decomposition $\lambda = \lambda_s + \lambda_a$, where λ_s is mutually singular with μ and λ_a is absolutely continuous with respect to μ. Then λ_s is called the *singular part* of λ. *See* Lebesgue decomposition.
(2.) A meromorphic function $f(z)$ with a pole at $z = z_0$ can be written in the form $f(z) = g(z) + h(z)$, where $g(z)$ is analytic at z_0 and $h(z) = \sum_{j=1}^{n} a_j(z - z_0)^{-j}$. Then $h(z)$ is called the *singular part* of f at $z = z_0$.

singular point *See* regular singular point, simple singular point.

singular positive harmonic function A positive harmonic function $u(z)$ in $|z| < 1$ has a (Poisson) representation

$$u(z) = \frac{1}{2\pi}\int_{-\pi}^{\pi} \frac{1 - |z|^2}{|e^{it} - z|^2} d\mu(t),$$

where μ is a (uniquely determined) positive measure on $[-\pi, \pi]$. The functon $u(z)$ is called a *singular* positive harmonic function if μ is a singular measure.

singular spectrum *See* spectral theorem.

singular support For a distribution T, the smallest closed set in the complement of which T is a $\mathbf{C}^{\infty}$ function.

singularity of linear map Let $A : E \to F$ be a linear map between locally convex, topological vector spaces, defined on a dense subspace $D[A] \subset E$. Let $\mathcal{U}$ be the filter of all neighborhoods of 0 in $D[A]$ and $A(\mathcal{U})$ the filter in F, generated by the images of the neighborhoods in $\mathcal{U}$. The *singularity* of A is the set $S[A]$ of all adherent points of $A(\mathcal{U})$. A is called *singular* if $S[A] \neq 0$ and *regular* otherwise.

sinusoid A curve with the shape of the graph of $y = \sin x$.

skeleton The *skeleton* of a simplex is the set of all vertices of the simplex. The skeleton of a simplicial complex K is the collection of all simplices in K which have dimension less than that of K.

skew surface A ruled surface which is not developable. Hyperboloids of one sheet and hyperbolic paraboloids are skew surfaces.

A ruled surface is one which is generated by straight lines, which are themselves called generators. A developable surface is a ruled surface on which "consecutive generators intersect."

skew-Hermitian form A bilinear form $a(u, v)$, for u, v belonging to a vector space X, satisfying $a(v, u) = -\overline{a(u, v)}$, for $u, v \in X$. *See* bilinear form.

skew-symmetric multilinear mapping A mapping $\Phi : V \times \cdots \times V \to W$, where V and W are vector spaces, such that $\Phi(v_1, \ldots, v_n)$ is linear in each variable and satisfies

$$\Phi(v_1, \ldots, v_i, \ldots, v_j, \ldots, v_n) = -\Phi(v_1, \ldots, v_j, \ldots, v_i, \ldots, v_n).$$

skew-symmetric tensor *See* antisymmetric tensor.

slope For a straight line in the plane, it is the ratio $(y_1 - y_0)/(x_1 - x_0)$, where (x_0, y_0) and (x_1, y_1) are any points on the line. *See also* slope-intercept equation of line.

slope function Suppose $y = \phi(x, \alpha)$ is a one-parameter family of non-intersecting curves, so that any point (x, y) on a curve lies on a unique curve of the family. The slope of the tangent to the unique curve passing through (x, y) is a function $p(x, y)$, referred to as the *slope function*.

slope-intercept equation of line An equation of the form $y = mx + b$, for a straight line in $\mathbf{R}^2$. Here m is the slope of the line and b is the y-intercept; that is, $y = b$, when $x = 0$.

slowly decreasing function A real-valued function $f(x)$, defined for $x > 0$ and with $\liminf[f(y) - f(x)] \geq 0$, as $x \to \infty$, $y > x$, $y/x \to 1$.

slowly decreasing sequence A sequence $\{a_j\}$ such that $a(x) = a_{[x]}$ is a slowly decreasing function (where $[x]$ is the greatest integer in x). *See* slowly decreasing function.

slowly increasing function A real-valued function $f(x)$, defined for $x > 0$ and with $\limsup[f(y) - f(x)] \leq 0$, as $x \to \infty$, $y > x$, $y/x \to 1$.

slowly increasing sequence A sequence $\{a_j\}$ such that $a(x) = a_{[x]}$ is a slowly increasing function (where $[x]$ is the greatest integer in x). *See* slowly increasing function.

slowly oscillating function A function $f(x)$, defined for $x > 0$, and such that $f(y) - f(x) \to 0$, as $x \to \infty$, $y > x$, $y/x \to 1$.

slowly oscillating sequence A sequence $\{a_j\}$ such that $a(x) = a_{[x]}$ is a slowly oscillating function (where $[x]$ is the greatest integer in x). *See* slowly oscillating function.

smooth function A function with a continuous first derivative.

smooth function in the sense of A. Zygmund A real-valued function $f(x)$ of a real variable, such that, at the point $x = x_0$, we have $\Delta^2 f(x_0, h)/h \to 0$, as $h \to 0$, where $\Delta^2 f(x_0, h) = f(x_0 + h) + f(x_0 - h) - 2f(x_0)$.

smooth manifold A connected Hausdorff space X such that every point $x \in X$ has a neighborhood U_x and a homeomorphism ϕ_x from U_x onto an open set in the complex plane with the property that, whenever two neighborhoods U_x and U_y overlap, $\phi_x \circ \phi_y^{-1}$ is of class $\mathbf{C}^1$, as a map from an open set in $\mathbf{C}$ into $\mathbf{C}$.

Sobolev space For D an open set in $\mathbf{C}^n$ and s a nonnegative integer, the class $W^s(D)$ of functions $f(z_1, \ldots, z_n)$, with weak derivatives satisfying

$$\frac{\partial^{\mu_1}}{\partial z_1^{\mu_1}} \cdots \frac{\partial^{\mu_n}}{\partial z_n^{\mu_n}} \frac{\partial^{\nu_1}}{\partial \bar{z}_1^{\nu_1}} \cdots \frac{\partial^{\nu_n}}{\partial \bar{z}_n^{\nu_n}} f \in L^2(D),$$

for all $\{\mu_j\}, \{\nu_j\} \in \mathbf{Z}_+^n$, such that $\sum \mu_j + \sum \nu_j \leq s$.

solution by quadrature *See* separation of variables.

solution operator Let $\{c_n\}$ be an N-periodic sequence, denoted by C the corresponding $N \times N$ circulant matrix given by

$$(C)_{nm} = c_{n-m}.$$

Consider the linear equation

$$Cx = b \quad (b \text{ a known vector}).$$

By the convolution theorem, the solution can be written as

$$x_n = (s \star b)_n = \sum_{m=1}^{N} s_{n-m} b_m,$$

where

$$(s \star b)_n = \sum_{m=1}^{N} a_{n-m} b_m,$$

$$s_n = \frac{1}{N}\sum_{m=1}^{N} \tilde{c}_m W_{mn},$$
$$\tilde{c}_m = (e_m, b)$$

and

$$W_{mn} = e^{\frac{2\pi i m n}{N}}, \quad n = 0, \pm 1, \pm 2, \ldots,$$
$$m = 1, 2, \ldots, N.$$

For this reason s_n is called the *solution operator* (or the fundamental solution).

Sommerfield's formula There are (at least) two equations known as Sommerfeld's formula. The first is

$$J_\nu(z) = \frac{1}{2\pi}\int_{-\eta+i\infty}^{2\pi-\eta+i\infty} e^{iz\cos t} e^{i\nu(t-\pi/2)}\, dt,$$

where $J_\nu(z)$ is a Bessel function of the first kind.

The second states that under appropriate restrictions,

$$\int_0^\infty J_0(\tau r) e^{-|x|\sqrt{\tau^2-k^2}} \frac{\tau\, d\tau}{\sqrt{\tau^2-k^2}} = \frac{e^{ik\sqrt{\tau^2+k^2}}}{\sqrt{r^2+x^2}}.$$

See Bessel function.

Sonine polynomial *See* Laguerre polynomial.

source of jet *See* jet.

Souslin's Theorem [Suslin's Theorem] If E is a subset of a Polish space, and E is both analytic and co-analytic, then E is a Borel set. Hence the set of subsets that are both analytic and co-analytic is exactly the Borel sets.

space *See* topological space.

space coordinates Coordinates in $\mathbf{R}^n$, for some $n \geq 3$. *See* coordinate.

There are three commonly-used coordinates in three-dimensional space: Cartesian coordinates (x, y, z), cylindrical coordinates (r, θ, z), and spherical coordinates (ρ, ϕ, θ). These coordinates are related by the equations $x = r\cos\theta$, $y = r\sin\theta$, $r^2 = x^2 + y^2$, $r = \rho\sin\phi$, $x = \rho\sin\phi\cos\theta$, $y = \rho\sin\phi\sin\theta$, $z = \rho\cos\phi$, $\rho^2 = r^2 + z^2 = x^2 + y^2 + z^2$.

space curve A continuous mapping $\gamma(t)$ from a real integral $[a, b]$ into a Euclidian space $\mathbf{R}^n$, for some $n \geq 3$.

space of complex numbers The set $\mathbf{C}$ of elements of the form $a+bi$, where a and b are real, $a, b \in \mathbf{R}$, and $i = \sqrt{-1}$. The distance function $d(a+bi, c+di) = |a+bi-(c+di)| = [(a-c)^2 + (b-d)^2]^{\frac{1}{2}}$ makes $\mathbf{C}$ a metric space.

space of imaginary numbers An imaginary number is a complex number of the form bi, where b is real and $i = \sqrt{-1}$. The set of imaginary numbers forms a metric space, as a subspace of the complex numbers $\mathbf{C}$.

space of irrational numbers The set of real numbers that are not rational. A metric space, considered as a subspace of the real numbers $\mathbf{R}$.

space of rational numbers The set $\mathbf{Q}$ of quotients m/n of integers, with $n \neq 0$ and with m/n and j/k identified whenever $mk = jn$. The usual distance function on $\mathbf{Q}$ is $d(p, q) = |p - q|$, under which $\mathbf{Q}$ becomes a metric space.

space of real numbers The set $\mathbf{R}$ of all infinite decimals or the set of all points on an infinite straight line, made into a metric space by defining the distance function $d(x, y) = |x - y|$. A construction of $\mathbf{R}$ can be achieved using Cauchy sequences of rational numbers or Dedekind cuts into the rational numbers $\mathbf{Q}$.

span of set For a set S in a vector space X, over a field $\mathbf{F}$, the set of all linear combinations $\{\sum \lambda_j x_j$, for $\lambda_j \in \mathbf{F}$ and $x_j \in S\}$.

Spectral Mapping Theorem Any of a number of theorems concerning a linear operator T on a complex Banach space and concluding that, for a certain class of functions $f(z)$, defined on the spectrum $\sigma(T)$ of T, one has $\sigma(f(T)) = f(\sigma(T))$. Examples of *Spectral Mapping Theorems* include the case in which T is bounded and f is analytic on (a neighborhood of) $\sigma(T)$ and the case where T is a self-adjoint operator on a Hilbert space and f is continuous on $\mathbf{R}$. *See also* functional calculus. Spectral Mapping Theorems in several variables, involving notions of *joint spectrum* of several operators, also hold.

spectral multiplicity *See* spectral theorem.

spectral radius For a bounded operator T on a complex Banach space, the number

$$r_\sigma(T) = \lim_{n\to\infty} \|T^n\|^{1/n}.$$

The spectrum of T is contained in the disk $\{z : |z| \le r_\sigma(T)\}$ and the series

$$\sum_{n=1}^{\infty} \frac{1}{\lambda^n} T^{n-1}$$

converges to the resolvent $(\lambda I - T)^{-1}$ for all $|\lambda| > r_\sigma(T)$.

spectral resolution *See* resolution of the identity.

spectral set (1.) A compact planar set S is a *K-spectral set* for a bounded linear operator T on a Banach space, if $\|p(T)\| \le K \sup_{z\in S} |p(z)|$, for every polynomial p. S is a *spectral set* for T if it is a K-spectral set, with $K = 1$.
(2.) A connected component of the spectrum of a linear operator T on a Banach space is sometimes referred to as a *spectral set of T*.
(3.) For a bounded, measurable function ϕ, the spectral set $\sigma(\phi)$ is the set of complex numbers λ such that $f \in L^2(\mathbf{R})$ and $f * \phi \equiv 0$ imply $\hat{f}(\lambda) = 0$. This is the set that *ought to be* the closed support of $\hat{\phi}$, if it existed.

spectral synthesis Let $\mathcal{M}$ be the maximal ideal space of a commutative Banach algebra B. The *kernel* $\mathbf{k}(E)$ of a subset $E \subset \mathcal{M}$ is the ideal $\cap_{M\in E} M$. A functional $\nu \in B^*$ *vanishes* on an open set $O \subset \mathcal{M}$ if $(x, \nu) = 0$, for every $x \in B$ such that the support of $\hat{x}$ is contained in O. The complement of the union of all the open sets on which ν vanishes is called the *support* of ν, denoted $\Sigma(\nu)$. For $M \in \mathcal{M}$, define the functional δ_M by $(x, \delta_M) = \hat{x}(M)$, for $x \in B$.

A functional $\nu \in B^*$ *admits spectral synthesis* if ν belongs to the weak-* closure of the span of $\{\delta_M : M \in \Sigma(\nu)\}$ in B^*.

A closed set $E \subset \mathcal{M}$ *is a set of spectral synthesis* if every $\nu \in B^*$ such that $\Sigma(\nu) \subseteq E$ is orthogonal to $\mathbf{k}(E)$.

Malliavin proved that if G is a discrete LCA group, then spectral synthesis fails for $A(G)$.

spectral theorem Let T be a self-adjoint operator on a Hilbert space $\mathcal{H}$. There exists a measure μ on the real line $\mathbf{R}$, a μ-measurable function $n : \mathbf{R} \to \mathbf{Z}_+ \cup \infty$, $\mathbf{Z}_+$ the nonnegative integers, and a unitary map U from $\mathcal{H}$ to a continuous direct sum $\int \mathcal{H}_\lambda d\mu(\lambda)$ of Hilbert spaces $\mathcal{H}_\lambda, \lambda \in \mathbf{R}$, such that, for every bounded Baire function β and for $x, y \in \mathcal{H}$, we have

$$(\beta(T)x, y)$$
$$= \int \beta(\lambda)(Ux(\lambda), Uy(\lambda))_\lambda d\mu(\lambda),$$

where $(\cdot, \cdot)_\lambda$ is the inner product in $\mathcal{H}_\lambda$ and $n(\lambda)$ is the dimension of $\mathcal{H}_\lambda$.

If $\mu = \mu_a + \mu_s$ is the Lebesgue decomposition of μ into its absolutely continuous and purely singular parts, then the closed supports of μ_a and μ_s are called the *absolutely continuous spectrum* and the *singular spectrum* of T. The function $n(\lambda)$ is called the *spectral multiplicity function* of T. (Sometimes *spectral multiplicity* refers to the essential supremum of $n(\lambda)$.)

Analogous theorems hold for unitary and normal operators.

spectral theory The study of the properties of the spectrum of a linear operator on a Banach or Hilbert space. This may involve theorems about general operators, about special classes of operators such as compact or self-adjoint operators, or specific operators, such as the Schroedinger operator.

spectrum (1.) For a linear operator T on a Banach space X, the set of complex λ such that $T - \lambda I$ fails to have a bounded inverse. (2.) For an element x in a Banach algebra, the set of complex λ such that $x - \lambda$ is not invertible.

speed *See* acceleration.

sphere The locus of

$$x^2 + y^2 + z^2 = R^2,$$

or one of its translates, in $\mathbf{R}^3$.

Sphere Theorem A complete, simply connected, Riemannian manifold of even dimension which is δ-pinched ($\delta > \frac{1}{4}$) is homeomorphic to a sphere. If M is of odd dimension and $\frac{1}{4}$-pinched, then, also, M is homeomorphic to a sphere. (By definition, M is δ-pinched if its sectional curvature K satisfies $A\delta \leq K \leq A$, for some positive constant A.)

spherical coordinates A system of curvilinear coordinates which is natural for describing positions on a sphere or spheroid. Define θ to be the azimuthal angle in the xy-plane from the x-axis with $0 \leq \theta < 2\pi$ (denoted λ when referred to as the longitude), ϕ to be the polar angle from the z-axis with $0 \leq \phi \leq \pi$ (colatitude, equal to $\phi = 90° - \delta$ where δ is the latitude), and r to be distance (radius) from a point to the origin.

Unfortunately, the convention in which the symbols θ and ϕ are reversed is frequently used, especially in physics, leading to unnecessary confusion. The symbol ρ is sometimes also used in place of r. Arfken (1985) uses (r, ϕ, θ), whereas Beyer (1987) uses (ρ, θ, ϕ). Be very careful when consulting the literature.

Here, the symbols for the azimuthal, polar, and radial coordinates are taken as θ, ϕ, and r, respectively. Note that this definition provides a logical extension of the usual polar coordinates notation, with θ remaining the angle in the xy-plane and ϕ becoming the angle out of the plane.

$$r = \sqrt{x^2 + y^2 + z^2}$$
$$\theta = \tan^{-1}\left(\frac{y}{x}\right)$$
$$\phi = \sin^{-1}\left(\frac{\sqrt{x^2 + y^2}}{r}\right) = \cos^{-1}\left(\frac{z}{r}\right),$$

where $r \in [0, \infty)$, $\theta \in [0, 2\pi)$, and $\phi \in [0, \pi]$. In terms of Cartesian coordinates,

$$x = r\cos\theta\sin\phi, \quad y = r\sin\theta\sin\phi$$
$$z = r\cos\phi.$$

The scale factors are

$$h_r = 1, \quad h_\theta = r\sin\phi, \quad h_\phi = r,$$

so the metric coefficients are

$$g_{rr} = 1, \quad g_{\theta\theta} = r^2\sin^2\phi, \quad g_{\phi\phi} = r^2.$$

The line element is

$$ds = dr\,\hat{\mathbf{r}} + r\,d\phi\,\hat{\phi} + r\sin\phi\,d\theta\,\hat{\theta},$$

the area element

$$d\mathbf{a} = r^2\sin\phi\,d\theta\,d\phi\,\hat{\mathbf{r}},$$

and the volume element

$$dV = r^2\sin\phi\,d\theta\,d\phi\,dr.$$

The Jacobian is

$$\left|\frac{\partial(x, y, z)}{\partial(r, \theta, \phi)}\right| = r^2|\sin\phi|.$$

The position vector is

$$\mathbf{r} \equiv \begin{bmatrix} r\cos\theta\sin\phi \\ r\sin\theta\sin\phi \\ r\cos\phi \end{bmatrix},$$

so the unit vectors are

$$\hat{\mathbf{r}} \equiv \frac{\frac{d\phi}{dr}}{\left|\frac{d\phi}{dr}\right|} = \begin{bmatrix} \cos\theta\sin\phi \\ \sin\theta\sin\phi \\ \cos\phi \end{bmatrix}$$

$$\hat{\theta} \equiv \frac{\frac{d\phi}{d\theta}}{\left|\frac{d\phi}{d\theta}\right|} = \begin{bmatrix} -\sin\theta \\ \cos\theta \\ 0 \end{bmatrix}$$

$$\hat{\phi} \equiv \frac{\frac{d\phi}{d\theta}}{\left|\frac{dx}{d\phi}\right|} = \begin{bmatrix} \cos\theta\cos\phi \\ \sin\theta\cos\phi \\ -\sin\phi \end{bmatrix}.$$

spherical harmonic A homogeneous, harmonic polynomial (of a given degree) in n real variables. Hence, a function of the form

$$P(x_1, \ldots, x_n) = \sum_{m_1+\ldots+m_n=k} a_{m_1\ldots m_n} x_1^{m_1} \cdots x_n^{m_n}$$

such that $\Delta P = \sum_1^n \frac{\partial^2}{\partial x_j^2} P = 0$.

spherical polygon A closed figure on a sphere, bounded by three or more arcs of great circles.

spherical sector The solid generated by revolving a sector of a circle about any diameter which does not cut the sector.

spherical segment The solid formed as the union of a zone and the planes that define it. *See* zone.

spherical triangle A spherical polygon of three sides. *See* spherical polygon.

spherical wedge The solid formed as the union of a lune and the planes of its bases. *See* lune.

spheriodal wave functon A solution of the differential equation

$$\frac{d}{dz}\left((1-z^2)\frac{du}{dz}\right) + \left(l - k^2z^2 - m^2(1-z^2)^{-1}\right)u = 0.$$

spiral A critical point of an autonomous system

$$\frac{dx}{dt} = F(x, y),$$
$$\frac{dy}{dt} = G(x, y)$$

(i.e., a point where $F(x, y) = G(x, y) = 0$) that is approached in a spiral-like manner by a family of paths that wind around the point an infinite number of times, as $t \to \infty$.

spiral of Archimedes The plane curve with polar equation

$$r^m = a^m\theta.$$

Sometimes the term *Archimedes' spiral* refers specifically to the case $m = 1$.

Splitting Theorem (of Grothendieck) Every vector bundle on $\mathbf{P}^1$ splits into a direct sum of line bundles, unique up to order.

square integrable function A complex-valued function f, measurable with respect to a positive measure μ, such that

$$\int |f|^2 d\mu < \infty.$$

The vector space of all such functions is denoted $L^2(\mu)$.

stability of solution *See* asymptotic stability.

star-shaped set A subset S of a vector space X over $\mathbf{R}$ or $\mathbf{C}$ such that $x \in S$ implies $tx \in X$, for $t \in [0, 1]$.

stationary curve *See* stationary function.

stationary function Any admissible solution of Euler's equation,

$$\frac{d}{dx}\left(\frac{\partial f}{\partial y'}\right) - \frac{\partial f}{\partial y} = 0.$$

Admissible means of class $\mathbf{C}^2$ and satisfying the boundary conditions $y(x_1) =$

$y_1, y(x_2) = y_2$. Also called *stationary curve.* The corresponding value of the integral $\int_{x_1}^{x_2} f(x, y, y')dx$ is called a *stationary value.*

stationary phase method A technique that, in physics, is synonymous with the classical approximation itself. There is a real integrand function F of the real variable x involved. $F(x)$ is supposed to have exactly one stationary point, which can be either a minimum or maximum, and which lies at $x = a$. A Taylor-type expansion of $F(x)$ about $x = a$ takes place. Integrals are then evaluated asymptotically by being replaced by suitable finite sums. This procedure picks out the value of F at its stationary point, without the necessity of finding the value of a. The method can be extended to give values of other functions of x at the stationary point. A classical application of the stationary phase method has been applied to show the Ponzano–Regge form satisfies Racah's identity asymptotically.

stationary point A value of the independent variable at which the first derivative of a differentiable function is equal to 0.

stationary value (1.) The value of the dependent variable at a stationary point.
(2.) *See* stationary function.

Stein space A complex space X such that
(i.) every connected component of X has a (second axiom) countable topology;
(ii.) the holomorphic functions on X separate the points of X: for $x, y \in X$ with $x \neq y$, there is f, holomorphic on X such that $f(x) \neq f(y)$;
(iii.) if K is a compact subset of the underlying topological space of X, then the holomorphically convex hull

$$\hat{K} = \cap\{x \in X : |f(x)| \leq \|f\|_K\}$$

is compact, where the intersection is over all f holomorphic on X.

step-by-step method An method involving repeated applications of a transformation.

step-down operator A second order, ordinary differential operator can sometimes be decomposed, or factored, into first order factors and then solved by finding a recurrence formula. (*See* ladder method.) For example, for Legendre's equation

$$L_n y = (1-x^2)((1-x^2)y')' + n(n+1)y = 0,$$

we can write $L_n = S_n T_n + n^2$, where $T_n = (1-x^2)\frac{d}{dx} + nx$ and $S_n = (1-x^2)\frac{d}{dx} - nx$. Thus, if y is a solution of $L_n y = 0$, $T_n y$ is a solution of $L_{n-1}y = 0$ and $S_{n+1}y$ is a solution of $L_{n+1}y = 0$. T_n [resp. S_n] is called the *step-down* [resp. *step-up*], *operator* with respect to n.

step-up operator *See* step-down operator.

Stieltjes moment problem Suppose $m_0, m_1, m_2, \ldots$ is a sequence of real numbers. The *Stieltjes moment problem* is to determine a measure μ defined on the Borel sets of $[0, \infty)$ such that $m_n = \int_0^\infty t^n d\mu(t)$ for $n = 0, 1, 2, \ldots$. A necessary and sufficient condition for the existence of such a measure is that $\sum_{i,j=0}^n m_{i+j}\alpha_i\bar{\alpha}_j \geq 0$ and $\sum_{i,j=0}^n m_{i+j+1}\alpha_i\bar{\alpha}_j \geq 0$ for every finite set of complex numbers $\alpha_0, \alpha_1, \ldots, \alpha_n$.

Stieltjes transform Of a monotonic function α:

$$F(z) = \int_0^\infty \frac{d\alpha(t)}{z+t},$$

for $\Re z > 0$.

Stirling's formula

$$\sqrt{2n\pi}\left(\frac{n}{e}\right)^n < n! < \sqrt{2n\pi}\left(\frac{n}{e}\right)^n\left(1 + \frac{1}{12n-1}\right),$$

so that the quantity on the left is a good approximation of $n!$, if n is large.

Stokes's differential equation The second order, linear, ordinary differential equation with one (irregular) singular point, and having the form

$$\frac{d^2y}{dx^2} + (Ax + B)y = 0.$$

Stokes's equation can be transformed to a special case of Bessel's equation. *See* Stokes's phenomenon.

Stokes's formula *See* Stokes's Theorem.

Stokes's phenomenon If the Hankel functions, obtained as solutions of Bessel's equation

$$\frac{d^2w}{dz^2} + \frac{1}{z}\frac{dw}{dz} + \left(1 - \frac{\nu^2}{z^2}\right)w = 0,$$

are expanded in asymptotic series in a region such as $-\pi + \delta \leq \arg z \leq 2\pi - \delta$, the coefficients change discontinuously with a continuous change in the range of z.

A similar phenomenon holds for the solutions of other equations.

Stokes's Theorem Let S be a piecewise smooth, oriented surface in $\mathbf{R}^n$, whose boundary C is a piecewise smooth simple, closed curve, directed in accordance with the given orientation in S. Let $\mathbf{u} = L\mathbf{i}+M\mathbf{j}+N\mathbf{k}$ be a vector field with continuous differentiable components, in a domain D of $\mathbf{R}^n$ containing S. Then

$$\int_C u_T ds = \int_S\int (\text{curl } \mathbf{u} \cdot \mathbf{n}) d\sigma,$$

where $\mathbf{n}$ is the chosen unit normal vector on S, that is,

$$\int_C L dx + M dy + N dz$$
$$= \int_S\int \left(\frac{\partial N}{\partial y} - \frac{\partial N}{\partial z}\right) dydz$$
$$+\left(\frac{\partial L}{\partial z} - \frac{\partial N}{\partial x}\right)dzdx + \left(\frac{\partial M}{\partial x} - \frac{\partial L}{\partial y}\right)dxdy.$$

Stolz angle The subset of the unit disk $D \subset \mathbf{C}$, defined by

$$\Gamma_a(e^{i\alpha}) = \{z \in D : |z - e^{i\alpha}| < a(1 - |z|)\}.$$

Approach to the boundary point $e^{i\alpha}$, through a Stolz angle is called *non-tangential* approach.

Stone-Cech compactification Let X be a completely regular space, $\mathbf{C}(X)$ the bounded, continuous functions on X with sup norm and $\mathcal{M}(X)$ the dual of $\mathbf{C}(X)$. Identify X with $\hat{X} \subset \mathcal{M}(X)$, the set of atomic measures, concentrated at points of X. The *Stone-Cech compactification* of X is the closure of $\hat{X}$ in the weak topology of $\mathcal{M}(X)$.

Stone-Weierstrass Theorem Let A be an algebra of real-valued, continuous functions on a compact set K. If A separates the points of K and vanishes at no point of K, then A is uniformly dense in $\mathbf{C}(K)$, the real-valued, continuous functions on K. The hypothesis that the functions in A and $\mathbf{C}(K)$ are real-valued can be replaced by the assumption that they are complex-valued and A is self-adjoint: $f \in A$ implies $\bar{f} \in A$.

strictly concave function A real-valued function $f(x)$ on a vector space V satisfying

$$f(\lambda x + (1 - \lambda)y) > \lambda f(x) + (1 - \lambda) f(y)$$

for $x, y \in V$ $(x \neq y)$ and $0 < \lambda < 1$.

strictly convex domain A domain $\Omega \subset \mathbf{R}^n$ with $\mathbf{C}^2$ boundary and with a defining function ρ such that

$$\sum_{i,j=1}^{n} \frac{\partial^2 \rho}{\partial x_i \partial x_j}(P) w_i w_j > 0,$$

for all tangent vectors $\mathbf{w} = (w_1, w_2, \dots, w_n)$ to Ω at P, for all $P \in \partial\Omega$.

strictly convex function *See* strongly convex function.

strictly monotonic function A function which is monotonic (*see* monotonically increasing function, monotonically decreasing function) and not constant on any interval.

strictly pseudoconvex domain A domain $\Omega \subset \mathbf{C}^n$ with $\mathbf{C}^2$ boundary and a defining function ρ such that the (Levi) form

$$\sum_{j,k=1}^{n} \frac{\partial^2 \rho}{\partial z_j \partial z_k}(P) w_j \bar{w}_k$$

is positive definite on the complex tangent space at P (i.e., $\sum_{j=1}^{n}(\partial\rho/\partial z_j)(P)w_j = 0$) for every $P \in \partial\Omega$.

strong convergence (1.) For a sequence $\{x_n\}$ in a Banach space, norm convergence to a limit. That is, $x_n \to x$, strongly if $\|x_n - x\| \to 0$.
(2.) For a sequence of operators $\{T_n\}$ on a Banach space B, norm convergence of $\{T_n x\}$, for every $x \in B$. *See also* weak convergence, uniform operator topology.

Strong Convergence Theorem A sequence $\{T_n\}$ of operators $T_n \in B(X, Y)$, where X and Y are Banach spaces and $B(X, Y)$ denotes the space of bounded linear operators, is strongly convergent if and only if:
(i) The sequence $\{\|T_n\|\}$ is bounded.
(ii) The sequence $\{T_n x\}$ is Cauchy in Y for every x in a total subset M of X.

strong operator topology The topology on the set $L(X, Y)$ of continuous linear operators between two locally convex linear topological spaces X and Y defined by the family of seminorms

$$p(T) = \sup_{j=1,\dots,n} q(Tx_j)$$

where $x_1, \dots, x_n$ are elements of X and q is a seminorm on Y. When X and Y are Banach spaces, this is the topology of strong convergence of sequences of operators. *See* strong convergence.

strongly convex domain *See* strictly convex domain.

strongly convex function A $\mathbf{C}^2$ function $f(x_1, \dots, x_n)$, defined on a domain $\Omega \subset \mathbf{R}^n$, such that the matrix $(\partial^2 f/\partial x_i \partial x_j)$ is positive definite at every point of Ω. Positive semidefiniteness of the matrix is equivalent to *convexity* of f.

strongly elliptic operator A differential operator

$$L = \sum_{|s|=|t|\le m} (-1)^{|t|} D^t c_{s,t} D^s,$$

on a bounded domain $G \subset \mathbf{R}^n$, where the functions $c_{s,t}$ are of class $\mathbf{C}^m$ on the closure of G, is *strongly elliptic* if

$$\Re \sum_{|s|,|t|\le m} c_{s,t}\xi^s \xi^t \ge c_0 |\xi|^{2m},$$

for $x \in G$ and for all $\xi = (\xi_1, \dots, \xi_m) \in \mathbf{R}^m$.

Here $D^s = \frac{\partial^{s_1}}{\partial x_1^{s_1}} \cdots \frac{\partial^{s_n}}{\partial x_n^{s_n}}$ and $|s| = s_1 + \dots + s_n$, when $s = (s_1, \dots, s_n)$.

strongly pseudoconvex domain *See* strictly pseudoconvex domain.

structure equations If ω, Θ, Ω are, respectively, the connection form, torsion form, and curvature form of a linear connection Γ of M, then

$$d\theta(X, Y) = -\frac{1}{2}(\omega(X) \cdot \theta(Y) - \omega(Y) \cdot \theta(X)) + \Theta(X, Y),$$

and

$$d\omega(X, Y) = -\frac{1}{2}[\omega(X), \omega(Y)] + \Omega(X, Y),$$

for $X, Y \in T_u(P)$ and $u \in P$.

Structure Theorems (1.) In P-spaces and P-algebras.
A. The Reduction Theorem. Let $\hat{P}$ and $\tilde{P}$ denote the Samelson subspace and a Samelson complement for the P-algebra $(S; \sigma)$. Then multiplication defines an isomorphism

$$g : \wedge\tilde{P} \otimes \wedge\hat{P} \xrightarrow{\cong} \wedge P$$

of graded algebras.

Next, let $(S;\bar{\sigma})$ be the $\bar{P}$-algebra obtained by restricting σ to $\bar{P}$, and denote its Koszul complex by $(S \otimes \wedge\bar{P}, \nabla_{\bar{\sigma}})$. Then $\nabla_{\bar{\sigma}}$ is the restriction of ∇_σ to $S \otimes \wedge\bar{P}$. Moreover,

$$(S \otimes \wedge\bar{P} \otimes \wedge\hat{P}, \nabla_{\bar{\sigma}} \otimes \iota)$$

is a graded differential algebra.

Theorem 1 (Reduction Theorem). Suppose that $(S;\sigma)$ is an alternating connected P-algebra with Samelson space $\hat{P}$, and let $\bar{P}$ be a Samelson complement. Then there is an isomorphism

$$f\colon (S\otimes\wedge\bar{P}\otimes\wedge\hat{P}, \nabla_{\bar{\sigma}}\otimes\iota) \xrightarrow{\cong} (S\otimes\wedge P, \nabla_\sigma)$$

of graded differential algebras, such that the diagram

$$\begin{array}{ccccc} & & S \otimes \wedge\bar{P} & & \\ & \lambda_1 \swarrow & & \searrow \lambda_2 & \\ S \otimes \wedge\bar{P} \otimes \wedge\bar{P} & & \underset{\cong}{\overset{f}{\longrightarrow}} & & S \otimes \wedge P \\ \bar{\varphi}_S\otimes\iota \downarrow & & & & \downarrow \varphi_S \\ \wedge\bar{P} \otimes \wedge\bar{P} & & \underset{g}{\overset{\cong}{\longrightarrow}} & & \wedge P \end{array}$$

commutes. (λ_1 and λ_2 are the obvious inclusion maps.)

B. Simplification Theorem. The theorem of this section is, in some sense, a generalization of Theorem 1. Let $(S;\sigma)$ be an alternating graded connected P-algebra. Assume $\alpha\colon P \to S^+ \cdot \sigma(P)$ is a linear map, homogeneous of degree 1, and define a second P-algebra $(S;\tau)$ by setting $\tau = \sigma + \alpha$.

Theorem 2 (Simplification Theorem). With the hypotheses and notation above, there is an isomorphism

$$f\colon (S \otimes \wedge P, \nabla_\tau) \xrightarrow{\cong} (S \otimes \wedge P, \nabla_\sigma)$$

of bigraded differential algebras, such that the diagram

$$\begin{array}{ccccc} & & S \otimes \wedge P & & \\ \iota_S & \swarrow & \downarrow & \searrow & \varphi_S \\ S & & \cong \downarrow f & & \wedge P \\ \iota_S & \searrow & \downarrow & \swarrow & \varphi_S \\ & & S \otimes \wedge P & & \end{array}$$

commutes.

C. A third structure theorem. Theorem 3. Let $(\vee Q;\sigma)$ be a symmetric P-algebra with Samelson space $\hat{P}$ and a Samelson complement $\bar{P}$. Then the following conditions are equivalent:

(i.) $(l^{\#}_{\vee Q})^+ = 0$.
(ii.) $\sigma_\vee\colon \vee\mathbf{P} \to \vee Q$ is surjective.
(iii.) $\bar{\sigma}_\vee\colon \vee\bar{\mathbf{P}} \to \vee Q$ is an isomorphism.

(iv.) $\varrho^{\#}_{\vee Q}\colon H(\vee Q \otimes \wedge P) \to \wedge\hat{P}$ is an isomorphism.
(v.) The algebra $H(\vee Q \otimes \wedge P)$ is generated by 1 together with elements of odd degree.

If these conditions hold, then

$$\dim P = \dim \hat{P} + \dim Q.$$

These results have been extended to more general algebras.

(2.) For W-algebra. The Racah operator

$$\left\{\begin{matrix} & J+\Delta & \\ 2J & & 0 \\ & J+\Delta' & \end{matrix}\right\} \equiv R$$

is uniquely determined (to within ± 1) by the zeros associated with both the trivial and characteristic null spaces, by the reflection symmetry, and by the requirements of normalization.

The action of a Racah operator on the coupled angular momentum basis of $H \otimes H$ (Hölder space product) is given by

$$R|_{(j_1 j_2)jm)} = [(2j_1 + 2\Delta + 1)(2j_2 + 1)]^{1/2} \times W(j, j_1, j_2 + \Delta', j; j_1 + \Delta)|_{(j_1+\Delta, j_2+\Delta')jm)}.$$

Note that this type of pattern calculus was invented as a tool for writing out the matrix elements of certain Wigner operators.

Sturm-Liouville operator An unbounded operator on a dense domain in $L^2(a, b)$ of the form

$$Lu = (p(t)u')' + q(t)u.$$

The domain is of the form
$D(L) = \{u \in L^2(a, b) : u$ is differentiable, pu' is of bounded variation and u satisfies

$$u(a)\cos\alpha - p(a)u'(a)\sin\alpha = 0$$
$$u(b)\cos\beta - p(b)u'(b)\sin\beta = 0\}.$$

Sturm-Liouville problem A boundary value problem, consisting of (1.) a differential equation of the form

$$(pu')' + (q + \lambda)u = 0$$

where $p(t) > 0$ and $q(t)$ is real-valued and continuous on a real interval $[a, b]$ and λ is a complex number and (2.) a homogeneous boundary condition

$$u(a)\cos\alpha - p(a)u'(a)\sin\alpha = 0,$$
$$u(b)\cos\beta - p(b)u'(b)\sin\beta = 0.$$

subfamily A subset of a family.

subharmonic function An upper semicontinuous function $u : D \to \mathbf{R} \cup \{-\infty\}$, where D is an open subset of $\mathbf{C}$, such that u satisfies

$$u(\zeta) \leq \frac{1}{2\pi}\int_{-\pi}^{\pi} u(\zeta + re^{i\theta})d\theta,$$

whenever the disk $\{|z - \zeta| \leq r\} \subset D$. The above integrals are also assumed to be $> -\infty$.

submanifold A subset A of an n-dimensional manifold M is a *k-dimensional submanifold* ($k \leq n$) if for each $a \in A$ there is an open neighborhood U_a of a and a homeomorphism $\phi_a : U_a \to \mathbf{R}^k$ that is the restriction of a coordinate chart from M. *See* atlas.

submersion A map ϕ from a differentiable manifold to itself, such that the differential of ϕ has locally constant rank. *See* differential of differentiable mapping.

subnet Suppose that D and E are directed sets and $\{S_n : n \in D\}$ and $\{T_m : m \in E\}$ are nets. T is said to be a subnet of S if there is a function $N : E \to D$ such that $T_m = S_{N_m}$ for all $m \in E$, and for every $n \in D$ there exists an $m \in E$ such that $N_k \geq n$ whenever $k \geq m$.

subnormal operator A bounded operator on a Hilbert space that is (unitarily equivalent to) the restriction of a normal operator to one of its invariant subspaces.

subsequence A restriction of a sequence to a proper subset of the positive integers. That is, given a sequence $\{a_1, a_2, \ldots\}$, a subsequence is a sequence $\{a_{n_1}, a_{n_2}, \ldots\}$, where $n_1 < n_2 < \ldots$ are positive integers.

substitution Any change of variable. For example, the integral

$$\int \sec^3\theta d\theta$$

is equal to

$$\int (1 + t^2)^{\frac{1}{2}} dt$$

after the *substitution* $t = \tan\theta$.

subtend To be opposite to, as a side of a triangle or an arc or a circle, subtending the opposite angle.

successive approximations (1.) Any method of approximating a number or function by a sequence $\{y_n\}$, where y_n is defined in terms of y_{n-1}.
(2.) The sequence of functions defined by

$$y_0(t) = y_0,$$

$$y_{n+1}(t) = y_0 + \int_{t_0}^{t} f(s, y_n(s))ds$$

which can be shown to converge to a solution of the problem

$$y' = f(t, y), y(t_0) = y_0,$$

under suitable conditions on $f(t, y)$. This is the method usually used to prove the Picard-Lindelöf Theorem.

sum of series An infinite series is a formal sum of the form $\sum_{j=1}^{\infty} a_j$, where the a_j come from some normed vector space X. The sum of the series is defined to be the limit, in X, of the partial sums $s_n = \sum_{j=1}^{n} a_j$. That is, the sum is equal to $x \in X$ provided $\|s_n - x\| \to 0$.

sum of vectors The + operation in a vector space. Geometrically, if $\mathbf{v} = (v_1, v_2, v_3)$ and $\mathbf{w} = (w_1, w_2, w_3)$ are vectors in $\mathbf{R}^3$, then their sum is given by the diagonal of the parallelogram with two of its sides equal to $\mathbf{v}$ and $\mathbf{w}$. Algebraically,

$$\mathbf{v} + \mathbf{w} = (v_1 + w_1, v_2 + w_2, v_3 + w_3).$$

summability A generalization of convergence for a numerical sequence or series. Consider the infinite matrix $T = (a_{ij}), i, j = 1, 2, \dots$ and set $\sigma_n = \sum_{k=1}^{\infty} a_{nj}s_j$. We say that the sequence $\{s_n\}$ or the series with partial sums $\{s_n\}$ *is summable* T *to the limit* s provided $\sigma_n \to s$. Under suitable restrictions on T, convergence in the usual sense implies summability.

summation A formal sum of a countable number of terms. Notation for the summation of $\{a_n\}$ is

$$\sum_{j=1}^{\infty} a_j = a_1 + a_2 + \dots$$

Some notion of convergence of the partial sums is required to assign a numerical value to a summation.

superior limit *See* upper limit.

support *See* support of a function.

support of a differential form For a differential form $\omega = \sum \omega_j dx_j$, the support of ω is the union of the supports of ω_j.

support of a distribution A generalized function T vanishes on an open set $U \subset \Omega$ if $Tu = 0$ for all $u \in \mathbf{C}^\infty(\Omega)$ with support contained in U. The support of T is the smallest closed set $F \subset \Omega$ such that T vanishes on $\Omega \backslash F$.

support of a function The set of points in the domain of a real or complex valued function $f(t)$, where $f(t) \neq 0$. Sometimes the closure of that set.

supporting function Let M be an oriented regular surface in $\mathbf{R}^3$ with normal $\mathbf{N}$. Then the supporting function of M is the function $h: M \to \mathbf{R}$ defined by

$$h(\mathbf{p}) = \mathbf{p} \cdot \mathbf{N}(\mathbf{p}).$$

surface The locus of an equation $f(x_1, \dots, x_n) = 0$ in $\mathbf{R}^n$.

surface area: figure of revolution If the graph of a positive, continuously differentiable function $y = f(x), a \leq x \leq b$ is rotated about the x axis, the area of the surface of revolution is defined to be

$$S = \int_a^b 2\pi f(x)\sqrt{1 + \left(\frac{dy}{dx}\right)^2} dx.$$

Analogous definitions can be given for rotation about other lines.

surface integral Let S be a surface in $\mathbf{R}^3$, given, for example, by $z = f(x, y), (x, y) \in R = \{(x, y) : a \leq x \leq b; c \leq y \leq d\}$, and assume that S is smooth (the partial derivatives of $f(x, y)$ are continuous in R). Let $g(x, y, z)$ be continuous on S. Divide S into pieces; that is, define a partition P by $a = x_0 < x_1 < \dots < x_n = b, c =$

$y_0 < y_1 < \ldots < y_n = d$ and pieces of S by $\Delta_{ij} = \{(x, y, f(x, y)) : x_{i-1} \le x \le x_i, y_{j-1} \le y \le y_j\}$. Denote by $\Delta_{ij}\sigma$ the area of Δ_{ij}. Then if we choose, for each i, j a point $(x_{ij}^*, y_{ij}^*, z_{ij}^*) \in \Delta_{ij}$, a Riemann sum for the surface integral of $g(x, y, z)$ over S is $R_P = \sum_{i,j} g(x_{ij}^*, y_{ij}^*, z_{ij}^*)\Delta_{ij}\sigma$. If, for every $\epsilon > 0$, there is a number I and a $\delta > 0$ such that $|I - R_P| < \epsilon$, whenever $\max_{ij}\{|x_i - x_{i-1}|, |y_j - y_{j-1}|\} < \delta$, then we say the surface integral of g over S exists and is equal to I, and write

$$I = \int_S\int g(x, y, z)d\sigma.$$

If the surface S is given parametrically or by an implicit function, the above discussion can be modified accordingly.

surface integral with respect to surface element *See* surface integral. If the surface S is given as the graph $z = f(x, y)$ then the *element of area* is given by

$$d\sigma = \int\sqrt{1 + \left(\frac{\partial z}{\partial x}\right)^2 + \left(\frac{\partial z}{\partial y}\right)^2}dxdy,$$

and so the surface integral of $g(x, y, z)$ over S becomes

$$\int_S\int g(x, y, z)d\sigma = \int_a^b\int_c^d g(x, y, f(x, y)) \sqrt{1 + \left(\frac{\partial z}{\partial x}\right)^2 + \left(\frac{\partial z}{\partial y}\right)^2}dxdy.$$

If the surface S is given parametrically or by an implicit function, the above discussion can be modified accordingly.

surface of constant curvature A surface whose total curvature K is the same at all points. If $K > 0$, the surface is a spherical surface. If $K < 0$, the surface is a pseudospherical surface. If $K = 0$, the surface is a developable surface.

surface of revolution The 2-dimensional figure in $\mathbf{R}^3$ formed when an arc is rotated about an infinite straight line, which does not intersect the arc.

symbol of an operator A function associated with an operator on a Banach or Hilbert space.

For the partial differential operator $P(z, \partial/\partial z)$ where $z \in \mathbf{C}^n$ and $\partial/\partial z = (\partial/\partial z_1, \ldots, \partial/\partial z_n)$, the symbol is $P(z, \zeta)$.

See pseudo-differential operator, Toeplitz operator, and Hankel operator for other examples where the symbol plays an important role.

symmetric bilinear form A bilinear form $a(x, y)$ such that $a(x, y) = \overline{a(y, x)}$.

symmetric kernel A function $k(s, t)$ of two variables, satisfying $k(t, s) = \overline{k(s, t)}$, so that the integral operator

$$Tf(t) = \int k(s, t)f(s)ds$$

is a Hermitian operator on L^2.

symmetric multilinear mapping A mapping $\Phi : V \times \cdots \times V \to W$, where V and W are vector spaces, such that $\Phi(v_1, \ldots, v_n)$ is linear in each variable and satisfies

$$\Phi(v_1, \ldots, v_i, \ldots, v_j, \ldots, v_n) = \Phi(v_1, \ldots, v_j, \ldots, v_i, \ldots, v_n).$$

symmetric operator An operator T, usually unbounded and defined on a dense subset D of a Hilbert space H, satisfying $(Tx, y) = (x, Ty)$, for all $x, y \in D$. Thus the domain of the adjoint, T^*, of T is bigger than D and $T^*x = Tx$, for $x \in D$. *See* adjoint operator.

symmetric positive system A system of differential operators which, as an operator T, between direct sums of Hilbert spaces of functions, is symmetric:

$$(Tu, v) = (u, Tv),$$

for all u, v belonging to the domain of T, and positive:

$$(Tu, u) \geq 0,$$

for all u in the domain of T.

symmetric relation A relation $\sim$ on a set S satisfying

$$x \sim y \Rightarrow y \sim x,$$

for all $x, y \in S$.

symmetric tensor A covariant tensor Φ of order r is *symmetric* if, for each i, j, $1 \leq i, j \leq r$, we have $\Phi(\mathbf{v}_1, \ldots, \mathbf{v}_i, \ldots \mathbf{v}_j, \ldots, \mathbf{v}_r) = \Phi(\mathbf{v}_1, \ldots, \mathbf{v}_j, \ldots, \mathbf{v}_i, \ldots, \mathbf{v}_r)$.

symmetrization A function S defined on covariant tensors K by

$$(SK) = \frac{1}{r!}\sum_{\pi} K(X_{\pi(1)}, \ldots, X_{\pi(n)}).$$

For any K, SK is symmetric.

symmetrizer *See* symmetrizing mapping.

symmetrizing mapping The mapping $\mathcal{S}$, generally on the space of covariant tensors of order r on a vector space V, satisfying

$$\mathcal{S}\Phi(\mathbf{v}_1, \ldots, \mathbf{v}_r) = \frac{1}{r!}\sum_{\sigma} \operatorname{sgn}\sigma\, \Phi(\mathbf{v}_{\sigma(1)}, \ldots, \mathbf{v}_{\sigma(r)}),$$

where the sum runs over all permutations σ of the set $\{1, \ldots, r\}$.

symmetry Let σ be an involutive automorphism of a connected Lie group G. Let H be a closed subgroup lying between the subgroup of fixed points of σ and its identity component. Then σ induces an involutive diffeomorphism σ_0 of G/H, called the *symmetry* around 0, the origin of G/H.

symmetry in space A symmetry T is a map from the set of unit rays in a Hilbert space H onto the set of unit rays in a second (not necessarily distinct) Hilbert space H_1 such that the following properties hold.
(i.) T is defined for every unit ray in H.
(ii.) Transition probabilities are preserved, that is

$$(T(\hat{e}), T(\hat{e}_1)) = (\hat{e}, \hat{e}_1),$$

where $\hat{e}, \hat{e}_1$ are unit rays in H, and $T(\hat{e})$, $T(\hat{e}_1)$ are the corresponding unit rays in H_1.
(iii.) The mapping T between unit rays is one-to-one and onto.

symmetry relative to axis Unchanged by rotation about that axis. A planar set S is symmetric with respect to the x-axis if $(x, y) \in S \Leftrightarrow (x, -y) \in S$. S is symmetric relative to the y-axis if $(x, y) \in S \Leftrightarrow (-x, y) \in S$.

symmetry relative to line For a planar set, being equal to its conjugate (i.e., the set of conjugates of its points) about that line.

symmetry relative to plane For a surface S relative to a plane $P \subset \mathbf{R}^3$, every line perpendicular to P and intersecting S at a point q not on P, also intersects S at a point q' such that q and q' are on opposite sides of P and equidistant from P.

symplectic manifold A $2n$-dimensional manifold M with a closed 2-form, which is non-degenerate at each point of M. Also *Hamiltonian manifold.*

symplectic structure A closed, non-degenerate 2-form on an even dimensional manifold. *See* symplectic manifold.

system of differential equations of the first order A simultaneous set of equations of the form

$$y_1' = f_1(t, y_1, \ldots, y_d)$$
$$\vdots$$
$$y_d' = f_d(t, y_1, \ldots, y_d),$$

where $y_1, \ldots, y_d$ are unknown, differentiable functions and $' = \frac{d}{dt}$.

system of differential operators An operator which can be represented by a matrix, with entries which are differential operators, acting between linear spaces which are direct sums of spaces of functions. For example, the operator

$$T = \begin{bmatrix} D & I \\ I & D \end{bmatrix}$$

acting on (an appropriate domain of pairs of differentiable functions in) $L^2 \oplus L^2$, where $D = \frac{d}{dt}$. The action of T is given by $T(f_1, f_2) = (f_1' + f_2, f_1 + f_2')$, so that the operator equation $T(f_1, f_2) = (g_1, g_2)$ corresponds to the system

$$\frac{df_1}{dt} + f_2 = g_1$$
$$\frac{df_2}{dt} + f_1 = g_2.$$

system of linear differential equations of the first order A simultaneous set of equations

$$y_1' = \sum_{j=1}^{d} a_{1j}(t)y_j + f_1(t)$$
$$\vdots$$
$$y_d' = \sum_{j=1}^{d} a_{dj}(t)y_j + f_d(t)$$

where $y_1, \ldots, y_d$ are unknown, differentiable functions and $' = \frac{d}{dt}$. The system can be rewritten $\mathbf{y}' = A(t)\mathbf{y} + \mathbf{f}(t)$, where $\mathbf{y}$ and $\mathbf{f}$ are $d \times 1$ column vectors and A is a $d \times d$ matrix function.

system of Souslin [Suslin] Given sets A, X, any collection of subsets of X, indexed by all finite sequences of elements of A. The standard notation for the set of all finite sequences of elements of A is $A^{<\mathbf{N}}$, hence a (Souslin) system of sets has the form $\{A_s : s \in A^{<\mathbf{N}}\}$.

Szegö's kernel function Let Ω be a domain in $\mathbf{C}^n$ with $\mathbf{C}^1$ boundary and let $H^2(\partial\Omega)$ be the closure in $L^2(d\sigma)$ of the continuous functions on $\bar{\Omega}$ that are analytic in Ω, where $d\sigma$ is arc length measure on $\partial\Omega$. The Szegö kernel for Ω is the function $S(z, \zeta)$, analytic in z and coanalytic in ζ, such that, for $z \in \Omega$ and $f \in H^2(\partial\Omega)$, we have

$$f(z) = \int_{\partial\Omega} S(z, \zeta) f(\zeta) d\sigma(\zeta).$$

T

tangent bundle A *(principal) fiber bundle over a manifold M* consists of a manifold P on which a group G acts, such that:
(1.) G acts freely on P, on the right;
(2.) $M = P/G$ (the quotient of P by the equivalence relation induced by G) and the quotient map (projection) $\pi : P \to M$ is differentiable;
(3.) P is locally trivial, that is, every point $x \in M$ has a neighborhood U such that there is a diffeomorphism $\psi : \pi^{-1}(U) \to U \times G$ such that $\psi(u) = (\pi(u), \phi(u))$, where ϕ is a mapping of $\pi^{-1}(U)$ into G satisfying $\phi(ua) = (\phi(u))a$, for $u \in \pi^{-1}(U)$ and $a \in G$.

Now suppose F is a manifold on which G acts on the left; we construct the *fiber bundle $E = E(M, F, G, P)$ over M with standard fiber F and group G, associated with P.* On $P \times F$, let G act on the right, by $(u, \xi) \to (ua, a^{-1}\xi)$, for $a \in G$, and let $E = P \times_G F$ be the quotient space of $P \times F$ by this action. The projection $\pi_E : E \to M$ is induced from the mapping $(u, \xi) \to \pi(u)$ from $P \times F$ to M.

The *bundle of linear frames* over a manifold M is the case where $P = L(M)$ is the set of all linear frames at all points of M (a linear frame at $x \in M$ is an ordered basis of the tangent space $T_x(M)$), $\pi : L(M) \to M$ assigns a linear frame at x to x, and $G = \mathrm{GL}(n; \mathbf{R})$. The action of $\mathrm{GL}(n; \mathbf{R})$ on $L(M)$ is given by

$$(X_1, \dots, X_n)(a_{ij}) = (\sum_j a_{j1}X_j, \dots, \sum_j a_{jn}X_j).$$

The *tangent bundle* is the bundle

$$E(M, \mathbf{R}^n, \mathrm{GL}(n; \mathbf{R}), L(M))$$

associated with $L(M)$ with standard fiber $\mathbf{R}^n$.

tangent circles Two circles that agree at a point and share the same tangent at that point.

tangent curves Two $\mathbf{C}^1$ curves that meet and have the same tangent vector, at a point.

tangent line The tangent line to the graph of a function $y = f(x)$ at the point $(x_0, y_0) = (x_0, f(x_0))$ is the straight line with equation $y = mx + b$, where $m = f'(x_0)$ is the slope of the curve at the point and $b = y_0 - mx_0$.

tangent line to a conic section *See* tangent line.

tangent plane to a surface Given a curve $u = u(t)$, $v = v(t)$ on the surface, the vector

$$\frac{d\mathbf{x}}{dt} = \dot{\mathbf{x}} = \mathbf{x}_u\dot{u} + \mathbf{x}_v\dot{v}$$

is tangent to the surface. The plane generated by all such vectors is called the *tangent plane to the surface at P.*

tangent r-frame An ordered basis of an r-dimensional subspace of the tangent space $T_x(M)$, at a point x in a manifold M.

tangent r-frame bundle *See* tangent bundle. The principal bundle over a manifold M where P is the set of all tangent r-frames at all points of M (a tangent r-frame at $x \in M$ is an ordered basis of an r-dimensional subspace of $T_x(M)$). The projection $\pi : P \to M$ assigns a tangent r-frame at x to x, and the action of $\mathrm{GL}(r; \mathbf{R})$ on P is given by

$$(X_1, \dots, X_r)(a_{ij}) = (\sum_j a_{j1}X_j, \dots, \sum_j a_{jr}X_j).$$

tangent space Let Ω be a domain in $\mathbf{R}^n$ with a $\mathbf{C}^1$ boundary and let $P \in \partial\Omega$. The tangent space at P is the set of vectors $\mathbf{w} = (w_1, w_2, \dots, w_n)$ such that

$$\sum_{j=1}^{n} \frac{\partial\rho}{\partial x_j}(P) \cdot w_j = 0,$$

0-8493-0320-6/00/$0.00+$.50

where ρ is a defining function for Ω. The vectors $\mathbf{w}$ in the tangent space are called *tangent vectors.*

tangent surface The *tangent surface* of a curve in space is the envelope of the osculating planes of the curve.

tangent vector *See* tangent space.

tangent vector bundle *See* tangent bundle.

tangent vector space *See* tangent space.

target of jet *See* jet.

Tauberian Theorems Any of a group of theorems whose hypothesis includes the summability of an infinite series $\sum a_n$, with respect to some summability method, in addition to some other hypotheses, and which concludes that $\sum a_n < \infty$. For example, Tauber's Theorem states that if $\sum_{n=0}^{\infty} a_n$ is Abel summable to L and if $na_n \to 0$, then $\sum_{n=0}^{\infty} a_n$ converges to L.

tautochrone A piece of wire is bent into the shape of the graph of a monotonic function $y = f(x)$. If $T(y)$ is the time of descent, as a function of initial height y, Abel's mechanical problem is to find $f(x)$, from $T(y)$. When $T(y)$ is given as a constant, $T(y) = t_0$, we get the tautochrone problem. The curve obtained is $f(y) = (b/y)^{1/2}$ yielding $x = a(\theta + \sin\theta)$, $y = a(1 - \cos\theta)$, where $a = 2gt_0^2/\pi^2$. The word *tautochrone* is from the Greek *tauto* (same) + *chronos* (time), since the time of descent is independent of the starting point.

Taylor expansion The formal expression for a function $f(x)$ having derivatives of all orders at the point $a \in \mathbf{R}$:

$$\sum_{n=0}^{\infty} \frac{1}{n!} f^{(n)}(a)(x-a)^n.$$

Taylor series *See* Taylor expansion.

Taylor's formula *See* Taylor expansion.

Taylor's Theorem If $f(x)$ is a real-valued function on $[a, b]$ and $f, f', \ldots, f^{(n-1)}$ exist and are continuous on $[a, b]$ with $f^{(n-1)}$ differentiable on (a, b), then there is a number $c \in (a, b)$ such that

$$f(b) = \sum_{j=0}^{n-1} \frac{1}{j!} f^{(j)}(a)(x-a)^j + \frac{1}{n!} f^{(n)}(c)(b-a)^n.$$

Teichmuller space For a Riemann surface X, a *marked Riemann surface modeled on* X is a triple (X, f, X_1), where X_1 is a Riemann surface and $f : X \to X_1$ is a quasiconformal homeomorphism. Two marked surfaces, (X, f, X_1) and (X, g, X_2), are called *Teichmuller equivalent* if there is a biholomorphism $\sigma : X_1 \to X_2$ such that $g^{-1} \circ \sigma \circ f : X \cup \partial X \to X \cup \partial X$ is homotopic to the identity, by a homotopy that keeps each point of ∂X fixed. The *Teichmuller space* of X is the set of equivalence classes of marked Riemann surfaces modeled on X, under this equivalence relation.

temperate distribution *See* tempered distribution.

tempered distribution For $\Omega \subset \mathbf{R}^n$, a linear form T on $\mathbf{C}_0^{\infty}(\Omega)$, such that there exists a constant C such that, for any integers m, M,

$$|T(\phi)| \leq C \sup_{x \in \Omega}[(1 + |x|)^M \sum_{|\alpha| < m} |D^{\alpha}\phi(x)|],$$

for all $\phi \in \mathbf{C}_0^{\infty}(\Omega)$. Also *temperate distribution. See* generalized function.

tensor *See* tensor product.

tensor algebra Let $\mathbf{V}$ be a vector space over $\mathbf{R}$. For $p = 0, 1, \ldots$, let $\otimes_0 \mathbf{V} = \mathbf{R}, \otimes_1 \mathbf{V} = \mathbf{V}, \otimes_2 \mathbf{V} = \mathbf{V} \otimes \mathbf{V}, \ldots$ (*see* tensor product). Then $\otimes_* \mathbf{V} = \otimes_{p=0}^{\infty} \otimes_p \mathbf{V}$ becomes an algebra under the obvious addition and multiplication defined to send a product of an element of $\otimes_p \mathbf{V}$ by an element of $\otimes_q \mathbf{V}$ to an element of $\otimes_{p+q}\mathbf{V}$. Then $\otimes_*$ is called the tensor algebra of $\mathbf{V}$.

tensor field Let M be a differentiable manifold with $T_s^r(p)$ the space of tensors of type (r, s) over the tangent space at $p \in M$. A *tensor field of type* (r, s) on $N \subset M$ assigns a tensor in $T_s^r(p)$, to each $p \in N$.

tensor field of class $\mathbf{C}^l$ Suppose a tensor field of type (r, s) over a subset N of a manifold M assigns to a point $p \in N$ the tensor K_p. Choose a local coordinate system $\{x_1, \ldots, x_n\}$ in a coordinate neighborhood U, and let $X_i = \frac{\partial}{\partial x_i}, i = 1, \ldots, n$, be a basis for the tangent space T_x, for $x \in U$, and $\omega_i = dx_i, i = 1, \ldots, n$, the dual basis in T_x^*. Then we can write K_p as

$$K_p = \sum K_{j_1 \cdots j_s}^{k_1 \cdots k_r} X_{j_1} \otimes \cdots \otimes X_{j_s} \otimes \omega_{k_1} \otimes \cdots \otimes \omega_{k_r}.$$

The tensor field is *of class* $\mathbf{C}^l$ if the coefficient functions $K_{j_1 \cdots j_s}^{k_1 \cdots k_r}$ are of class $\mathbf{C}^l$ in U.

tensor product Let $\mathbf{V}_1, \ldots, \mathbf{V}_n$ be vector spaces over $\mathbf{R}$. The tensor product $\mathbf{V}_1 \otimes \ldots \otimes \mathbf{V}_n$ is the vector space generated by all n-tuples $(\mathbf{v}_1, \ldots, \mathbf{v}_n) \in \mathbf{V}_1 \times \ldots \times \mathbf{V}_n$, with the added requirements that
$(\mathbf{v}_1, \ldots, \mathbf{x}_j, \ldots, \mathbf{v}_n) + (\mathbf{v}_1, \ldots, \mathbf{y}_j, \ldots, \mathbf{v}_n)$
$= (\mathbf{v}_1, \ldots, \mathbf{x}_j + \mathbf{y}_j, \ldots, \mathbf{v}_n)$; and
$c(\mathbf{v}_1, \ldots, \mathbf{v}_j, \ldots, \mathbf{v}_n)$
$= (\mathbf{v}_1, \ldots, c\mathbf{v}_j, \ldots, \mathbf{v}_n)$.
The elements of $\mathbf{V}_1 \otimes \ldots \otimes \mathbf{V}_n$ are called *tensors*.

tensor space Let V be a vector space and r a positive integer. The vector space $\otimes_r V = V \otimes \cdots \otimes V$ (*see* tensor product) is called the *contravariant tensor space of degree r* and $\otimes^r V = V^* \otimes \cdots \otimes V^*$ is called the *covariant tensor space of degree r*.

tensorial form Let M be a differentiable manifold, $P(M, G)$ a principal fiber bundle, and ρ a representation of G on V, a finite dimensional vector space. Thus $\rho(a) : V \to V$ is a linear mapping, for $a \in G$, satisfying $\rho(ab) = \rho(a)\rho(b)$. A *pseudotensorial form of degree r on P of type* (ρ, V) is a V-valued r-form ϕ on P such that

$$R_a^* \phi = \rho(a^{-1}) \cdot \phi,$$

for $a \in G$, where R_a is right translation by $a \in G$. The form ϕ is also called a *tensorial form* if $\phi(X_1, \ldots, X_r) = 0$, whenever at least one of the tangent vectors $X_1, \ldots, X_r$ of P is vertical, i.e., tangent to a fiber.

term A single mathematical symbol or expression; usually one of several that are being added, multiplied, or otherwise acted upon in series.

terminal point of integration The upper limit of a real integral over an interval (*see* upper limit), or the final end point of a path over which a line integral is being taken (*see* line integral).

termwise integrable series A series of integrable functions $\sum_{n=1}^{\infty} f_n(x)$ for which it is true that the sum of the series is also integrable and

$$\int \sum f_n(x)dx = \sum \int f_n(x)dx.$$

A number of different hypotheses are sufficient for the interchange of summation and integration, depending upon the nature of the functions and the integral involved. For continuous functions and the Riemann integral over a compact real interval, uniform convergence of the series of functions implies that it is termwise integrable.

tetragamma function The third derivative of $\log\Gamma(x + 1)$, second derivative of the digamma function and derivative of the trigamma function. Sometimes $\log \Gamma(x)$ is used.

Theorem of Identity (for power series) If two power series $\sum_{n=0}^{\infty} a_n(z - z_0)^n$ and

$\sum_{n=0}^{\infty} b_n(z-z_0)^n$ have a positive radius of convergence and have the same sum in some neighborhood of z_0 then $a_n = b_n$ $(n \geq 0)$.

Theorem on Termwise Differentiation Suppose that $f_1(x), f_2(x), \ldots$ are real-valued, differentiable functions on an interval $[a, b]$, that the series of derivatives $\sum_{n=1}^{\infty} f_n'(x)$ converges uniformly on $[a, b]$, and that $\sum f_n(c)$ converges at some point $c \in [a, b]$. Then we have termwise differentiation of the series on $[a, b]$:

$$\frac{d}{dx}\sum_{n=1}^{\infty} f_n(x) = \sum_{n=1}^{\infty} f_n'(x),$$

for $x \in [a, b]$.

third derivative The derivative of the second derivative, when it exists:

$$\frac{d^3 f(x)}{dx^3} = \frac{d}{dx}\frac{d^2 f(x)}{dx^2}.$$

Also denoted $f'''(x)$, $D_{xxx}f(x)$ and $D^3 f(x)$.

third quadrant of Cartesian plane The subset of $\mathbf{R}^2$ of points (x, y), where $x < 0$ and $y < 0$.

three-space The set $\mathbf{R}^3 = \mathbf{R} \times \mathbf{R} \times \mathbf{R}$ of triples (x, y, z) of real numbers, realized as three-dimensional space, with three mutually perpendicular axes, each representing one of the copies of the real numbers $\mathbf{R}$.

Tissot-Pochhammer differential equation The equation

$$\sum_{j=0}^{n} \phi_j(z)\frac{d^j w}{dz^j} = 0,$$

where

$$\phi_j(z) = \frac{(-1)^{n-j-1}}{(h+n-2)\ldots(h+1)h}$$
$$\times \left(\binom{h+n+2}{n+1} P_1^{n-j-1}(z) + \binom{h+n+2}{n-j} P_0^{n-j}(z) \right)$$

and

$$P_0(z) = \prod_1^n (z - a_k),$$

$$P_1(z) = P_0(z)\sum_1^n \frac{\beta_k}{z - a_k}.$$

Toeplitz matrix (1.) An $(n+1) \times (n+1)$ matrix of the form

$$M = \begin{bmatrix} a_0 & a_1 & \ldots & a_n \\ a_{-1} & a_0 & \ldots & a_{n-1} \\ \cdot & \cdot & \cdot & \\ a_{-n+1} & a_{-n+2} & \ldots & a_1 \\ a_{-n} & a_{-n+1} & \ldots & a_0 \end{bmatrix}.$$

That is, M is constant along the main diagonal and all parallel diagonals.
(2.) An infinite matrix of the form (a_{j-k}), $j, k = \ldots, -1, 0, 1, \ldots$, $j - k \geq 0$. *See* Toeplitz operator.
(3.) An infinite matrix of the form (a_{j-k}), $j, k = \ldots, -1, 0, 1, \ldots$ Although this bi-infinite operator is now usually called a *Laurent* matrix, this was what Hilbert originally called a *Toeplitz matrix*.

Toeplitz operator The operator $T_\phi : H^2 \to H^2$, defined as follows. ϕ is an L^∞ function on $[-\pi, \pi]$ and $P : L^2 \to H^2$ is the orthogonal projection. Then

$$T_\phi f = P(\phi f),$$

for $f \in H^2$. The matrix of T_ϕ with respect to the orthonormal basis $\{e^{int}, n = 0, 1, 2, \ldots\}$ of H^2 is the infinite Toeplitz matrix formed of the Fourier coefficients of ϕ, which is called the *symbol*. *See* Toeplitz matrix. Toeplitz operators can also be similarly defined with H^2 replaced by other H^p spaces, by several-variables H^2, by H^2 of other domains or by the Bergman space.

topological manifold A connected Hausdorff space X such that every point $x \in X$ has a neighborhood that is homeomorphic to an open set in $\mathbf{R}^n$ [resp. $\mathbf{C}^n$]. Such a manifold is said to have real [resp. complex] dimension n. *See also* atlas.

topological manifold with boundary A connected Hausdorff space X such that every point $p \in X$ has a neighborhood U_p and a homeomorphism ϕ_p from U_p into either $\mathbf{R}^n$ or $\mathbf{R}^n_+ = \{(x_1, \ldots, x_n) : x_j \geq 0, \text{for} j = 1, \ldots, n\}$. *See also* atlas.

topological space A set X, together with a collection τ of subsets of X, such that
(1.) $S_1, S_2 \in \tau \Rightarrow S_1 \cap S_2 \in \tau$, and
(2.) $S_\alpha \in \tau$, for $\alpha \in A \Rightarrow \cup_{\alpha \in A} S_\alpha \in \tau$.
The sets in τ are called *open* sets and τ is called a *topology on* X.

If X is a metric space, the usual topology is defined by $S \in \tau \Leftrightarrow$ for every $x \in S$ there is an $\epsilon > 0$ such that $\{t : d(t, x) < \epsilon\} \subset S$.

topological vector space A vector space X over a field $F = \mathbf{R}$ or $\mathbf{C}$, with a topology on X such that the maps
(i.) $(\alpha, x) \to \alpha \cdot x$ from $F \times X$ to X and
(ii.) $(x, y) \to x + y$ from $X \times X$ to X
are continuous. Abbreviated *t.v.s.*, in many treatments.

Torelli's Theorem Let M and M' be compact Riemann surfaces of genus g with period matrices (Π_0, Π_1) and (Π'_0, Π'_1), with respect to arbitrary homology bases, having intersection matrices J, J', respectively, and arbitrary bases for their holomorphic 1-forms. Then M is biholomorphically equivalent to M' if and only if

$$(\Pi_0, \Pi_1) = A(\Pi'_0, \Pi'_1)\Sigma, \; J = \Sigma^t J' \Sigma,$$

for some $A \in GL(g, \mathbf{C})$, $\Sigma \in GL(2g, \mathbf{Z})$.

torsion The limit of the ratio of the angle between the osculating planes at two neighboring points of a curve and the length of arc between these two points. If $\mathbf{t}$ is the tangent vector, $\mathbf{n}$ the normal, and $\mathbf{b} = \mathbf{t} \times \mathbf{n}$ the binormal, then $d\mathbf{b}/ds = -\tau\mathbf{n}$, where τ is the torsion. *See also* torsion tensor.

torsion form Let P denote the bundle of linear frames of a differentiable manifold M and θ the canonical form of P. A connection in P is called a *linear connection* and the *torsion form* Θ of a linear connection is the exterior covariant differential of θ:

$$\Theta = D\theta.$$

torsion tensor For a differentiable manifold M with bundle of linear frames $L(M)$ and tangent space $T_p(M)$ at p, the *torsion tensor field* or *torsion* T is defined by

$$T(X, Y) = u(2\Theta(X^*, Y^*))$$

for $X, Y \in T_p(M)$, where $u, X^*, Y^* \in L(M)$, with $\pi(u) = p, \pi(X^*) = X$ and $\pi(Y^*) = Y$.

torus (1.) The surface in $\mathbf{R}^3$ obtained by rotating a circle about a line, lying in the same plane of the circle, and not intersecting the circle. An example is parameterized by $x = [R - \cos v] \cos u$, $y = [R - \cos v] \sin u$, $z = \sin v$, $-\pi \leq u, v \leq \pi$ with $R > 1$.
(2.) The set $\{(e^{is}, e^{it})\}$ in $\mathbf{C}^2$, also called the *distinguished boundary of the bidisk* $\mathbf{T}^2$.

total curvature (1.) For a curve, the quantity $\sqrt{\kappa^2 + \tau^2}$, where κ is the curvature and τ is the torsion. Also *third curvature.*
(2.) For a surface, the product of the principal curvatures, $K = \kappa_1 \kappa_2$. Also *Gaussian curvature.*
(3.) For a surface, the integral of the Gaussian curvature $\int \int_A K dA$ is also referred to as the total curvature. Also *integral curvature.*

total differential A differential form

$$\omega = dp = \frac{\partial p}{\partial x_1} dx_1 + \ldots + \frac{\partial p}{\partial x_n} dx_n.$$

total subset A subset of a normed vector space X whose span is dense in X.

total variation (1.) For a real-valued function $f(x)$, on an interval $[a, b]$, the quantity

$$V = \sup_P \sum_{j=1}^{n} |f(x_j) - f(x_{j-1})|$$

where the supremum is over all partitions $P = \{x_j\}, a = x_0 < x_1 < \cdots < x_n = b$ of $[a, b]$. When the variation of f over $[a, b]$ is finite, we say f is *of bounded variation.* Any monotonic function on $[a, b]$ has $V = |f(b) - f(a)|$ and so is of bounded variation. *See also* positive variation.
(2.) For a complex measure μ, the set function $|\mu|$ defined on measurable sets E by

$$|\mu|(E) = \sup_{\{E_j\}} \sum_{j=1}^{n} |\mu(E_j)|,$$

where the supremum is over all partitions of the set E. *See also* positive variation.

totally bounded set A subset M of a metric space X such that, for every $\epsilon > 0$, there is a finite set of points $\{m_1, m_2, \ldots, m_n\} \subset M$ with the property that every point $m \in M$ is at a distance $< \epsilon$ from some m_j.

trace class The class of operators T on a separable Hilbert space H satisfying

$$\sum_{j=1}^{\infty} |(Tx_j, y_j)| < \infty,$$

for some (and hence all) complete orthonormal sets $\{x_j\}$ and $\{y_j\}$ in H.

trace of operator For an operator T on a separable Hilbert space H, the sum

$$\sum_{j=1}^{\infty} (Tx_j, x_j) < \infty,$$

where $\{x_j\}$ is a complete orthonormal set in H. The sum is finite for every operator of trace class and is independent of the orthonormal set. *See* trace class.

trace of surface The set of points that lie on a given surface.

trace operator (1.) $\mathbf{R}^2$. Let a patch be given by the map $\mathbf{x}: U \to \mathbf{R}^n$, where U is an open subset of $\mathbf{R}^2$, or more generally by $\mathbf{x}: A \to \mathbf{R}^n$, where A is any subset of $\mathbf{R}^2$. Then $\mathbf{x}(U)$ (or more generally, $\mathbf{x}(A)$) is called the *trace operator* of $\mathbf{x}$.
(2.) $(L(E; E))$. A symmetric p-linear operator in the space $L(E; E)$ of bounded linear operators on a Banach space E denoted by T_{r_p} and given by

$$T_{r_p}(x_1, x_2, \ldots, x_p) = \frac{1}{p!} \sum_{\sigma} \mathrm{tr}(x_{\sigma(1)} \circ \cdots \circ x_{\sigma(p)}), \quad p \geq 1.$$

(3.) Hilbert space. Let T denote a trace class operator defined on a Hilbert space E, and $\{e_n\}$ $n \geq 1$ denote an orthonormal basis on E. The trace operator $\mathrm{Tr}(T)$ is defined by

$$\mathrm{Tr}(T) = \sum_n (T \cdot e_n, e_n).$$

Note that the above sum is independent of the chosen orthonormal basis $\{e_n\}$ $n \geq 1$.

tractrix The curve with parametric equations

$$x = a \log(\sec t + \tan t) - a \sin t,$$
$$y = a \cos t,$$

where $-\frac{\pi}{2} < t < \frac{\pi}{2}$, or polar equation

$$\rho = a \tan \phi.$$

transcendental curve The graph of a transcendental function.

transcendental function Any function that is not a polynomial, an algebraic function, or in the field they generate.

transcendental singularity Any singularity other than a pole or an algebraic singularity.

transformation (1.) A function; usually a linear function from one vector space to

another. Also *operator, linear map.*
(2.) A *linear-fractional* transformation is a function $f : \mathbf{C} \to \mathbf{C} \cup \{\infty\}$, of the form $f(z) = (az + b)/(cz + d)$.
(3.) An *affine* transformation is a function of the form $f(x) = ax + b$.

transformation of coordinates A mapping that transforms the coordinates of a point in one coordinate system on a space to the coordinates of the point in another coordinate system.

transformation of local coordinates *See* local coordinate system.

transition point *See* turning point.

transitive relation A relation $\sim$ on a set S satisfying

$$x \sim y, y \sim z \Rightarrow x \sim z,$$

for $x, y, z \in S$.

translate *See* translation.

translation In an additive group G, the map $t \to t - g$ is *translation.* So, for example, $f_g(t) = f(t - g)$ is the *translation* or *translate* of a function $f(t)$ on G.

translation number Let f be a complex valued function defined on the real numbers. Then for any $\epsilon > 0$, a number t is called a *translation number* of f corresponding to ϵ if $|f(x+t) - f(x)| < \epsilon$ for all real numbers x.

translation-invariant (1.) A function $f : G \to G$, on a group G which is unchanged by the change of variable $x \to x - \lambda$:

$$f(x - \lambda) = f(x).$$

(2.) A subset of a group G which is unchanged after the map $x \to x - \lambda, \lambda \in G$, is applied.

transposed operator *See* dual operator.

transversal *See* Transversality Theorem.

Transversality Theorem (Thom's Transversality Lemma) Let M and N be smooth manifolds of dimension m and n, respectively, and let S be a p-dimensional submanifold of N. Then the set of all $\mathbf{C}^\infty$-maps of M to N which are transversal to S is a dense open subset of $\mathbf{C}^\infty(M, N)$.

By definition, a smooth map $f : M \to N$ is *transversal* to S if, for each $x \in M$ with $f(x) \in S$, we have

$$df(TM)_x + (TS)_{f(x)} = (TN)_{f(x)}.$$

transverse axis Of the axes of symmetry of a hyperbola, one intersects the hyperbola. The *transverse axis* is the segment of this axis between the two points of intersection with the hyperbola.

trapezoid rule An approximation to the Riemann integral of a nonnegative function $f(x)$ over a real interval $[a, b]$. Starting with a partition $a = x_0 < x_1 < \ldots < x_n = b$ one takes (in the interval $[x_{j-1}, x_j]$) not the area of the rectangle $f(c_j)[x_j - x_{j-1}]$, where $x_{j-1} \le c_j \le x_j$, but the area of the trapezoid

$$\frac{f(x_{j-1} + f(x_j)}{2}[x_j - x_{j-1}].$$

triangle inequality For a distance function $d(x, y)$:

$$d(x, y) \le d(x, z) + d(z, y).$$

So called because of the case where the three distances represent the lengths of the sides of a triangle and the inequality states that the sum of the lengths of any two sides is not less than the length of the third.

trigamma function The second derivative of $\log\Gamma(x + 1)$. The derivative of the digamma function. Sometimes $\log\Gamma(x)$ is used.

trigonometric function Any of the six functions:

$$\text{sine} : \sin t = \frac{e^{it}-e^{-it}}{2i},$$
$$\text{cosine} : \cos t = \frac{e^{it}+e^{-it}}{2},$$
$$\text{tangent} : \tan t = \frac{\sin t}{\cos t},$$
$$\text{cotangent} : \cot t = \frac{1}{\tan t},$$
$$\text{secant} : \sec t = \frac{1}{\cos t},$$
$$\text{cosecant} : \csc t = \frac{1}{\sin t}.$$

trigonometric polynomial A function of the form

$$p(t) = \sum_{j=-n}^{n} a_j e^{ijt},$$

where the a_j are complex numbers and $t \in [-\pi, \pi]$.

trigonometric series A formal series of the form

$$\sum_{n=-\infty}^{\infty} a_n e^{int},$$

where the a_n are constants and $t \in [-\pi, \pi]$.

trigonometric system The set of functions $\{e^{int} : n = \ldots, -1, 0, 1, \ldots\}$, for $t \in [-\pi, \pi]$. The set forms an orthonormal basis in $L^2[-\pi.\pi]$.

trihedral The figure formed by the union of three lines that intersect at a common point and which are not in the same plane. In this definition, the three lines can be replaced by three non-coplanar rays with a common initial point. In a polyhedron, the pairwise intersections of three faces at a vertex form a trihedral. The three faces are said to form a trihedral angle of the polyhedron.

trihedral angle The opening of three planes that intersect at a point.

trilinear coordinates Given a triangle ΔABC, the trilinear coordinates of a point P which, with respect to ΔABC, form an ordered triple of numbers, each of which is proportional to the directed distance from P to one of the sides. *Trilinear coordinates* are denoted $\alpha : \beta : \gamma$ or (α, β, γ) and are also known as barycentric coordinates. *See* barycentric coordinates.

triple integral An integral over a Cartesian product $A \times B \times C$ in $\mathbf{R}^3$, parameterized so as to be evaluated in three integrations. Hence any integral of the form

$$\int_a^b \int_{c(z)}^{d(z)} \int_{e(y,z)}^{f(y,z)} g(x, y, z)dxdydz.$$

trochoid *See* roulette.

tubular neighborhood A subset U of an n-dimensional manifold N is a *tubular neighborhood* of an m-dimensional submanifold M of N provided U has the structure of an $(n - m)$-dimensional vector bundle over M with M as the zero section.

turning point A zero of the coefficient $f(x)$ in the equation

$$\frac{d^2w}{dx^2} = f(x)w.$$

At such a point (if the variables are real and the zero of odd order) the nature of the solution changes from exponential type to oscillatory. Also *transition* point.

twisted curve A space curve that does not lie in a plane.

two-point equation of line The equation

$$y - y_0 = [\frac{y_1 - y_0}{x_1 - x_0}](x - x_0),$$

which represents the straight line passing through (x_0, y_0) and (x_1, y_1).

U

UHF algebra A unital $\mathbf{C}^*$ algebra $\mathcal{A}$ which has an increasing sequence $\{A_n\}$ of finite-dimensional $\mathbf{C}^*$ subalgebras, each containing the unit of $\mathcal{A}$ such that the union of the $\{A_n\}$ is dense in $\mathcal{A}$.

ultrafilter A filter that is maximal; such that there exists no finer filter.

umbilical point A point on a surface where the coefficients of the first and second fundamental forms are proportional. Also *umbilic, navel point.*

uncountable set An infinite set that *cannot* be put into one-to-one correspondence with the integers.

underlying topological space *See* complex space.

unicellular operator An operator, from a Banach space to itself, whose invariant subspace lattice is totally ordered by inclusion.

unicursal curve A curve of genus 0.

uniform asymptotic stability Consider the system of autonomous differential equations of the form

$$\dot{x} = F(x; M). \qquad (1)$$

Here, F is a vector field which does not explicitly depend on the independent variable t and is represented by a map $F: R^n \times R^m \to R^n$; $M \in R^m$ is a vector of control parameters and $x = x(t)$ is finite dimensional, $x \in R^n, t \in R$. A solution $u(t)$ of (1) is Lyapunov stable if, given a small number $\varepsilon > 0$, there exists a number $\delta = \delta(\varepsilon) > 0$ such that any other solution $v(t)$ for which $\|u - v\| < \delta$ at time $t = t_0$ satisfies $\|u - v\| < \varepsilon$ for all $t > t_0$. Lyapunov stability is also called uniform stability because δ is independent of the initial time t_0. A solution $\{u_k\}$ of (1) is said to be uniform asymptotic stable if it is uniform stable and

$$\lim_{k\to\infty} \|u_k - v_k\| \to 0.$$

uniform boundedness principle *See* Banach-Steinhaus Theorem.

uniform continuity The property of being uniformly continuous, for a complex-valued function.

uniform convergence A sequence of bounded, complex-valued functions $\{f_n\}$ on a set D converges uniformly to f provided that, for every $\epsilon > 0$, there is an integer n_0 such that

$$n \geq n_0 \Rightarrow |f_n(x) - f(x)| < \epsilon,$$

for $x \in D$.

uniform function A single-valued function. Technically, a function must be single-valued, but the term may be used for an analytic function of a complex variable, where some notion of multi-valued function is sometimes unavoidable.

uniform norm The norm on a set of bounded, complex-valued functions on a set D, given by

$$\|f\| = \sup_{x\in D} |f(x)|.$$

uniform operator topology The norm topology on the algebra of all bounded operators on a Banach space. Thus a sequence of operators $\{T_n\}$ converges uniformly to T, provided $\|T_n - T\| \to 0$, as $n \to \infty$.

uniformization For a multiple-valued function g on a Riemann surface S, the process of replacing $g(z)$ by $g(F(w))$, where F is a conformal map to S of one of the

0-8493-0320-6/00/$0.00+$.50

three canonical regions (sphere, sphere minus point, or sphere minus ray) to which the universal covering surface of S is conformal. The resulting function $g(F(w))$ is no longer multiple-valued.

uniformly continuous function A function $f : D \to \mathbf{C}$, satisfying: for every $\epsilon > 0$ there is a $\delta > 0$ such that

$$x, y \in D, |x-y| < \delta \Rightarrow |f(x)-f(y)| < \epsilon.$$

unilateral shift *See* shift operator.

unique continuation property The property of a function, analytic at a point z_0, that any two analytic continuations to a point z_1, possibly along different paths, lead to functions that agree in a neighborhood of z_1.

uniqueness The existence of at most one. For example, an equation with *uniqueness of solutions* possesses at most one solution.

uniqueness theorem Any theorem asserting uniqueness.

One famous uniqueness theorem is the Picard-Lindelöf Theorem of ordinary differential equations. *See* Picard-Lindelöf Theorem.

Another theorem, sometimes referred to as *The Uniqueness Theorem* comes from classical Fourier series on the unit circle $\mathbf{T}$: If $f \in L^1(\mathbf{T})$, $\mathbf{T}$ the unit circle, and $\hat{f}(j) = 0$, for all j, then $f = 0$.

Uniqueness Theorem of the Analytic Continuation Any two analytic continuations of the same analytic function element along the same curve result in equivalent analytic function elements. That is, if the curve γ is covered by two chains of disks, $\{A_1, A_2, \dots, A_n\}$, $\{B_1, B_2, \dots, B_m\}$, where $A_1 = B_1$, and $f(z)$ is a power series convergent in A_1, which can be analytically continued along the first chain to $g_n(z)$ and along the second chain to $h_m(z)$, then $g_n(z) = h_m(z)$ in $A_n \cap B_m$.

unital (1.) Having a unit. For example, a $\mathbf{C}^*$ algebra is *unital* if it contains the identity operator I, which satisfies $IT = TI = T$ for every operator T.
(2.) Sending unit to unit. For example, a unital representation of a *unital* algebra $\mathcal{A}$ on $\mathcal{L}(H)$, the bounded linear operators on a Hilbert space H, sends the unit of $\mathcal{A}$ to the identity operator on H.

unitarily equivalent operators Two operators T_1, T_2, on a Hilbert space H satisfying $T_1 = UT_2U^*$, where $U : H \to H$ is unitary. *See* unitary operator, similar linear operators.

unitary dilation Given a (contraction) operator T on a Hilbert space H, a *unitary dilation* U of T is a unitary operator on a Hilbert space H' containing H, such that

$$T^n x = \mathbf{P}U^n x, \quad n = 1, 2, \dots$$

for all $x \in H$, where $\mathbf{P} : H' \to H$ is the orthogonal projection.

unitary operator An isomorphism (or automorphism) of Hilbert spaces. That is, a bounded (onto) operator $U : H_1 \to H_2$, satisfying $(Ux, Uy) = (x, y)$, for $x, y \in H_1$.

universal set (1.) Given a class S of subsets of a set X, a set $U \subseteq \mathbf{N}^{\mathbf{N}} \times X$ is *universal for* S if, for every $A \in S$, there is $a \in \mathbf{N}^{\mathbf{N}}$ such that $A = U_a$.
(2.) When considering sets of objects, the *universal set* is the set of all objects that appear as elements of a set. Any set that is being considered is a subset of the universal set.

upper bound For a subset S of an ordered set X, an element $x \in X$ such that $s \leq x$, for all $s \in S$.

upper envelope Given two commuting, self-adjoint operators A, B on Hilbert space, their *upper envelope*, denoted $\sup\{A, B\}$, is the operator $\frac{1}{2}(A + B + |A - B|)$, as it is the smallest self-adjoint operator that majorizes A and B and commutes with both of them. Similarly, $\frac{1}{2}(A + B - |A + B|)$ is the lower envelope of A and B, denoted $\inf\{A, B\}$.

upper limit (1.) For a sequence $S = \{a_1, a_2, \ldots\}$, the limit superior of the sequence. It can be described as the supremum of all limit points of S, denoted $\overline{\lim}_{n\to\infty} a_n$. (2.) The right endpoint of a real interval over which an integrable function $f(x)$ is being integrated. In $\int_a^b f(x)dx$, b is the upper limit of integration.

upper semicontinuous function at point x_0 A real valued function $f(x)$ on a topological space X is *upper semicontinuous at* x_0 if $x_n \to x_0$ implies

$$\limsup_{n\to\infty} f(x_n) \le f(x_0).$$

upper semicontinuous function in E For a real valued function $f(x)$ on a topological space X, $\{x : f(x) < \alpha\}$ is open, for every real α. Equivalently, f is upper semicontinuous at every point of X. *See* upper semicontinuous function at point x_0.

V

value of variable A specific element of the domain of a function, when the independent variable is equal to that element.

vanish Become equal to 0, as applied to a variable or a function.

vanishing theorem A theorem asserting the vanishing of certain cohomology groups. For example, for a finite dimensional manifold X such that $H^q(X, \mathcal{F}) = 0$, for every $q \geq 1$ and for every coherent analytic sheaf $\mathcal{F}$ near X, we have $H^{\dim X + j}(X, \mathbf{C}) = 0$, for every $j \geq 1$.

variable A quantity, often denoted x, y, or t, taking values in a set, which is the domain of a function under consideration. When the function is $f : X \to Y$, we write $y = f(x)$ when the function assigns the value $y \in Y$ to $x \in X$. In this case, x is called the independent variable, or abscissa, and y the dependent variable, or ordinate.

variation *See* negative variation, positive variation, total variation.

variational derivative Of a function $F(x, y, y')$, the quantity

$$[F]_y = F_y - \frac{d}{dx} F_{y'}.$$

variational principle A class of theorem where the main hypothesis is the positive definiteness or an extremal property of a linear functional, often defined on the domain of an unbounded (e.g., differential) operator. Classically, such a hypothesis may appear as an integral condition.

For example, the largest eigenvalue of a Hermitian matrix H is the maximum value of

$$(Hu, u)/(u, u)$$

for u, a non-zero vector.

vector (1.) A line segment in $\mathbf{R}^2$ or $\mathbf{R}^3$, distinguished by its length and its direction, but not its initial point.
(2.) An element of a vector space. *See* vector space.

vector analysis A more advanced version of vector calculus. *See* vector calculus.

vector bundle Let $\mathbf{F}$ be one of the fields $\mathbf{R}$ or $\mathbf{C}$, M a manifold, G a group, and P a principal fiber bundle over M (*See* tangent bundle). We let $GL(m; \mathbf{F})$ act on $\mathbf{F}^m$ on the left by $a(\xi_1, \dots, \xi_m) = (\sum_j a_{j1}\xi_j, \dots, \sum_j a_{im}\xi_j)$ and let ρ be a representation on G on $GL(m; \mathbf{F})$. The associated bundle

$$E(M, \mathbf{F}^m, G, P)$$

with standard fiber $\mathbf{F}^m$ on which G acts through ρ is called a *real* or *complex vector bundle* over M, according as $\mathbf{F} = \mathbf{R}$ or $\mathbf{C}$.

vector calculus The study of vectors and vector fields, primarily in $\mathbf{R}^3$. Stokes' Theorem, Green's Theorem, and Gauss' Theorem are the principal topics of the subject.

vector field A vector-valued function V, defined in a region D (usually in $\mathbf{R}^3$). The vector $V(p)$, assigned to a point $p \in D$ is required to have its initial point at p. An example would be the function which assigns to each point in a region containing a fluid, the velocity vector of the fluid, at that point.

vector field of class C^r A vector field defined by a function V of class C^r. *See* vector field.

vector product of vectors For two vectors $\mathbf{x} = (x_1, x_2, x_3)$ and $\mathbf{y} = (y_1, y_2, y_3)$ in $\mathbf{R}^3$,

0-8493-0320-6/00/$0.00+$.50

the vector

$$\mathbf{x}\times\mathbf{y} = (x_2y_3 - x_3y_2, x_3y_1 - x_1y_3, x_1y_2 - x_2y_1),$$

which is orthogonal to both $\mathbf{x}$ and $\mathbf{y}$, has length $|\mathbf{x}||\mathbf{y}|\sin\theta$, where θ is the angle between $\mathbf{x}$ and $\mathbf{y}$, and is such that the triple $\mathbf{x}, \mathbf{y}, \mathbf{x}\times\mathbf{y}$ is right-handed. In matrix notation,

$$\mathbf{x}\times\mathbf{y} = \begin{vmatrix} \mathbf{i} & \mathbf{j} & \mathbf{k} \\ x_1 & x_2 & x_3 \\ y_1 & y_2 & y_3 \end{vmatrix}.$$

vector quantity Any quantity determined by its direction and magnitude, hence describable by a vector or vector function.

vector space A set V, together with a field F (often the real or complex numbers and referred to as the field of scalars) satisfying
(1) there is an operation $+$ defined on V, under which V is an Abelian group;
(2) there is a multiplication on V, by elements of F, called scalar multiplication. This operation satisfies $\lambda\mathbf{v} = \mathbf{v}\lambda$, for $\lambda \in F$ and $\mathbf{v} \in V$ and
(a.) $\lambda(\mu\mathbf{v}) = (\lambda\mu)\mathbf{v}$, for $\lambda, \mu \in F, \mathbf{v} \in V$,
(b.) $0\mathbf{v} = \mathbf{0}$, for $0 \in F, \mathbf{v}, \mathbf{0} \in V$,
(c.) $\lambda\mathbf{0} = \mathbf{0}$, for $\lambda \in F, \mathbf{0} \in V$,
(d.) $(\lambda + \mu)\mathbf{v} = \lambda\mathbf{v} + \mu\mathbf{v}$, for $\lambda, \mu \in F, \mathbf{v} \in V$, and
(e.) $\lambda(\mathbf{v}_1 + \mathbf{v}_2) = \lambda\mathbf{v}_1 + \lambda\mathbf{v}_2$, for $\lambda \in F, \mathbf{v}_1, \mathbf{v}_2 \in V$.

velocity The derivative of a curve $t \to f(t)$ from $\mathbf{R}$ to $\mathbf{R}^n$, assumed to represent the position of an object at time t. That is, the vector function

$$\mathbf{v}(t) = \lim_{h\to 0} \frac{f(t+h) - f(t)}{h}.$$

See also acceleration.

vertical space Of a Riemannian manifold (M, g) at a point $p \in M$: the subspace of T_pM,

$$\ker D\pi(p) = T_p\pi^{-1}(\pi(p))$$

where $\pi : M \to N$ is a submersion. A vector field tangent to the vertical space is called a *vertical vector field.* The orthogonal complement of the vertical space is called the *horizontal space* at p. A vector field tangent to the horizontal space is called a *horizontal vector field.*

vertical vector field *See* vertical space.

visibility manifold A Hadamard manifold (a complete, simply connected, Riemannian manifold of nonpositive sectional curvature) X such that, if x and y are points at ∞ of X and $x \neq y$, then there exists a geodesic $\gamma : \mathbf{R} \to X$ such that $\gamma(\infty) = x$ and $\gamma(-\infty) = y$.

Vitali Covering Theorem Let E be a subset of $\mathbf{R}$ with finite outer measure. Suppose $\mathcal{I}$ is a Vitali covering of E; that is, $\mathcal{I}$ is a collection of intervals such that each point in E is in an interval $I \in \mathcal{I}$ of arbitrarily small length. Then, for each $\epsilon > 0$, there is a finite, disjoint subcollection $\{I_1, \dots, I_n\} \subset \mathcal{I}$ such that the outer measure of $E \setminus \cup_{j=1}^n I_j$ is $< \epsilon$.

Volterra integral equation of the first kind An integral equation, in an unknown function $x(t)$, having the form

$$\int_0^t k(t,s)x(s)ds = f(t),$$

where $0 < t < \infty$. In case the kernel has the form $k(t,s) = k(t-s)$ the equation is called a *convolution* Volterra equation of the first kind.

Volterra integral equation of the second kind An integral equation, in an unknown function $x(t)$, having the form

$$x(t) + \int_0^t k(t,s)x(s)ds = f(t),$$

where $0 < t < \infty$. In case the kernel has the form $k(t,s) = k(t-s)$, the equation is called a *convolution* Volterra equation of the second kind.

Volterra operator An operator on $L^2[a, b]$ having the form

$$Vf(t) = \int_t^b v(t,s)f(s)ds,$$

for $f \in L^2[a, b]$.

volume element Any nonvanishing form μ on a manifold M. Hence, if f is a continuous function with compact support on M, we can form the n-form $\omega = f\mu$ and integrate f as $\int_{M^n} f\mu = \int_{M^n} \omega$.

volume of figure of revolution For the graph of a continuous function $y = f(x), a \leq x \leq b$, the volume of revolution about the x axis is defined by the integral

$$V = \pi \int_a^b [f(x)]^2 dx.$$

von Neumann algebra *See* W* algebra.

von Neumann inequality If T is a contraction operator on a Hilbert space and $p(z)$ is a polynomial, then

$$\|p(T)\| \leq \sup_{|z|=1} |p(z)|.$$

von Neumann's Selection Theorem Let X and Y be Polish spaces, A an analytic subset of $X \times Y$, and $\mathcal{A}$ the σ-algebra generated by the analytic subsets of X. Then there is an $\mathcal{A}$-measurable map $u : A_1 \to Y$, where $A_1 = \{x : (x, y) \in A, \text{ for some } y \in Y\}$ satisfying $(x, ux) \in A$, for every $x \in A_1$.

W* algebra An *abstract W* algebra* is an abstract C* algebra which is a dual Banach space.

Also called *von Neumann algebra.*

Watson transform The unitary operator $U : L^2(a, b) \to L^2(a, b)$, defined by

$$Uf = \frac{d}{dx}\int_a^b \frac{\chi(xt)}{t} f(t)dt$$

where the function χ is chosen so that $\chi(t)/t \in L^2(a, b)$ and

$$\int_a^b \frac{\overline{\chi(xt)}\chi(yt)}{t^2}dt = \begin{cases} \min[|x|, |y|], & \text{if } xy \geq 0, \\ 0, & \text{if } xy \leq 0, \end{cases}$$

for every $x, y \in (a, b)$.

wave equation The partial differential equation

$$\frac{\partial^2 u}{\partial x^2} = \frac{\partial^2 u}{\partial t^2}.$$

weak L^1 The class of measurable functions $f(t)$, with respect to a positive measure μ, such that $\mu\{t : |f(t)| > \lambda\} \leq C/\lambda$, for some constant C.

weak * convergence A sequence $\{x_n^*\}$ in the dual B^* of a Banach space B converges weak * to $x^* \in B^*$, provided $x_n^*(x) \to x^*(x)$, for every $x \in B$. *See also* weak * topology, weak convergence.

weak * topology In the dual B^* of a Banach space B, the weakest topology that makes all the functionals $x^* \to x^*(x)$ on B^*, (for $x \in B$) continuous.

weak convergence A sequence $\{x_n\}$ in a Banach space B converges weakly to $x \in B$ provided $x^*(x_n) \to x^*(x)$, for all $x^* \in B^*$. *See also* weak topology, weak * convergence.

weak operator topology The topology on $L(X)$, the algebra of bounded operators on a Hilbert space X, that is the weakest topology making all the functions $T \to (Tx, y)$, from $L(X)$, to $\mathbf{C}$ continuous.

weak solution A continuous function $u(x)$ is a weak solution of a partial differential equation $P(D)u = 0$ on $\Omega \subseteq \mathbf{R}^n$ provided

$$\int u(x)P(D)\phi(x)dx = 0$$

for all $\mathbf{C}^\infty$ functions ϕ with compact support in Ω.

weak topology In a Banach space B, the weakest topology that makes all the functionals in B^* continuous.

weak type (1,1) A linear operator T from $L^1(\mu)$ into the μ-measurable functions is weak type (1,1) if it is continuous as a map from $L^1(\mu)$ into weak L^1. That is, if

$$\mu\{t : |Tf(t)| > \lambda\} \leq C\|f\|_{L^1}/\lambda.$$

Weber function The function

$$W_n(z) = \frac{Y_n(z)\cos n\pi}{\pi e^{n\pi i}},$$

where

$$Y_n(z) = 2\pi e^{in\pi}\frac{J_n(z)\cos n\pi - J_{-n}(z)}{\sin 2n\pi},$$

with J_n a Bessel function. Also called *Bessel function of the second kind.*

Weber's differential equation The equation $u''+tu'-2\lambda u = 0$, where λ is a constant.

wedge *See* Edge of the Wedge Theorem.

Weierstrass canonical form The function

$$\prod_{n=1}^{\infty}(1 - \frac{z}{z_n})\exp[\frac{z}{z_n} + \frac{z^2}{z_n^2} + \ldots + \frac{z^{p_n}}{z_n^{p_n}}].$$

0-8493-0320-6/00/$0.00+$.50

Every entire function $f(z)$ with $f(0) \neq 0$ has a representation as $e^{g(z)}$ times such a product, where $z_1, z_2, \ldots$, are the zeros of $f(z)$, included with their multiplicities, and $g(z)$ is an entire function.

Weierstrass elliptic function *See* Weierstrass $\wp$ function.

Weierstrass $\wp$ function An elliptic function of order 2 with double pole at the origin, normalized so that its singular part is z^{-1}. For the period lattice $\omega = n_1\omega_1 + n_2\omega_2$, such a function has the expansion

$$\wp(z) = \frac{1}{z^2} + \sum_{\omega \neq 0} (\frac{1}{(z-\omega)^2} - \frac{1}{\omega^2}).$$

Weierstrass point A point P on a Riemann surface S, where $i[P^g] > 0$, where $i[P^g]$ is the dimension of the vector space of Abelian differentials that are multiples of P^g and g is the genus of S.

Weierstrass σ function The entire function represented by the infinite product, taken over the period lattice $\{n_1\omega_1 + n_2\omega_2\}$:

$$\sigma(z) = z \prod_{\omega \neq 0} (1 - \frac{z}{\omega}) e^{z/\omega + \frac{1}{2}(z/\omega)^2}.$$

The function satisfies $\sigma'(z)/\sigma(z) = \zeta(z)$. *See* Weierstrass zeta function.

Weierstrass zeta function The negative of the (odd) antiderivative of the Weierstrass $\wp$ function. Thus, the function

$$\zeta(z) = \frac{1}{z} + \sum_{\omega \neq 0} (\frac{1}{z-\omega} + \frac{1}{\omega} + \frac{1}{\omega^2}).$$

Weierstrass's Preparation Theorem Suppose $f(z)$ is analytic in a neighborhood of 0 in $\mathbf{C}^n$ and $f(0, \ldots, 0, z_n)$ has a zero of multiplicity m at $z_n = 0$. Then there is a polydisk $\Delta = \Delta' \times \Delta_n$, with center at 0, such that, for each $(z_1, \ldots, z_{n-1}) \in \Delta'$, $f(z_1, \ldots, z_{n-1}, \cdot)$ has m zeros in Δ_n and we can write

$$f(z) = W(z)h(z)$$

where h is analytic and nonzero in Δ, and

$$W(z) = b_0 + b_1 z_n + \ldots + b_{m-1} z_n^{m-1} + z_n^m$$

where $b_j = b_j(z_1, \ldots, z_{n-1})$ is analytic in Δ' and $b_j(0, \ldots, 0) = 0$, for $j = 1, \ldots, n-1$.

Weierstrass's Theorem **(1.)** A continuous, real-valued function on a real interval $[a, b]$ is the uniform limit of a sequence of polynomials on $[a, b]$. *See also* Stone-Weierstrass Theorem.

(2.) If $f_n(x)$ satisfies $|f_n(x)| \leq M_n$, for $n = 1, 2, \ldots$, with $\sum M_n < \infty$, then the series $\sum_{n=1}^{\infty} f_n$ converges uniformly. Also known as the Weierstrass M-test.

(3.) Let $D \subseteq \mathbf{C}$, let $\{z_n\} \subset D$ have no limit point in D and let $\{n_j\}$ be integers. Then there is a meromorphic function f on D such that $f(z)/(z - z_j)^{n_j}$ is analytic and nonzero at each z_j.

Weierstrass-Stone Theorem *See* Stone-Weierstrass Theorem.

Weingarten surface A surface with the property that each of its principal radii is a function of the other.

Weingarten's formula

$$\delta z = -\varphi_*(\psi),$$

where: $\varphi: M \to \mathbf{R}^{n+1}$ is an immersion of an oriented n-manifold in an oriented $(n+1)$-space, equipped with the induced Riemannian metric; $Z: M \to S^n$ is a unique smooth map such that

$$(h, t) \to (d\varphi)_x(h) + tz(x),$$

$h \in T_x(M)$, $t \in \mathbf{R}$ defining an orientation preserving isometry from $\tau_M \oplus \varepsilon$ to $M \times \mathbf{R}^{n+1}$; $\psi \in A^1(M; \tau_M)$.

Weyl's conformal curvature tensor Let g be a pseudo-Riemannian metric in a smooth

n-manifold M. There is a unique pseudo-Riemannian connection ∇ in τ_M with torsion zero (Levi–Civita connection). Let R be its curvature. Let R have components $R^{\,k}_{ij\ell}$ with respect to a local coordinate system. The expressions

$$(\mathrm{Ric})_{ij} = \sum_{\alpha} R^{\alpha}_{\alpha ij},$$
$$(\widehat{\mathrm{Ric}})^{i}_{j} = \sum_{\alpha} g^{\alpha i}(\mathrm{Ric})_{\alpha j},$$

define tensor fields Ric and $\widehat{\mathrm{Ric}}$ on M, called *Ricci tensors.* Ric is symmetric. Define $\phi \in L(M)$ by $\phi = \sum_{\alpha}(\widehat{\mathrm{Ric}})^{\alpha}_{\alpha}$. ϕ is called the *Ricci scalar curvature.* Assume $n \geq 3$. We can show that a 2-form $C \in A^2(M; S_{k_{\ell m}})$ is defined by

$$\begin{aligned} C^{\,h}_{\ell mk} &= R^{\,h}_{\ell mk} \\ &-\frac{1}{n-2}\{\delta^{h}_{\ell}(\mathrm{Ric})_{mk} - \delta^{h}_{m}(\mathrm{Ric})_{\ell k} \\ &+g_{km}(\widehat{\mathrm{Ric}})^{h}_{\ell} - g_{k\ell}(\widehat{\mathrm{Ric}})^{h}_{m}\} \\ &-\frac{\phi}{(n-1)(n-2)}\{\delta^{h}_{m}g_{k\ell} - \delta^{h}_{\ell}g_{km}\}. \end{aligned}$$

C is called the *Weyl conformal curvature tensor.*

Weyl's Lemma If $\Omega \subseteq \mathbf{R}^n$ and u is continuous on Ω and is a weak solution of Laplace's equation, i.e.,

$$\int u(x)(\frac{\partial^2}{\partial x_1^2} + \ldots + \frac{\partial^2}{\partial x_n^2})\phi dx = 0$$

for all $\phi \in \mathbf{C}_c^{\infty}(\Omega)$, then u is harmonic in Ω.

Weyl-Stone-Titchmarsh-Kodaira theory A theory dealing with pattern calculus. This particular branch of calculus was invented as a tool for writing out the matrix elements of certain Wigner operators.

Whitney's Theorem If M is a compact, smooth manifold of dimension n then there is an immersion $f : M \to \mathbf{R}^{2n}$ and an embedding $g : M \to \mathbf{R}^{2n+1}$.

Whittaker's differential equation The equation

$$\frac{d^2W}{dx^2} + \left\{-\frac{1}{4} + \frac{k}{x} + \frac{1/4 - m^2}{x^2}\right\}W = 0,$$

which has two singularities: a regular singular point at 0 and an irregular singular point at ∞. It is satisfied by the confluent hypergeometric function $W_{k,m}(x)$.

Whittaker's function The solution $W_{k,m}(x)$ of Whittaker's differential equation. (*See* Whittaker's differential equation.) When $\Re(k - \frac{1}{2} - m) \leq 0$, we have

$$W_{k,m}(x) = \frac{e^{-x/2}x^k}{\Gamma(1/2 - k - m)} \cdot \int_0^{\infty} t^{-k-1/2+m}\left(1 + \frac{t}{x}\right)^{k-1/2+m} e^{-t}dt.$$

Also called *confluent hypergeometric function.*

Wiener's formula

$$W(f) = 2\int \psi(\tau)\cos 2\pi f\tau\, d\tau,$$

where $W(f)$ is the power spectrum (spectrum of squared amplitudes) and $\psi(\tau)$ is the autocorrelation function for large τ. This formula appears in cybernetics in the study of methods of minimizing interference (separating signals from noise).

Wiener-Hopf equation An eigenvalue problem on $(0, \infty)$, involving an integral operator with a kernel depending on the difference of the arguments:

$$\int_0^{\infty} k(t - s)f(s)ds = \lambda f(t).$$

Wiener-Levy Theorem If $f : [-\pi, \pi] \to (\alpha, \beta)$ is a function with an absolutely convergent Fourier series and $F(z)$ is analytic at every point of (α, β), then $F(f(t))$ has an absolutely convergent Fourier series.

winding number If γ is a closed path in the complex plane and z is a point off γ, the winding number of γ about z is the quantity $\frac{1}{2\pi i}\int_\gamma \frac{d\zeta}{\zeta - z}$, which is an integer and represents the number of times γ winds in a counterclockwise direction about z.

witch of Agnesi The locus of a point P located as follows: A circle of radius r is centered at $(0, r)$. If S is any ray from the origin, cutting the circle at Q and the line $y = 2a$ at A, then P is at the intersection of the horizontal line through Q and the vertical line through A. The parametric representation for the witch of Agnesi is

$$x = 2a \tan t$$
$$y = 2a \cos^2 t.$$

WKB method Wentzel-Kramers-Brillouin method for applying the Liouville-Green approximation by relating exponential and oscillatory approximations around a turning point. Sometimes called WKBJ (Jeffreys) method. *See* Liouville-Green approximation.

Wronskian determinant For functions $u_1(t), \ldots, u_k(t)$, of class $\mathbf{C}^{k-1}$, the determinant $W(t, u_1, \ldots, u_k) = \det(u_i^{j-1}(t))$.

X

x axis (1.) In $\mathbf{R}^3$, one of the three mutually perpendicular axes, usually horizontal and pointing west to east.
(2.) In $\mathbf{R}^2$, one of the two mutually perpendicular axes, usually horizontal.

x coordinate When points in $\mathbf{R}^3$ are designated by triples of real numbers, the *x coordinate* is usually the first entry in the triple and denotes the number of units (positive or negative) to be traveled parallel to the x axis to reach the point denoted.

x intercept A point at which a curve or graph in $\mathbf{R}^3$ crosses the x axis. That is, a point of the form (x, 0, 0), for some $x \in \mathbf{R}$, lying on the curve or graph.

xy-plane The plane in $\mathbf{R}^3$ containing the x and y axes. It is the set of points of the form $(x, y, 0)$, for $x, y \in \mathbf{R}$.

xyz-space The space $\mathbf{R}^3$ of triples of real numbers, where the three coordinates of a point (p_1, p_2, p_3) are referred to as the x coordinate (p_1), y coordinate (p_2), and z coordinate (p_3).

xz-plane The plane in $\mathbf{R}^3$ containing the x and z axes. It is the set of points of the form $(x, 0, z)$, where $x, z \in \mathbf{R}$.

0-8493-0320-6/00/$0.00+$.50

y axis (**1.**) In $\mathbf{R}^3$ one of the three mutually perpendicular axes, usually horizontal and pointing south to north.
(**2.**) In $\mathbf{R}^2$, one of the two mutually perpendicular axes, usually vertical.

y coordinate When points in $\mathbf{R}^3$ are designated by triples of real numbers, the *y coordinate* is usually the second entry in the triple and denotes the number of units (positive or negative) to be traveled parallel to the y axis to reach the point denoted.

y intercept A point at which a curve or graph in $\mathbf{R}^3$ crosses the y axis. That is, a point of the form $(0, y, 0)$, for some $y \in \mathbf{R}$, lying on the curve or graph.

Yang-Mills equation Given by

$$\frac{\partial}{\partial \overline{x}_1}\left(\Omega^{-1}\frac{\partial \Omega}{\partial x_1}\right) + \frac{\partial}{\partial \overline{x}_2}\left(\Omega^{-1}\frac{\partial \Omega}{\partial x_2}\right) = 0.$$

(Also called *anti-self dual Yang–Mill equation.*)

Yosida approximation Let B be a closed linear operator and let $R(\lambda, B) = (\lambda I - B)^{-1}$ be the resolvent of B. A *Yosida approximation* to B is $B_\lambda = \lambda^2 R(\lambda, B) - \lambda I$. B is the generator of a semigroup if and only if $|R(\lambda, B)| \leq 1/\lambda$ for all $\lambda > 0$; this can be proved by generating the approximating semigroup e^{tB_λ}.

Young's inequality Let $f \in L^p(\mathbf{R}^n)$, $g \in L^r(\mathbf{R}^n)$ and $h = f * g$, then $h \in L^q(\mathbf{R}^n)$, where $1/q = 1/p + 1/r - 1$ and

$$\|h\|_q \leq \|f\|_p \|g\|_r.$$

yz-plane The plane in $\mathbf{R}^3$ containing the y and z axes. It is the set of points of the form $(0, y, z)$, for $y, z \in \mathbf{R}$.

0-8493-0320-6/00/$0.00+$.50

Z

z axis One of the three mutually perpendicular axes, usually the vertical axis, in $\mathbf{R}^3$.

z coordinate When points in $\mathbf{R}^3$ are designated by triples of real numbers, the *z coordinate* is usually the third entry in the triple and denotes the number of units (positive or negative) to be traveled parallel to the z axis to reach the point denoted.

z intercept A point at which a curve or graph in $\mathbf{R}^3$ crosses the z-axis. That is, a point of the form $(0, 0, z)$, for some $z \in \mathbf{R}$, lying on the curve or graph.

zero element An element, denoted 0, in a set S, with an addition operation $+$, satisfying $x + 0 = 0 + x = x$, for all $x \in S$.

zero of function For a function $f : X \to \mathbf{C}$, a point $x_0 \in X$ satisfying $f(x_0) = 0$.

zero point *See* zero point of the kth order, zero point of the $-k$th order.

zero point of the $-k$th order A pole of order $k > 0$ of a function $f(z)$. *See* pole. That is, a point z_0 such that $f(z)$ is analytic in $D \backslash \{z_0\} \subseteq \mathbf{C}$ (or $D \backslash \{z_0\} \subseteq M$, for an analytic manifold M), and

$$0 < |\lim_{z \to z_0} (z - z_0)^k f(z)| < \infty.$$

zero point of the kth order For a function $f(z)$, analytic in an open set $D \subseteq \mathbf{C}$ (or $D \subseteq M$, for M an analytic manifold), a point $z_0 \in D$ such that $f(z_0) = f'(z_0) = \cdots = f^{(k-1)}(z_0) = 0$ and $f^{(k)}(z_0) \neq 0$.

zero set *See* set of measure 0.

zone The portion of the surface of a sphere lying between two parallel planes. *See also* spherical segment.

Zygmund class The class of real, measurable functions on a measure space (S, Ω, μ) such that

$$\int_S |f(t)| \log^+ f(t) d\mu(t) < \infty$$

where $\log^+(t) = \max(\log t, 0)$.

0-8493-0320-6/00/$0.00+$.50